109 Topics in Current Chemistry

Fortschritte der Chemischen Forschung

Managing Editor: F. L. Boschke

W0106846

Wittig Chemistry

Dedicated to Professor Dr. G. Wittig

With Contributions by
H. J. Bestmann, D. Hellwinkel, A. Krebs,
H. Pommer, U. Schöllkopf, P. C. Thieme,
O. Vostrowsky, J. Wilke

With 34 Figures and 22 Tables

Springer-Verlag
Berlin Heidelberg GmbH 1983

This series presents critical reviews of the present position and future trends in modern chemical research. It is addressed to all research and industrial chemists who wish to keep abreast of advances in their subject.

As a rule, contributions are specially commissioned. The editors and publishers will, however, always be pleased to receive suggestions and supplementary information. Papers are accepted for "Topics in Current Chemistry" in English.

ISBN 978-3-662-15316-1   ISBN 978-3-540-39508-9 (eBook)
DOI 10.1007/978-3-540-39508-9

Library of Congress Cataloging in Publication Data. Main entry under title: Wittig chemistry. (Topics in current chemistry; v. 109)
Bibliography: p. Includes index.
1. Chemistry, Physical organic — Addresses, essays, lectures. 2. Wittig, Georg, 1897 —.
I. Wittig, Georg, 1897 —. II. Bestmann, H. J. (Hans Jürgen), 1925 —. III. Series.
QD1.F58   vol. 109 [QD476]   540s [547.1'3]   82-19662

© by Springer-Verlag Berlin Heidelberg 1983
Originally published by Springer-Verlag Berlin Heidelberg New York in 1983
Softcover reprint of the hardcover 1st edition 1983

Table of Contents

Penta- and Hexáorganyl Derivatives of the
Main Group Five Elements
D. Hellwinkel   1

Enantioselective Synthesis of Nonproteinogenic
Amino Acids
U. Schöllkopf 65

Selected Topics of the Wittig Reaction
in the Synthesis of Natural Products
H. J. Bestmann, O. Vostrowsky. 85

Industrial Applications of the Wittig Reaction
H. Pommer, P. C. Thieme 165

Angle Strained Cycloalkynes
A. Krebs, J. Wilke 189

Author Index Volumes 101–109 235

Penta- and Hexaorganyl Derivatives of the Main Group Five Elements

Dieter Hellwinkel

Organisch-Chemisches Institut der Universität Heidelberg, Im Neuenheimer Feld 270, D-6900 Heidelberg 1, West-Germany

Table of Contents

1 Introduction 2

2 Early Developments 3

3 The Discovery of the Wittig Reaction 7

4 Further Investigations in the Field of Penta- and Hexaaryl
  Phosphorus Compounds 9

5 Crystal Structures of Spirocyclic Pentaorganylphosphoranes 23

6 Bis-2,2'-biphenylylenehydridophosphoranes 26

7 A Nitrogen Interlude 29

8 Alkylphosphoranes 30

9 Newer Developments in the Area of Penta- and
  Hexaorganyl Arsenic Compounds 35

10 Recent Developments in the Field of Pentaaryl Antimony Compounds 44

11 Pentaalkylstiboranes and Related Compounds 52

12 Penta- and Hexaaryl Bismuth Compounds 54

13 Concluding Remarks 55

14 Acknowledgements 56

15 Bibliography 56

16 References 58

1 Introduction

The main goals of experimental chemistry have always been the generation of new compounds on the one hand and the development of new reactions on the other. But, as one can learn from more historically orientated studies, for many areas of chemistry these two objectives are most oftenly closely interrelated so that they stimulate and even necessitate each other in various, though sometimes quite incidental ways.

A demonstrative example for this may be seen in the "Wittig-Reaction" — the well-known carbonyl olefination by phosphorus ylides — whose origin can very precisely be traced back to a fortuitous outcome of a routine trapping experiment in the course of fundamental investigations in the field of phosphorus pentacoordination. In other words, the discovery of the Wittig reaction was not so much the result of a specifically designed search for new reactions, than a rather unexpected but of course highly welcomed offshoot from a totally different main stream of activities dedicated to the generation and study of pentacoordinated organic derivatives of the nitrogen/phosphorus group elements. Because of its high synthetic importance, the Wittig reaction quickly attracted the general attention of the chemical community, perhaps even at the expense of the interest in the underlying basic research program on new pentacoordinated molecules. But this did not really mean that the latter topic had now become less appealing and would therefore have been pursued with less endeavor. Quite the contrary: the continuation of work on pentaorganylphosphorus and homologous derivatives did in fact lead to a wealth of fascinating new compounds and concepts which were certainly as interesting, and perhaps even more interesting — at least from a more fundamental point of view — than the growing importance of the practical applications of the Wittig reaction to synthesis.

In this article I shall present a coherent insight into that framework of research activities that led from very early ideas and experiments on nitrogen pentacoordination quite logically and, as I think, also stringently first to that important branching point where the Wittig reaction took its origin, and then via a plethora of interesting new compounds to the actual conceptions of the stereochemistry of pentacoordination. I will concentrate, however, on those compounds containing five or six carbon groups linked directly by covalent bonds to a nitrogen/phosphorus group elemental center, which mostly were investigated in the Heidelberg laboratories. Pertinent general results obtained with derivatives containing less than five element-carbon bonds will of course be included where necessary.

Before entering the material part proper of this review, a note on nomenclature has to be made. Although not really needed, the only partially IUPAC-approved terms "pentaorganylphosphorane, -arsorane, -stiborane and -bismuthorane" have asserted themselves widely in the literature instead of the less systematic but equally unambiguous names "pentaphenylphosphorus, -arsenic, -antimony and -bismuth". Chemical Abstracts, however, has only adopted the first two systematic terms, that is, "phosphorane" and "arsorane" as parent compound entries, whereas pentacoordinated antimony and bismuth derivatives have to be looked for under "antimony and bismuth, pentaorganyl" respectively. To stay consistent I shall use the systematic "-orane" set of names throughout. A similar naming dichotomy exists for the trivalent derivatives of these elements where the older and still IUPAC-approved and

CA-used terms "triorganylphosphine, -arsine, -stibine and -bismuthine" seem to become progressively replaced by the newer parent hydride-based names "triorganylphosphane, -arsane, -stibane and -bismuthane". Again I will exclusively use the first set of names which prevails by far in the works discussed in this paper. According to a recent IUPAC proposal which has already been adopted by some journals, parent hydrides with non-standard valence states can be designated by the λ^n symbol in front of the name of the standard hydride, where the bonding number n gives the total number of classical valence bonds originating from the central atom in question. Correspondingly, compound types PR_5, AsR_5, SbR_5, BiR_5 would be named as λ^5-phosphanes, λ^5-arsanes, λ^5-stibanes and λ^5-bismuthanes.

2 Early Developments

The history in this field goes very far back indeed. As early as 1862 Cahours had claimed [1] to have prepared pentamethylarsorane $(CH_3)_5As$ (*1*) by the reaction of tetramethylarsonium iodide with dimethylzinc. First attempts to synthesize pentaalkyl nitrogen compounds, e.g. pentaethylnitrogen (*2*) have been reported by Lachman in 1896 who, however, in reacting tetraethylammonium iodide and the triethylamine/bromine adduct with diethylzinc could only identify tetraethylammonium triiodide and the reduction product triethylamine respectively [2]. It may be noted that at this time the imagination of the experimentalist was not yet limited (or spoiled?) by too restrictive bonding theory considerations [3] as is clearly indicated in Lachman's own words: "*There is a priori no reason why nitrogen in its pentavalent state cannot combine with five like atoms or radicals . . .*" [2]

More conclusive results were obtained by Schlenk and Holz as much as twenty years later. In reactions of tetramethylammonium chloride with triphenylmethylsodium [4] and, later, benzylsodium [5], bright red compounds were formed which according to elemental analyses did contain five organic groups per nitrogen atom. Although the formulation of these compounds, e.g.

$$(C_6H_5)_3 \equiv C - N \equiv (CH_3)_4$$

still reflected the "whishful thinking" of that time, their properties (intense color, electrical conductivity and high susceptibility to polar agents and oxygen) made it evident that they had to be considered as salt-like [6] ion pair structures such as

$$[C_6H_5CH_2]^\ominus\ [N(CH_3)_4]^\oplus \quad \textit{4}$$

At about the same time the famous octet principle based on the work of Kossel [7], Lewis [8] and Langmuir [9] was developed. It explained and, moreover, even postulated the nonexistence of five covalent bonds around a nitrogen center in a very stringent way. Notwithstanding these theoretically imposed restrictions, the conception of pentacoordinated nitrogen persisted in the minds of dedicated experimentalists. Thus, Staudinger and Meyer resumed the experiments for synthesizing pentamethylnitrogen and pentaethylphosphorane by the reaction of the corresponding onium salts with dimethyl- and diethylzinc, respectively, but again with negative results [10].

In the twenties, another series of experiments aimed at the synthesis of pentacoordinated nitrogen compounds became known through Hager and Marvel. They dealt with the reactions of the much more reactive lithium alkyls with tetraalkylammonium salts. But again nitrogen proved its inability or at least great reluctance to take on a fifth ligand, in that no pentaalkyl nitrogen derivatives, but only mixtures of amines and hydrocarbons, could be isolated [11]. In another series of experiments with the same aim, tetraethylphosphonium iodide was used together with triphenyl-methylsodium as nucleophile, but again no pentacoordinated structure was produced [12]. When, however, triphenylalkylphosphonium salts were treated with tritylsodium or butyllithium in diethyl ether, orange solutions were obtained which were shown to contain triphenylphosphinemethylenes 5, a compound class, in principle already known through the work of Staudinger and Meyer [13]. Besides hydrolysis no further chemistry was done with these compounds at that time [12].

$$Ph_3\overset{\oplus}{P}-CHR_2 \xrightarrow[\text{or BuLi}]{Ph_3CNa} Ph_3P=CR_2 \xrightarrow{H_2O} Ph_2P\overset{CHR_2}{\underset{O}{\diagdown}}$$

$$5, R = H, Me, Ph$$

At the end of the paper dealing with this work a casual but highly interesting remark is made (l.c. [12] p 3498). It states, that by studying compounds not having "*one group which carries a hydrogen atom on the carbon attached to the phosphorus atom . . . pentaalkyl phosphorus compounds may (thus) be obtained*". In a way this remark anticipates the further development in this field by more than two decades. Analogous experiments were then performed with tetraalkylarsonium salts where again no pentaalkyl derivatives of arsenic could be isolated [14].

The investigations reported so far consisted of more or less limited sets of experiments which evidently did not lead to any more systematic studies. Probably, the octet rule stifled any initiative to do so. Looking back it appears now that only a more general and more broadly designed research program could instigate such coherent systematic investigations as would be necessary to bring about real progress in the field.

Such a comprehensive and fundamental program was started in the laboratories of G. Wittig when, in the early thirties, he systematically began to explore the synthetic potential of organyllithium compounds [15], particularly of phenyllithium [16]. In the course of these investigations many interesting new research areas were disclosed and a first climax was reached with the formulation of the dehydrobenzene concept at the beginning of the forties [17]. The encouraging and highly stimulating experiences thus made with the now readily accessible phenyllithium led naturally back to the question of pentacoordinate nitrogen, since it could be hoped that this strongly nucleophilic reagent would eventually be able to overcome the inherent resistence of nitrogen in ammonium salts to accepting a fifth covalently bonded ligand.

In this respect the experiments were again a total failure since no pentacoordinated nitrogen compound could be identified. But in contrast to all previous researchers, Wittig did not acknowledge this failure as such, instead, he considered it merely as an unexpected result which nevertheless had to be explored thoroughly. In doing so, he laid the foundations for the ultimate success of his early and original ideas on pentacoordination if not of nitrogen, but of the higher elements of main group five. At the same time, this perseverance opened up the way to a quite different line of research which very soon would prove even further reaching. And this all came about as follows.

Contrary to the resonance stabilized triphenylmethyl- [4], benzyl- [5] and fluorenyl-anions [8], phenyllithium did not leave the tetramethylammonium ion unaffected. Instead, it removed a proton to form trimethylammonium-methylide (6) [19], the prototype of that interesting class of zwitter-ionic compounds for which Wittig coined the name "ylides" [20].

$$[Me_3\overset{\oplus}{N}-CH_3]Cl^{\ominus} \xrightarrow[-PhH,-LiCl]{PhLi} Me_3\overset{\oplus}{N}-\overset{\ominus}{\overset{..}{C}}H_2 \xrightarrow{Ph_2CO} Me_3\overset{\oplus}{N}-CH_2-\overset{\overset{\overset{..}{\overset{\ominus}{O}}}{|}}{C}Ph_2$$

$$\begin{matrix} 6 & & 7 \end{matrix}$$

$$\left[Me_3\overset{\oplus}{N}-CH_2\overset{\overset{OH}{|}}{C}Ph_2\right] X^{\ominus} \xleftarrow{HX}$$

$$8$$

The constitution of 6 was proved primarily by trapping it with benzophenone to give another zwitter-ion 7 which easily yielded the adducts 8 with various acids. Since it was now perfectly clear that even the most nucleophilic agents would attack only the periphery, not the nitrogen center of tetraalkylammonium salts, extension of the investigations to the next heavier element of the group, phosphorus, was only logical. This was especially so, since it was known, in principal at least, that this element could combine covalently with five like ligands, as in phosphorus-pentachloride and -penta-fluoride [21].

In a similar sequence of experiments as in the case of the ammonium analogue, tetramethylphosphonium iodide reacted with phenyl- or methyllithium first to give trimethylphosphoniummethylide (Wittig also used the name trimethylphosphine-methylene to account for the doubly-bound resonance form 9a), which again was trapped with benzophenone to yield 10 and its hydroiodic acid adduct 11 respectively [22].

$$[Me_3\overset{\oplus}{P}-CH_3]I^{\ominus} \xrightarrow[or\ MeLi]{PhLi} Me_3\overset{\oplus}{P}-\overset{\ominus}{\overset{..}{C}}H_2 \longleftrightarrow Me_3P=CH_2$$

$$\begin{matrix} 9 & & 9a \end{matrix}$$

$$\Big\downarrow Ph_2CO$$

$$Me_3\overset{\oplus}{P}-CH_2-\overset{\overset{\overset{..}{\overset{\ominus}{O}}}{|}}{C}Ph_2 \xrightarrow{HI} [Me_3\overset{\oplus}{P}-CH_2\overset{\overset{OH}{|}}{C}Ph_2]I^{\ominus}$$

$$\begin{matrix} 10 & & 11 \end{matrix}$$

Thus, in alkylphosphonium salts too the C—H functions adjacent to the positively charged key atom were too acidic to let the attacking nucleophile proceed on its way towards the phosphorus center. A way out of this dilemma should consist in the application of a phosphonium salt with less acidic neighboring C—H functions. Such a salt was tetraphenylphosphonium iodide which in the reaction with phenyl-lithium did indeed lead to the first compound containing five covalent carbon-element bonds, pentaphenylphosphorane, $(C_6H_5)_5P$ (12) [23]. Its covalent nature was clearly demonstrated by its stability towards water, its relatively low melting point (124 °C under decomposition), its solubility in cyclohexane and its thermal decomposition into triphenylphosphine and biphenyl in a radical reaction. Polar reagents such as hydrohalogenic acids led to regeneration of the tetraphenylphosphonium cation. Later it was shown by Russian workers that many of the thermal reactions of pentaphenylphosphorane in the absence or presence of other reaction partners can be explained on the basis of initial phenyl radical formation [24].

After this decisive breakthrough, the preparation of the higher homologues of pentaphenylphosphorane was only a question of time. Thus, by similar methods of synthesis pentaphenylarsorane (13) [25], pentaphenylstiborane (14) [25] and pentaphenylbismuthorane (15) [26] were obtained in rapid succession. The reactions of all these new compounds are in principle very similar, as can be seen from the following scheme:

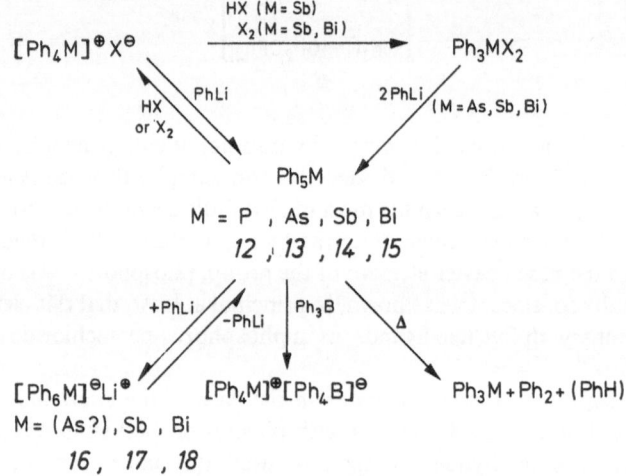

Several points, though, need special consideration. Whereas pentaphenylphosphorane, -arsorane and -stiborane are colorless to slightly yellowish, pentaphenylbismuthorane forms violet crystals. On the basis of the decomposition points of substituted derivatives it was argued that the increasing stability in going from the severly congested pentaarylphosphorus to the much more relaxed pentaaryl-antimony compounds is primarily due to the increase of the central atom radius. Along this line pentaphenylbismuthorane should be even more stable than the quite resistent pentaphenylstiborane. In fact it is much less stable than even pentaphenylphosphorane [27, 28]. In anticipation of later research it might be mentioned here that under mass spectroscopical conditions all the pentaphenyl derivatives showed very poor stability. The molecular ions are either very weak ((C_6H_5)$_5$P, As, Sb) or not observable at all ((C_6H_5)$_5$Bi), and the most intense fragments nearly all belong to a mass level below $M - 2 C_6H_5$ which corresponds to the triphenyl derivatives [29].

Also, the inability of pentaphenylphosphorane to form a hexaphenylphosphate complex with phenyllithium can be explained solely by steric overcrowding. In the light of this, the hexaphenylbismuthate ion 18 should be more stable than the hexaphenylantimonate complex 17, but again this is not the case. As could be shown many years later, pentaphenylbismuthorane (*15*) reacts with excess phenyllithium only at $-70\,°C$ to give the yellow lithium hexaphenylbismuthate (*Li-18*) which readily regenerates the violet starting material *15* on warming to room temperature. [30] Pentaphenylarsorane also forms an unstable yellowish precipitate with phenyllithium in tetrahydrofurane at $-70\,°C$ but it is not clear whether this really does consist of lithium hexaphenyl arsenate (*Li-16*) or of a rather labile adduct, $(C_6H_5)_5As \cdot C_6H_5Li$ [30].

With this wealth of experimental experience gained in the study of the above pentaphenyl derivatives as background, renewed efforts to obtain similar pentaalkyl derivatives were made. When dibromotrimethylstiborane or tetramethylstibonium bromide were treated with methyllithium, pentamethylstiborane (*19*) was obtained as a distillable, water-sensitive liquid [31] whose reactions obeyed the same principles as established for pentaphenylstiborane.

$$[Me_4Sb]^{\oplus}\,Br^{\ominus} \underset{Br_2}{\overset{MeLi}{\rightleftharpoons}} \underset{\underset{\downarrow Ph_3B}{19}}{Me_5Sb} \overset{2\ MeLi}{\longleftarrow} Me_3SbBr_2$$

$$[Me_4Sb]^{\oplus}[BPh_3Me]^{\ominus}$$

Similar experiments with analogous arsenic precursors did not allow the isolation of pentamethylarsorane (*1*), whose existence, however, could be made probable by some trapping reactions [31].

With these experiments, a somewhat clearer picture relating to main group five element pentacoordination began to emerge. Thus, in the form of the pentaaryl derivatives, more or less stable model compounds for all the elements except nitrogen had become accessible, which were to particularly facilitate the study of the stereochemical questions pertinent to this field. For the pentaalkyl derivatives, however, the situation was much less satisfying and pentamethylstiborane remained the sole clearly characterized representative at that time. A compromise was therefore sought, which would also make the preparation of alkyl derivatives of pentacoordinate phosphorus possible. Such a compromise could, for example, consist in the generation of mixed aryl-alkylphosphoranes [32], an idea, which turned out to be very successful, but in a totally unexpected and yet most significant way.

3 The Discovery of the Wittig Reaction

That the "carbonyl olefination by phosphorus ylides" was in fact discovered rather incidentally in the course of experiments aimed at the study of pentaphenylphosphorane and the syntheses of new derivatives thereof is clearly demonstrated in the title of the original publication by Wittig and Geissler: "Zur Reaktionsweise des Pentaphenylphosphors und einiger Derivate" [33].

When, with the objective of synthesizing methyltetraphenylphosphorane (*20*), methyltriphenylphosphonium iodide was reacted with phenyllithium, the nucleophile was once again caught by the acidified hydrogens of the methyl group and thus prevented from reaching the phosphorus center. Instead, the known triphenylphosphinemethylene *5* (R = H) [12] was formed. So far, everything was quite normal and could have been predicted from earlier investigations [12]. The crucial point, however, was reached when the routine trapping experiment with benzophenone was performed. In this case the reaction went straight beyond the potential initial zwitterionic addition product *21* to the cyclic phosphorane *22* which then collapsed to 1,1-diphenylethylene and triphenylphosphine oxide [33]. It must be noted here that a precursor to that type of reaction had already been described as early as 1919 by Staudinger and Meyer [13b] but was not explored any further. A proper evaluation of this and other relevant earlier work has been given by Wittig in a special review dealing with the more historical aspects of the phosphororganic carbonyl olefination reaction [34].

$$Ph_4P-CH_3 \xleftarrow{\;PhLi\;} [Ph_3\overset{\oplus}{P}-CH_3]I^{\ominus} \xrightarrow{\;PhLi\;} Ph_3P=CH_2 \xrightarrow{\;Ph_2CO\;} Ph_3\overset{\oplus}{P}-CH_2$$
$$20 \qquad\qquad\qquad\qquad 5$$

$$\downarrow Ph_2CO \qquad\qquad \overset{\ominus}{O}\diagdown^{CPh_2}_{\;\;21}$$

$$Ph_3P + Ph_2C=CH_2 \longleftarrow\qquad\quad Ph_3P-CH_2$$
$$\qquad\qquad\qquad\qquad\qquad O-CPh_2$$
$$\qquad\qquad\qquad\qquad\qquad 22$$

In the first report on this famous reaction, Wittig expressed a clear preference for the cyclic phosphorane intermediate *22* such as had already been formulated by Staudinger and Meyer in their earlier work. This view became generally accepted only many years later [35]. In this respect, the Wittig reaction did indeed still reflect some prominent features of the underlying research program on penta-coordinated molecules, from which it now evolved rapidly to a very active life of its own [36].

In conclusion of this line of experiments, the first mixed arylalkylphosphorane *23* could finally be synthesized by the reaction of triphenyl(triphenylmethyl)phosphonium chloride with phenyllithium.

$$[(C_6H_5)_3P^{\oplus}-C(C_6H_5)_3]\,Cl^{\ominus} + C_6H_5Li \rightarrow (C_6H_5)_4P-C(C_6H_5)_3 \qquad 23$$

Unlike the formally analogous, but in reality ionic, tetramethyl(triphenylmethyl)-nitrogen compound *3* of Schlenk and Holtz [4], tetraphenyl(triphenylmethyl)phosphorane (*23*) was unambiguously proven to be a covalent compound [33]. Alkinyl-tetraphenylphosphoranes *24* were prepared only many years later [37] and shown to be stable only below −20 °C.

$$[Ph_4P]^{\oplus}Cl^{\ominus} + KC\equiv\!\equiv CR \xrightarrow[-KCl]{(NH_3)\,liquid} Ph_4P-C\equiv\!\equiv CR \xrightarrow{H^{\oplus}} Ph_4P^{\oplus} + HC\equiv\!\equiv CR$$
$$24$$
$$R = Me,\ OMe,\ p\,(MeO)C_6H_4$$

$$[Ph_4P]^{\oplus}Br^{\ominus} + LiCH=CH_2 \xrightarrow{Et_2O} Ph_4P-CH=CH_2 \rightarrow Ph_3P + PhHC=CH_2$$
$$24a$$

Similarly, a detailed study of the reaction of tetraphenylphosphonium bromide with vinyl- and isopropenyllithium gave evidence of the intermediacy of tetraphenyl-vinylphosphorane derivatives *24a* [38].

4 Further Investigations in the Field of Penta- and Hexaaryl Phosphorus Compounds

With the accessibility of the fairly stable pentaphenylphosphorane (*12*) and its even more stable homologues, there arose the intriguing question of the stereochemistry of such species. The first speculations were of course based on a trigonal bipyramidal model, since this structure type had already been proved for phosphorus penta-halides [21]. It was not possible, however, to differentiate in chemical reactions between the unlike ligands in compounds such as tetraphenyl(triphenylmethyl)-phosphorane (*23*) and tetraphenyl-p-tolylphosphorane [33] or in pentaphenylphos-phorane derivatives appropriately labeled with deuterium or ^{14}C [39].

To my knowledge, the first detailed discussion dealing with the stereochemical implications of five unlike ligands around a pentacoordinated phosphorus center was presented in 1959 in the doctoral thesis of Kochendoerfer [40]. The later general development of these questions, especially under consideration of skeletal nonri-gidity, can be followed in a more recent review by Gielen [41]. At the beginning, in the late fifties, however, the conception of a potential inner flexibility of a penta-coordinate molecular framework did not yet play a major role in stereochemical discussions. Thus, for a rigid trigonal bipyramidal phosphorane with five different ligands, 20 stereoisomers, that is, 10 diastereomeric pairs of enantiomers could be pre-dicted, a number much too high to allow reasonable experimental studies. The ut-most reduction of stereochemical possibilities for such compounds which still left a system from which relevant stereochemical information could potentially be expected consisted in the introduction of two bidentate ligands. Thus, the spiro-cyclic bis-2,2′-biphenylylenephenylphosphorane (*25*) was designed, for which, on the basis of a trigonal-bipyramidal structure with axial-equatorial linking of the chelate groups to the central atom, only one pair of enantiomeres was possible [40, 42]. At the same time, such spirophosphoranes were expected to be more stable than pentaphenyl-phosphorane itself because of significantly reduced steric overcrowding.

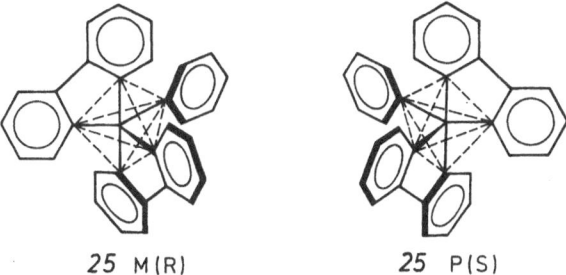

25 M(R)   25 P(S)

Here again a note on nomenclature is needed. If compounds of type *25* are considered as substituted phosphoranes, names such as the one given above are appropriate. If, on the other hand, their heterocyclic nature stays in the foreground, the correct names (as used by Chemical Abstracts) are such as: 5-phenyl-5,5′-spirobi[5H-dibenzophosphole]. Both naming procedures are consistent with IUPAC rules and are applicable to similar arsenic, antimony- and bismuth-heterocycles as well (see later chapters).

Since the earlier methods could not be used for preparing *25* and similar compounds, a more general and also more convenient synthesis for pentaarylphosphoranes was developed. This consisted in the reaction of quaternized triarylphosphine imines with two equivalents of aryllithium [40, 42].

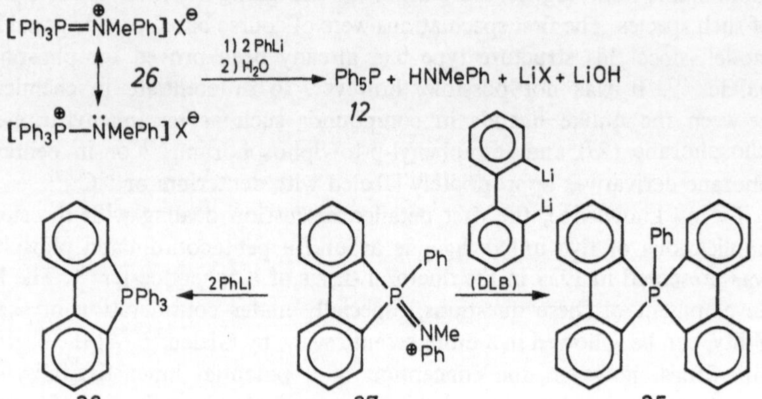

Obviously, in the precursor immonium compounds *26*, *27* the phosphorus bears enough partial positive charge to allow easy attack by aryl anions. Along similar lines of argument an even simpler synthetic procedure could be found in using analoguous p-toluenesulfonylimine (tosylimine) derivatives *29* [43], which had the appropriate polarization of the P=N double bond already built in [44].

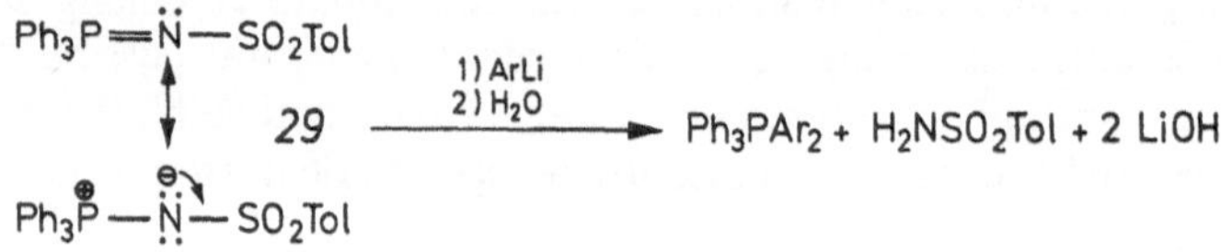

With these methods a number of substituted spirophosphoranes, e.g. *30–32*, was easily obtainable [42, 45].

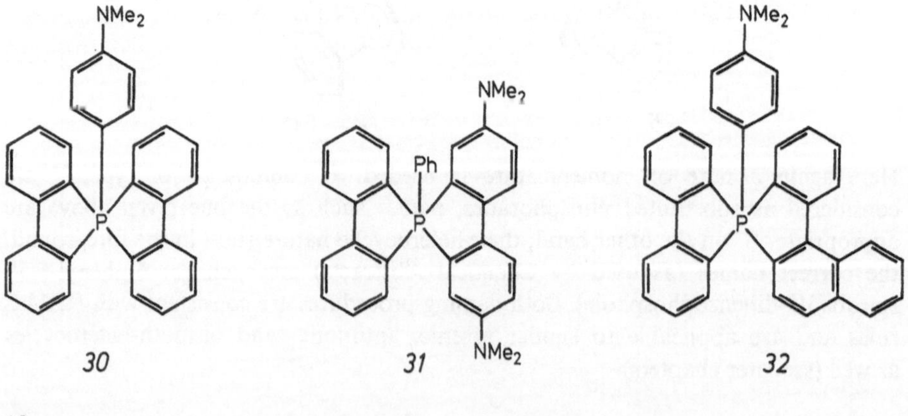

A modification of the tosylimine method finally led to a simple one-pot synthesis of several substituted pentaphenylphosphoranes [46].

$$(PhO)_3P = NSO_2Tol + 5 \ ArLi \longrightarrow Ar_5P + Li_2NSO_2Tol + 3 \ LiOR$$

$$Ar = C_6H_5, \ p-CH_3C_6H_4, \ p-ClC_6H_5, \ p-C_6H_5C_6H_4$$

Of special interest with respect to the stereochemical questions was the yellow spirophosphorane *31* since according to the new synthetic procedure it could formally be obtained by two ways, either from the tosylimine of 5-phenyldibenzo-phosphole and 4,4'-bis(dimethylamino)-2,2'-dilithiobiphenyl or, the other way round, from 3,7-bis(dimethylamino)-5-phenyldibenzophosphole tosylimine and 2,2'-dilithio-biphenyl.

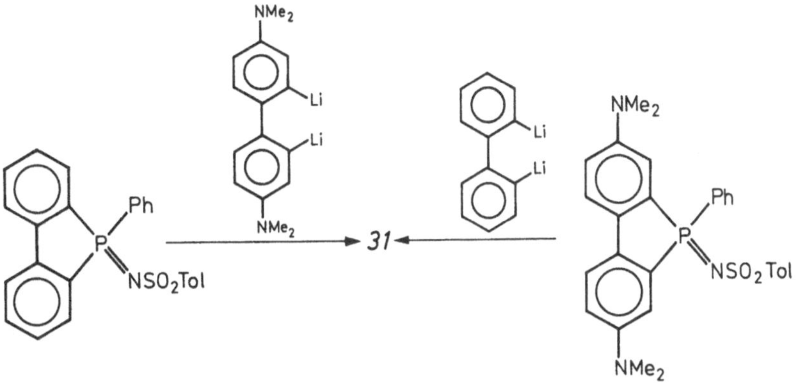

Both alternatives gave the same compound *31*, thus demonstrating the equivalency of the two chelate group arrangements, which is only in accordance with chiral trigonal-bipyramidal structures of type *25M, P.*

The chemistry of such spirophosphoranes seems to be determined by an inherent tendency to attain lower valence states and the not negligible ring strain of the dibenzophosphole systems. Thus, with acids and other electrophiles, spirophospho-ranes of type *25* react to give phosphonium salts *33* in which one of the di-benzophosphole units has been cleaved; removal of the phenyl group to yield the most important bis-2,2'-biphenylylenephosphonium ion *34* (5,5'-spirobi[5H-dibenzophos-pholium]) could never be achieved [42, 45].

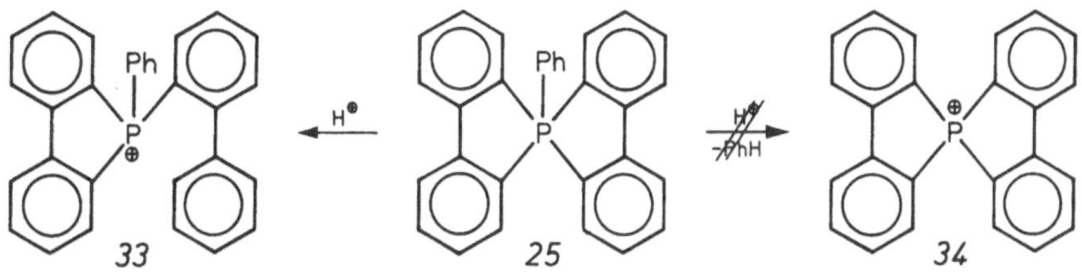

When the above spirophosphoranes are heated above their melting points rearrangements to isomeric phosphines occur [45].

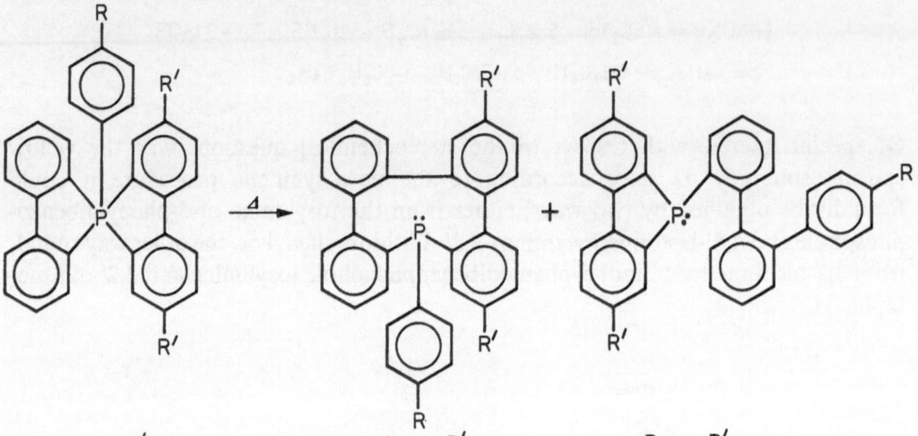

	R	R'		R	R'		R	R'
25	H	H	*35*	H	H	*38*	H	NMe$_2$
30	NMe$_2$	H	*36*	NMe$_2$	H			
31	H	NMe$_2$	*37*	H	NMe$_2$			

In a detailed study with tetramethyl substituted skeletons it could be shown, that the constitution of the isomers so formed depended characteristically on the size of the single fifth ligand, R. If R is relatively small (CH$_3$, C$_6$H$_5$, 3-(i-C$_3$H$_7$) C$_6$H$_5$) only 1-organyl-1H-tetrabenzo[b,d,f,h]phosphonines *40* are formed. With a medium-sized R (2-biphenylyl), in addition to *40*, 5-organyl-5H-dibenzophosphole *41* is formed, and with very spacious groups R (8-quinolinyl, 9-anthryl) only the latter phosphine type is isolated [47]. The isomerizations can be rationalized [47] in terms of cheletopic reactions involving allowed axial-axial and/or equatorial-equatorial bond rupture-bond regeneration mechanisms [48] with "aromatic" transition states [49].

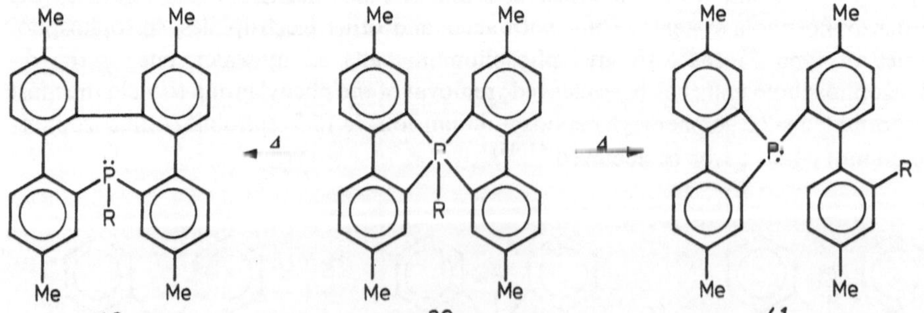

From a stereochemical point of view, tetrabenzophosphonine type systems *40* can be analyzed in terms of chiral bridged biphenyl systems, whose racemization barriers depend on the size of the single ligand R and whose racemization transition states closely correspond to pictorial representations such as *40* [47)].

Whereas the above constitutional rearrangements involve diminuation of the coordination number at phosphorus from 5 to 3, in special cases pentaorganyl-phosphoranes can also undergo constitutional isomerizations with perservation of coordination level five [50)]. When spirophosphonium salt *42-I* (see later) was treated with 8-(lithiooxy)-1-naphthyllithium in tetrahydrofurane, the unhydrolyzed reaction mixture showed a ^{31}P-NMR-signal at $\delta = -89$, indicating that the initially formed phosphoranyl-substituted phenolate *43* did not close the ring to the ate complex *44*.

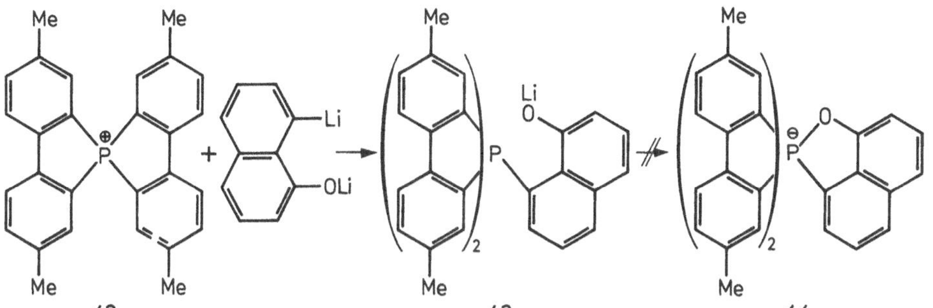

For the latter, on the basis of earlier investigations with mixed aryl-aryloxy-phosphates, e.g. *45*, one could expect a much more shielded phosphorus having a δ^{31}P of at least -150 [51)]. It should be noted, however, that when the underlying phosphorane precursor is substituted with more electronegative groups, such ate-complexes containing an additional five-membered ring are formed quite readily, e.g. *46* [52)] and *47* [53)].

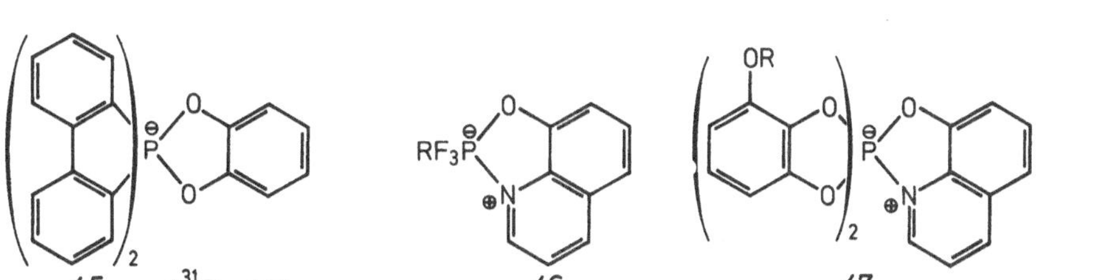

When hydrolyzed, *43* gave not the expected pentaarylphosphorane *48*, but instead a mixture of two isomeric phosphoranes *49*, *50* each containing a P—O bond. In the ^{1}H-NMR-spectrum, the four pairs of methyl singnals for *49* + *50* coalesce at about 70 °C to give four singlets, thus indicating reversible equilibration via rotation of the 2-biphenylyl group with an activation barrier of $\Delta G_c^{\neq} \sim 18(75)$ kcal(kJ)/mol [50)].

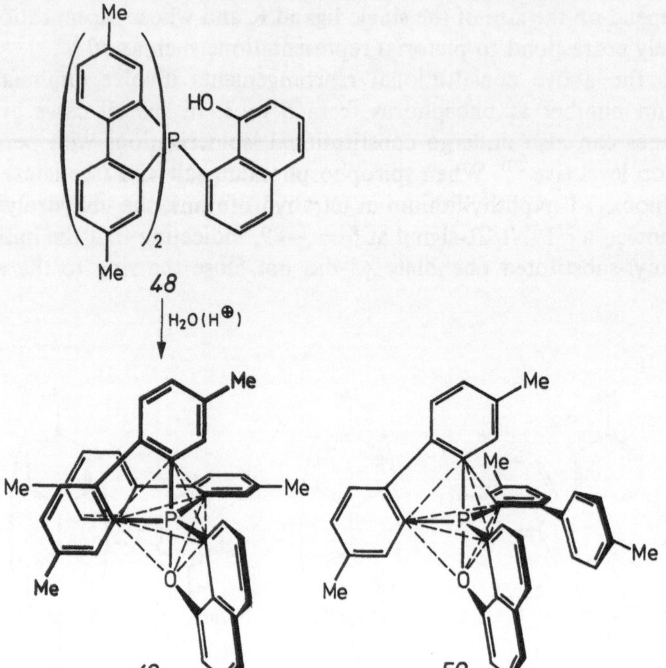

Obviously, the acidic phenolic group of 48 attacked here one of the slightly strained phosphafluorene systems under cleavage of a P—C and formation of a stronger P—O bond, which can be arranged very conveniently in an axial direction, satisfying thus the high apicophilicity of oxygen [54]. This reaction is an intramolecular equivalent of the decomposition of pentaphenylphosphorane with phenol, leading to phenoxytetraphenylphosphorane [24a].

A double rearrangement within coordination level five at phosphorus can be achieved with phosphoranes of type 51, containing the less acidic hydroxymethyl groups. Here, heating above 250 °C is needed [50] to produce the dialkoxytriaryl-phosphorane type 52 [55]. When only boiled in methanol, even addition of a few drops of concentrated hydrochloric acid leaves the original skeleton 51 unaffected. Obviously, the bulky 2,6-bis(hydroxymethyl)phenyl group blocks the PC$_5$ center efficiently against ring-cleaving attack by protons. [50]

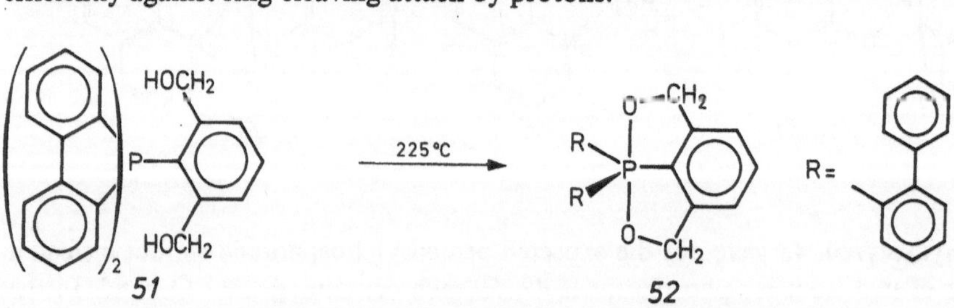

In nucleophilic substitutions with organyllithium compounds the spirocyclic skeleton of phosphorane type 25 remains essentially intact. In principle, such reactions can proceed either as one step S_N2- or two step addition-elimination-processes involving intermediate hexaccordinate ate-complexes 53 [56].

$$Ar_5P + RLi \quad \underset{\longleftrightarrow}{\overset{[Ar_5PR]^{\ominus} Li^{\oplus}}{\diagup \quad 53 \quad \diagdown}} \quad Ar_4PR + ArLi$$

In contradistinction to pentaphenylstiborane [26] and pentaphenylbismuthorane [30] and also other main group element aryls [57], pentaphenylphosphorane and phenyllithium never gave an isolable or at least directly observable hexaphenylphosphate complex $(C_6H_5)_6P^{\ominus}$. Only when steric congestion around the relatively small phosphorus center was considerably reduced, complexes of such a type could finally be isolated [58] as will be discussed in some detail later on. On the basis of ligand exchange experiments with pentaphenylphosphorane derivatives and alkyl-[46, 59] and aryllithium compounds [46, 59, 60] it was concluded, however, that hexaorganylphosphate complexes 53 could at any rate be short lived intermediates here. In this context it may also be noted that potassium amide in liquid ammonia at 120 °C displaces four phenyl groups from pentaphenylphosphorane to give benzene and a complex salt, $[C_6H_5P(= NK) (NHK)] NH$, still containing one P—C bond [61].

When analogous exchange reactions were then achieved with various spirophosphoranes of type 25 and 25', respectively, it was found that in all cases only the single fifth substituent R, R' was replaced and never any product stemming from the ring opened carbanion 54 was observed.

R = Alkyl , Aryl , Alkinyl
R'= Alkyl , Aryl

Evidently, under nucleophilic conditions the potential ring cleavage reaction $25 \rightleftharpoons 54$ is extremely reversible because the incipient carbanionic function of 54 can never totally leave the influential sphere of the phosphorus center [46, 59]. Again, no hexacoordinate ate-complexes 55 could be identified in these reactions.

All these chemical studies had shown correspondingly that bis-2,2'-biphenylyleneorganylphosphoranes were indeed considerably more stable than their simple pentaphenylphosphorane counterparts. However, all attempts to seperate them into optical antipodes were in vain. Consequently, all discussions pertinent to their stereochemistry rested on unsafe ground [45], although a crystal structure determination had shown, that, at least for solid pentaphenylphosphorane, the trigonal-bipyramidal model conception was valid [62,63]. Pentaphenylphosphorane crystallizes as an only slightly distorted trigonal bipyramid with mean axial P—C bond lengths of 1.987 and mean equatorial bond lengths of 1.850 Å. The axial C—P—C bond angle is 176.9° whereas the equatorial C—P—C angles are found to lie between 117.8° and 122.6°, and the axial-equatorial ones between 86.3 and 92.7° [63].

Since all conventional ways of obtaining optically active forms of spirophosphoranes 25 by direct or indirect resolution procedures had failed [42,45], other means for getting information on the stereochemistry of those compounds had to be looked for. One of them consisted in the detour over a field which yet had to be explored: the hexaarylphosphate complexes. As mentioned before, there existed only indirect evidence for the intermediacy of such ions in nucleophilic substitutions on penta-arylphosphoranes, and their obvious instability was attributed mainly to steric overcrowding. Therefore, similar to the pentaphenylphosphorane case, where the combination of each two phenyl groups to a 2,2'-biphenylylene ligand had led to perfectly stable spirophosphoranes of type 25, the attachment of three 2,2'-biphenylylene ligands to a phosphate center appeared very promising. And, indeed, on reacting a suspension of phosphorus pentachloride with 2,2'-dilithiobiphenyl in diethyl ether a yellow salt was obtained which was shown to be the fascinating bis-2,2'-biphenylylenephosphonium-tris-2,2'-biphenylylenephosphate (56) [58,64].

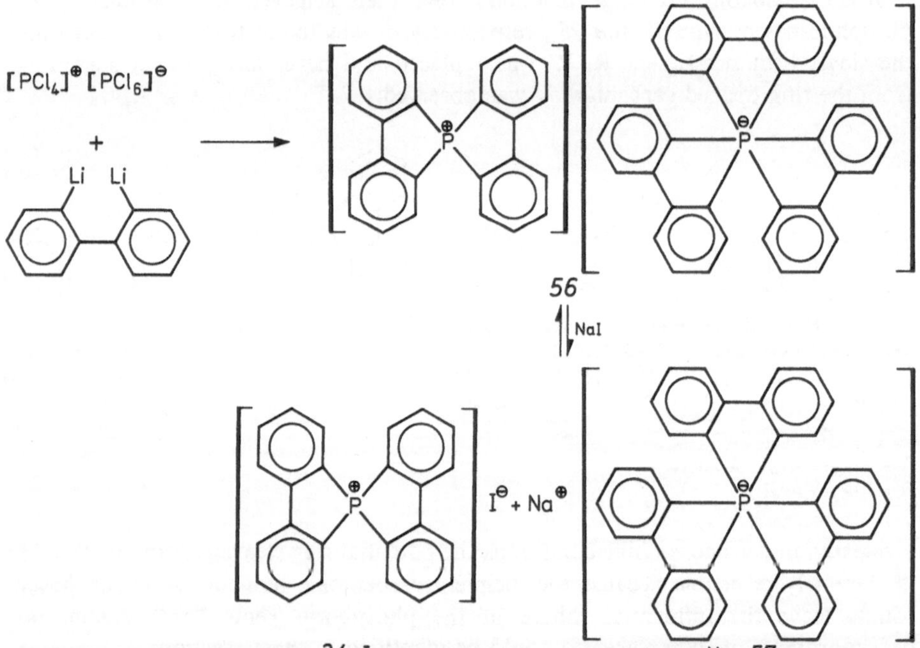

With excess sodium iodide, onium-ate complex 56 was readily cleaved into its component ions, which were essential for all further studies in this field. The preparative problems involved here became easily controllable when it was found that spirophosphonium iodide 34-I was generated in high yields by treating the reaction product of triphenylphosphate and 2,2'-dilithiobiphenyl with acid and potassium iodide. Further treatment of *34* with 2,2'-dilithiobiphenyl in tetrahydrofurane gave excellent yields of the lithium salt *Li-57* of the phosphate anion *57*.

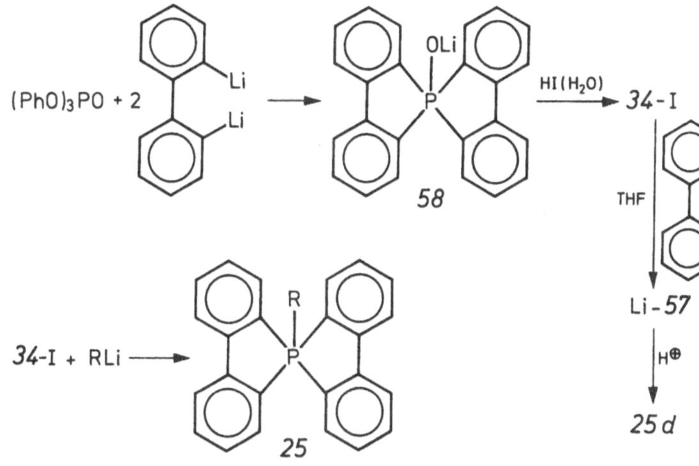

	a	*b*	*c*	*d*	*e*	*f*
R =	Me	Bu	Ph	2- Biph	1-Np	8-(Me$_2$N)-1- Np

Also, with monovalent lithio compounds, a variety of alkyl-, aryl- [58], alkynyl- [46, 59], alkenyl- and hetaryl-bis-2,2'-biphenylylenephosphoranes [65] *25* became very readily accessible.

On the basis of these findings a new level had been reached, from which a very efficient pursuit of the stereochemical and other questions pertinent to those classes of compounds became possible. In this respect, the key reaction consisted in the protolytic cleavage of phosphate ion *57* to give 2-biphenylyl-bis-2,2'-biphenylylene-phosphorane *25d*. On the assumption that *57* very probably had an octahedral structure, it was then hoped that the fate of the optical activity on protolytic cleavage of optically active forms of this ion would also enable some conclusions on the stereochemistry of the resulting phosphorane *25d* to be made. The resolution of *57* could readily be achieved with methylbrucinium iodide and finally yielded the optically active potassium salts *K-57* with $[\alpha]_{578}^{24} = \pm 1930$ and $[M]_{578}^{24} = \pm 10164°$ [66]. On the basis of CD and UV-spectra it could later be shown that the (—)-isomer had the M(C$_3$)-configuration of a left-handed screw [67]. An attempt to influence the further reaction of spirophosphonium ion *34* with 2,2'-dilithiobiphenyl in an asymmetric way (by starting from the onium-ate-complex *56* with an optically active phosphate part) was unsuccessful within the limits of experimental error [68].

As can be seen from the following scheme, cleavage of the optically active ate-complex *57* with protons should lead to a chiral, optically active phosphorane, *B*, provided this has a rigid trigonal-bipyramidal configuration.

17

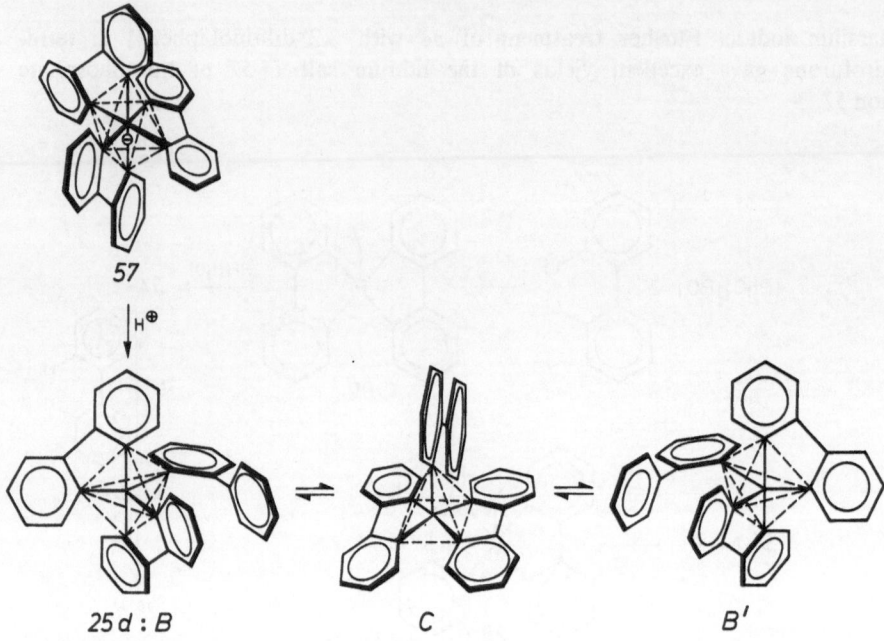

57

25 d : B   *C*   *B'*

When the experiment was performed, however, not the least trace of optical activity was found for the resulting phosphorane [66]. A similar result was obtained in the decomposition with bromine [68], where the optically inactive phosphorane *59* was formed.

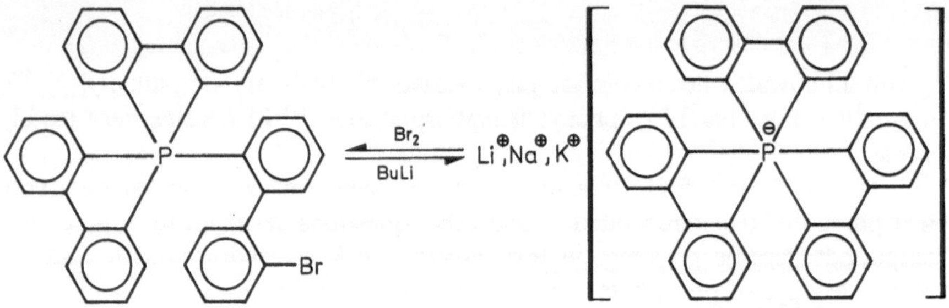

59
inactive

57
optically active

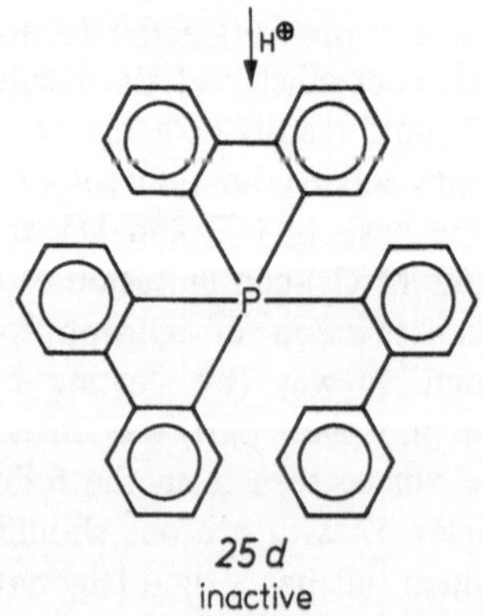

25 d
inactive

This could only be explained in terms of an intramolecular ligand reorganization process which led from B through a tetragonal-pyramidal transition state C to the enantiomer B' and therefore to racemization. Such a process, in which essentially the axial bonds of a trigonal bipyramid are transformed into equatorial ones and two equatorial bonds into two axial ones, was first proposed by Berry [69] to explain the apparent non-rigidity of phosphorus pentafluoride on the NMR-time scale, and soon accepted by others [70,71]. Fully analogous results were obtained with the hexamethyl substituted phosphate ion 60, which with acids again produced the inactive phosphorane 61 [72].

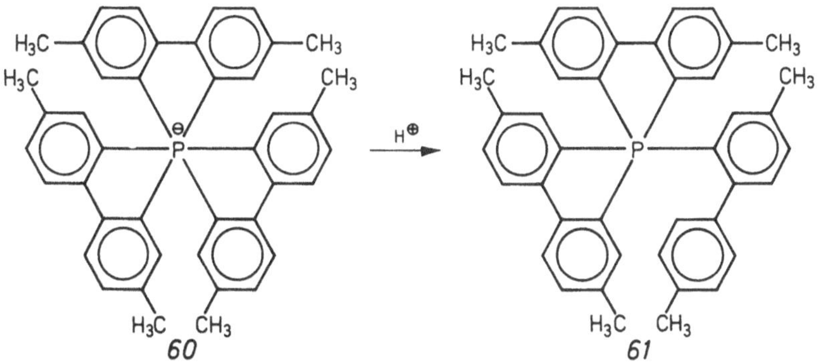

| 60 | 61 |
| optically active | inactive |

According to these findings it was clear that an optically active spirophosphorane of that type was only possible if at least one unsymmetrically substituted 2,2'-biphenylylene ligand was present. Only then, as is shown in the next scheme, the Berry process involving tetragonal-pyramidal transition state C will not lead to racemization, since now the trigonal bipyramids B, B' are no longer enantiomers but diastereomers.

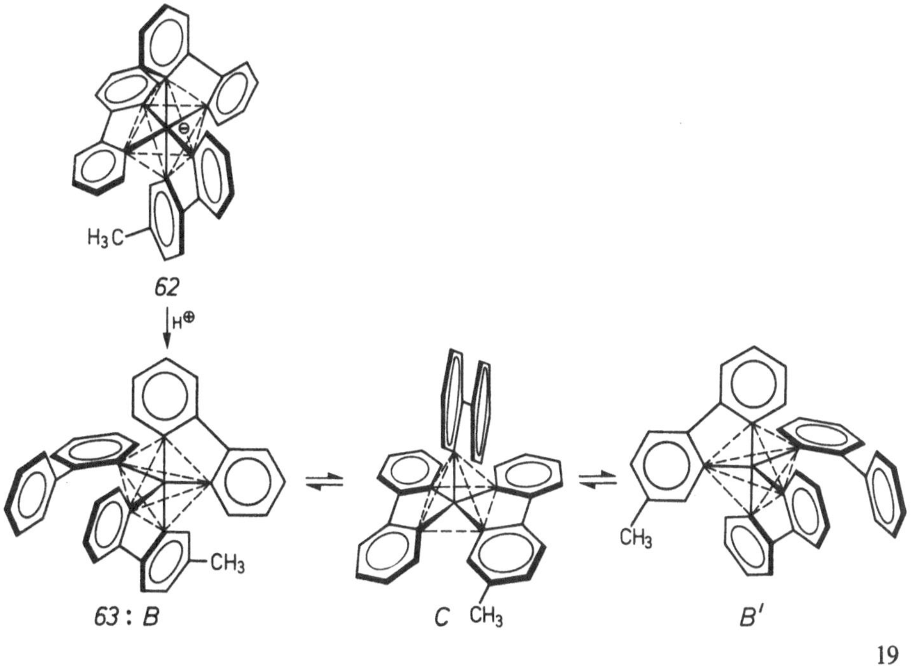

62

63 : B   C   B'

19

To verify this idea the methyl substituted phosphate complex *62* was synthesized and separated into its optical antipodes. The reaction with protons was not regiospecific but to a certain extent stereoselective, and the isolation of optically active phosphorane *63* was greatly helped by fortunate incursion of a spontaneous racemate separation [73].

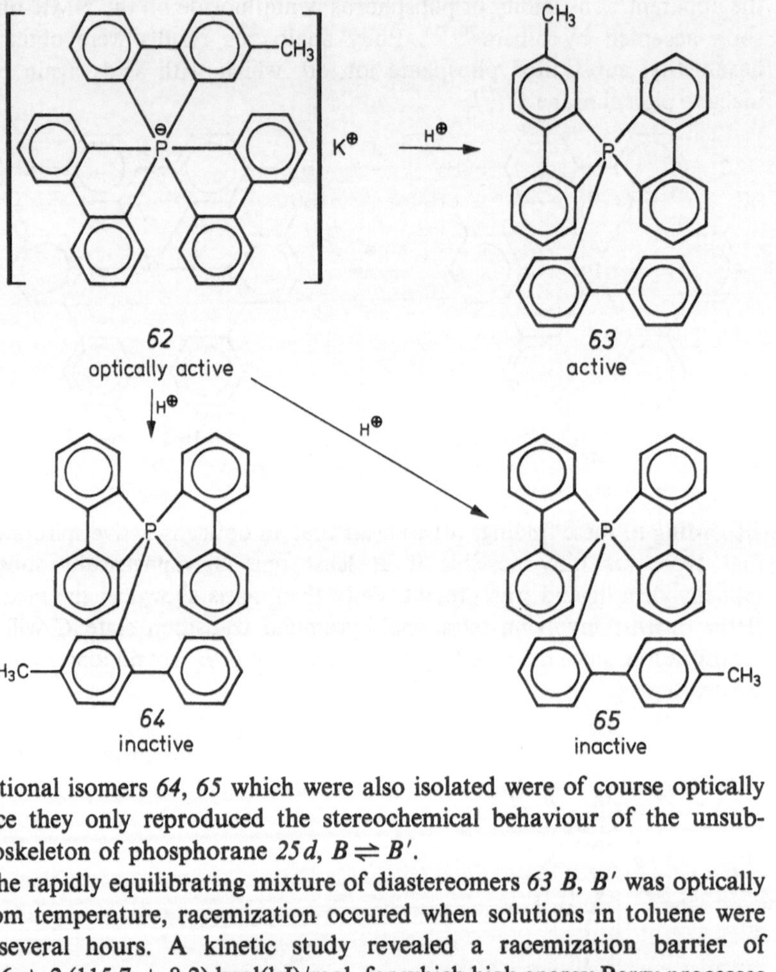

62
optically active

63
active

64
inactive

65
inactive

The constitutional isomers *64, 65* which were also isolated were of course optically inactive, since they only reproduced the stereochemical behaviour of the unsubstituted spiroskeleton of phosphorane *25 d*, $B \rightleftharpoons B'$.

Whereas the rapidly equilibrating mixture of diastereomers *63 B, B'* was optically stable at room temperature, racemization occured when solutions in toluene were heated for several hours. A kinetic study revealed a racemization barrier of $\Delta G_{298}^{\neq} = 27.6 \pm 2\,(115.7. \pm 8.2)$ kcal(kJ)/mol, for which high energy Berry processes running through the decisive, achiral trigonal-bipyramidal transition states *66, 67* with diequatorial 2,2'-biphenylylene ligands were made responsible [74]. Equivalent

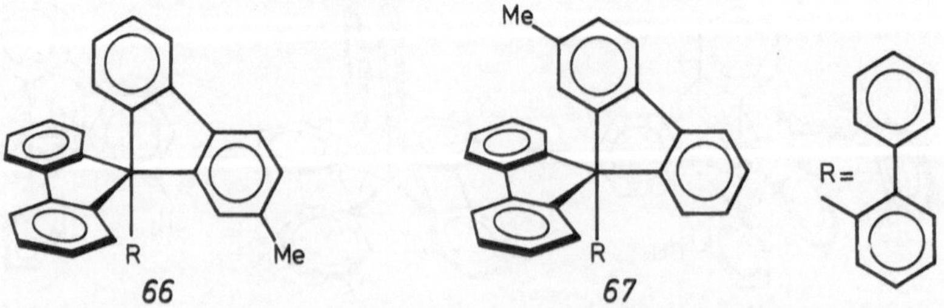

66

67

$R =$

transition states can be formulated in the framework of the turnstile rotation model [75], for which relative rotations of a trio of ligands formed by the unsubstituted 2,2′-biphenylylene group and the single ligand R (= 2-biphenylyl) against the ligand duo formed by the substituted 2,2′-biphenylylene group have to be assumed here [74]. At any rate, both modes of rearrangement are experimentally indifferentiable, but from theoretical considerations it follows that the Berry process has generally lower energy requirements than the turnstile rotation process [75, 76]. The fact that 63 still shows optical activity (although a certain intramolecular ligand exchange process, namely, the low energy Berry pseudorotation, is fully excited) represents another manifestation of the general principles of "*residual stereoisomerism*" which were established in the course of stereochemical studies with orthosubstituted triphenylmethane derivatives [76a] and subsequently applied to analogous triphenylamine derivatives [76b]. Depending on the substitution pattern, such compounds can exhibit "*residual diastereotopism*", "*residual diastereoisomerism*" and "*residual enantiomerism*" when only a limited set of ring flip processes is operating.

Accordingly for 63, residual stereoisomerism expresses itself as residual enantiomerism — observable here in the form of "*residual optical activity*" — as long as the intramolecular motions are restricted to the low energy Berry process. However, once the high energy Berry modes, passing through conformations 66 and 67, come into existence, any residual stereoisomerism will be annihilated, concomitant, of course, with total racemization.

Parallel to the polarimetric studies, ¹H-NMR-investigations were initiated which were intended particularly to provide information pertinent to the lower energy Berry processes. The first experiments with penta-p-tolylphosphorane were undecisive since they gave only one single methyl signal even at —60 °C [77]. More conclusive results were obtained for a series of bis-(4,4′-dimethyl-2,2′-biphenylylene)-organylphosphoranes 39 with widely varying groups R, which were easily accessible by reactions of the spirophosphoniumsalt 42-I with the corresponding lithium

a	*b*	*c*	*d*	*e*	*f*	*g*	*h*	*i*
R = Me	Et	PhCH₂	2-Biph	2-Np	1-Np			

21

organyls. Here, depending on the nature of R, at lower temperatures two to four methyl resonances were observed which at higher temperatures coalesced to a single line. This again was entirely compatible with the usual trigonal-bipyramidal ground state structures undergoing intramolecular ligand exchanges at elevated temperatures according to the Berry pseudorotation mechanism [78,79]. The barriers for these processes, obtained as usual by the coalescence temperature approximation [80], were 12–13 (50–54.5) kcal (kJ)/mol when R was a simple alkyl or aryl group, and 15–21 (63–88) kcal (kJ)/mol when R was a sterically more demanding 2-biphenylyl or 8-substituted 1-naphthyl group. For the latter compound types total equilibration of the four different methyl positions can only occur when pseudorotation of the spirophosphorane skeleton and rotation of the axially unsymmetrical groups R are intimatly coupled if not synchronous processes involving two different tetragonal-pyramidal transition states *68, 69* as is shown for the 2-biphenylyl case *39d* [79].

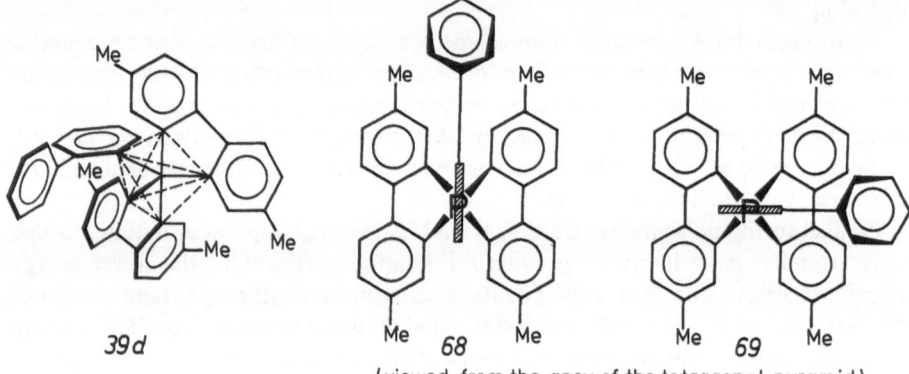

39d　　　　　　　*68*　　　　　　　*69*

(viewed from the apex of the tetragonal pyramid)

Similar conclusions were drawn from detailed ^{1}H-NMR-analyses of the dynamic behaviour of the isopropylphenyl derivatives *39h, i* where even the (very improbable) possibility of a tetragonal-pyramidal *intermediate* with freely rotating isopropylphenyl substituents was taken into consideration [81].

When in such spirophosphoranes another benzo group is annelated across the 3,4-positions of one of the 2,2'-biphenylylene groups, as in *70*, the usual Berry pseudorotation over the now very overcrowded tetragonal-pyramidal transition state *71* should be strongly hindered, if not strictly forbidden. Nevertheless, coalescence of the two methyl signals of *70* was observed at 89 °C, corresponding to an activation barrier of $\Delta G_c^{\neq} = 18.2$ (76.0) kcal (kJ)/mol [82]. This can be satisfactorily explained with a Berry process leading now via the tetragonal pyramids *72, 72'* through the strained trigonal bipyramid *73* with a diequatorial 2,2'-biphenylylene group which here marks most probably the transition state.

That the observed barrier is significantly higher than that for comparable spirophosphoranes of type *25, 39* [$\Delta G_c^{\neq} \sim 12$–13 (50–54.5) kcal (kJ)/mol] is understandable in view of the considerably congested intermediate or transitionary conformations *72, 73, 72'*.

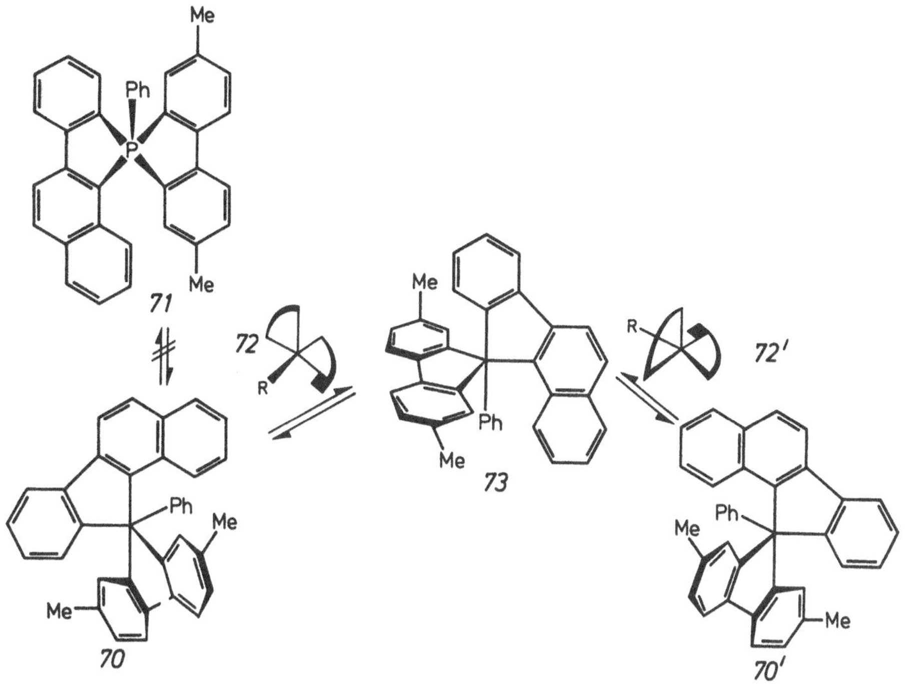

5 Crystal Structures of Spirocyclic Pentaorganylphosphoranes

Very recently Holmes et al. conducted crystal structure determinations for some typical representatives of the bis-2,2'-biphenylylene-organylphosphorane type, namely 25a, c [83] and 25e, f [84], which in principle confirmed but also supplemented the conclusions drawn earlier from solution studies.

Grossly viewed, all four phosphoranes investigated adhere to the trigonal-bipyramidal structure pattern, although several more or less significant deviations therefrom may be noted, depending on the reference features considered. The most prominent characteristics of a trigonal-bipyramidal arrangement of ligands around a central atom are undoubtedly the two longer coaxial bonds and the three shorter equatorial bonds which span a trigonal plane orthogonal to the axis [85]. If one concentrates on and compares just these parameters for the above spirophosphoranes (see Table 1), no doubt about their overall trigonal-bipyramidal shape remains. In all cases, though somewhat less pronounced for the 8-dimethylamino-1-naphthyl compound 25f, two longer axial P—C bonds are clearly discernible from three shorter equatorial bonds. Also, the axial C—P—C bond angles do not deviate substantially from the ideal 180° value.

For the equatorial C—P—C bond angles, however, larger deviations from their ideal 120° values are found. Whether these alone, without concomitant comparable alterations of the axial C—P—C angles, can justify the definition of a gradual distortion towards a specific tetragonal-pyramidal structure might be questioned. On the basis of a proper evaluation of more latent structural features, as done in

Table 1. Significant structural data for Spirophosphoranes *25a, b, d, e* as related to a trigonal bipyramidal structure

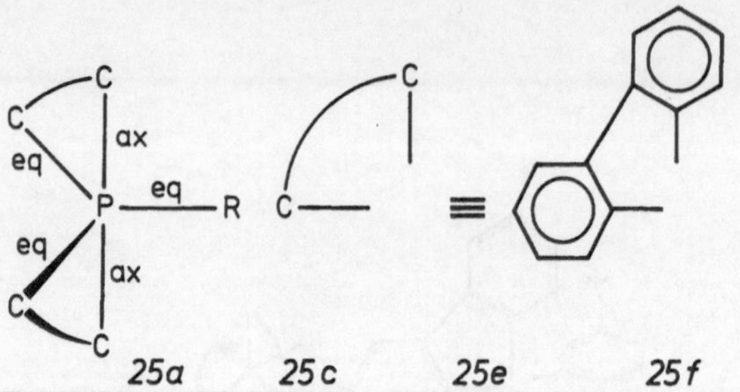

	25a	25c	25e	25f
R	Me	Ph	1 - Np	8-Me$_2$N-1-Np
P－C$_{ax}$	193.7 pm 194.1	193.4 pm 193.4	191.0 pm 193.6	194.1 pm 190.9
P－C$_{eq}$	186.0 186.4	186.2 186.2	185.2 187.5	184.6 188.6
P－R$_{eq}$	185.9	185.9	186.6	189.2
∢C$_{ax}$PC$_{ax}$	178.3°	179.3°	177.4°	175.5°
∢C$_{eq}$PC$_{eq}$	111.0°	111.5°	110.9°	101.8°
∢C$_{eq}$PR$_{eq}$	123.5° 125.6°	124.3° 124.3°	116.4° 132.6°	106.1° 151.9°

the dihedral angle method [86], such a discussion can become more meaningful. Here, the sum of the changes of the dihedral angles (as defined by the angles between the normals of the polytopal faces sharing a common edge) in going from the actual structure toward an idealized tetragonal pyramid or trigonal bipyramid, respectively, is translated into a percentual transition from the latter into the former. In a sense, this corresponds to an averaging of a more significant number of coordinates considered descriptive for a Berry pseudorotation mode.

In such a way, for spirophosphoranes *25a, c, e, f* gradual distortions of trigonal-bipyramidal towards (different kinds of) tetragonal-pyramidal structures of −16.9, −15, 25 and 64%, respectively, have been derived. Interestingly, for *25e*, and even more pronounced for *25f*, the tetragonal pyramids aimed at are of the unusual type containing the single ligand R not in the apical but in an basal position, thus leaving one of the 2,2'-biphenylylene groups with an apical-basal arrangement. From such a conformation, a continuous path to a trigonal bipyramid with one diequatorial 2,2'-biphenylylene group is easily imaginable, thus supporting earlier views that for sterically highly congested cases intramolecular ligand exchange may well take place through transient conformations with just that type of geometry [81,82,86].

After all, there remains the fact that in all four spirophosphoranes *25a, c, e, f* the axial C－P－C angle of the underlying trigonal-bipyramidal structure does not deviate significantly from its ideal 180° value. Hence, the assignment of a super-imposed percentual tetragonal-pyramidal character to a basically trigonal-bipyramidal skeleton still appears somewhat arbitrary. This can be simply visualized by displaying one and the same molecular graph of, for example *25f*, under different viewing angles; clearly, the left presentation insinuates a trigonal-bipyramidal, the right one a tetragonal-pyramidal structure, which sometimes is also named a square pyramid or a rectangular pyramid.

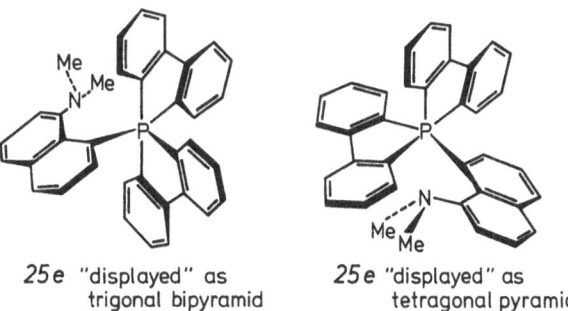

| 25e "displayed" as trigonal bipyramid | 25e "displayed" as tetragonal pyramid |

In passing, it might be noted that earlier predictions about structural details of such spirophosphoranes (such as the necessary pyramidality of the nitrogen atom in *25f*[79]) or the size of the intraannular angles of the phosphafluorene system, ~86° for C—P—C and ~113° for C—C—C, respectively [28,30a,66]) have been fully con-firmed by the present x-ray studies of Holmes and his group.

In the light of the above structure investigations and together with the vast amount of structural information obtained with other phosphorane types, it can be concluded that pentacoordinated phosphoranes in general prefer a trigonal-bipyramidal ground state geometry and that only in special cases can tetragonal-pyramidal characteristics attain predominance [86–89]. All attempts to realize other ground state geometries, namely of the turnstile type [75], with the help of specifically designed tridentate ligand systems have so far met with failure. Thus, even phosphoranes containing a trioxaadamantane- (74, 75)[90] or a still more rigid azaphosphatriptycene skeleton (76)[91,92] have to be described in terms of trigonal bipyramids which are not substantially distorted.

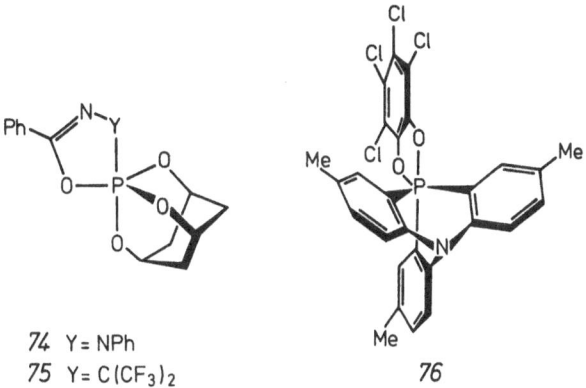

74 Y = NPh
75 Y = C(CF₃)₂

76

6 Bis-2,2'-biphenylylenehydridophosphoranes

The easy accessibility of all kinds of stable bis-2,2'-biphenylylene-organylphosphoranes by the spirosalt method led quite naturally to the question of the potential viability of bis-2,2'-biphenylylene-hydridophosphorane (*77*) whose simpler prototype, hydrido-tetraphenylphosphorane $(C_6H_5)_4PH$, could never be prepared [93]. However, reaction of spirophosphonium salt *34-I* with lithium aluminum hydride in ether or sodium borohydride in ethanol under carefully controlled conditions in fact gave good yields of crystalline phosphorane *77* which was fairly stable when kept as a solid under nitrogen [94]. In solution, on the other hand, a very unusual spontaneous dissociation into the violet bis-2,2'-biphenylylenephosphoranyl radical *78a* and hydrogen could be observed, which at room temperature took some 25 hours to produce the maximum radical concentration. The obvious weakness of the P—H bond in *77* was also reflected in its somewhat eccentric spectroscopical behaviour, indicated by a very lowfield chemical shift, ($\delta = 9.33$), small coupling constant $(J_{PH} = 482\ Hz)$ and low stretching frequency $(\nu_{PH} = 2096\ cm^{-1})$ for the proton bound to phosphorus.

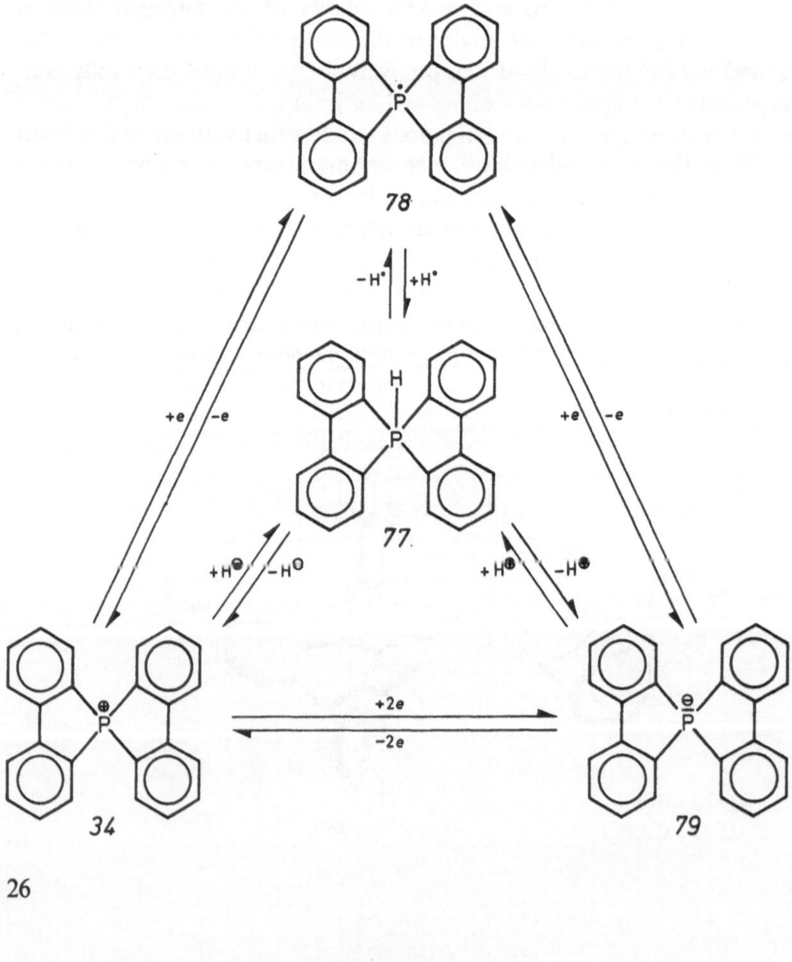

With strong bases, e.g. t-butyllithium, *77* loses a proton to give the phosphoranyl anion *79*; strong electrophiles $(H^\oplus, X^\oplus, CH_3I)$ remove a hydride ion to regenerate spirophosphonium ion *34*. The general chemistry of *77* and of its tetramethyl derivative bis-(4,4'-dimethyl-2,2'-biphenylylene)-hydridophosphorane [95)] as well can be abstracted according to the foregoing scheme, some points, though, needing special consideration.

First, radical *78* in inert solvents and under nitrogen decays only slowly during several weeks to give eventually the P—C dimer *80* and the ring-opened phosphine *81*. Interestingly, in the mass spectrum of dimer *80* the doubly charged parent ion is much more intense than the corresponding singly-charged ion. This can be explained if one assumes rapid decomposition of the singly-charged species, which still has the original constitution, and rearrangement of the doubly-charged ion to a significantly more stable isomer *83* [96)].

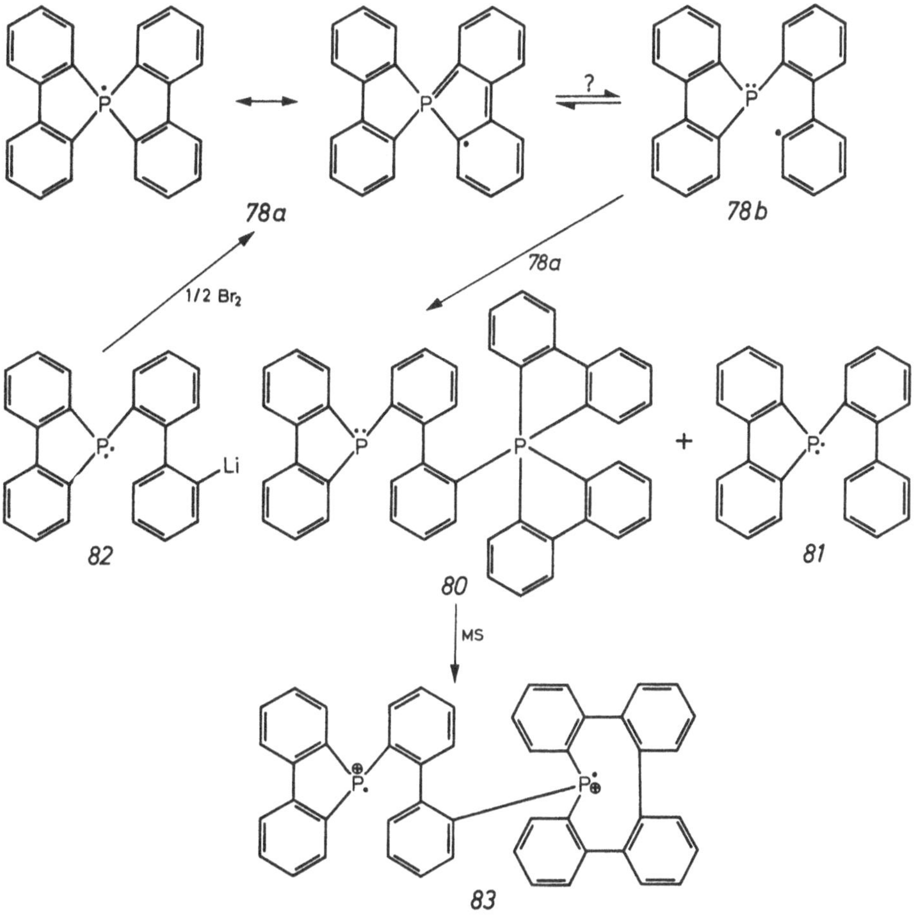

The formation of phosphine *81* and dimer *80* in the decomposition of *77* does not necessarily prove a radical equilibrium *78a*⇌*78b*, although phosphoranyl radical *78a* was also accessible by treating the open ring lithium compound *82* with less than equivalent amounts of bromine [97)]. Excess bromine leads to spirophosphonium salt *34-Br* which can again be reduced to radical *78* with potassium or sodium metal or electrolytically. The investigation of a series of substituted spirophosphoranyl radicals of that sort containing in all four 4,4'- [95)] or 5,5'-positions [98)] methyl, dimethylamino, or nitro groups revealed relatively small phosphorus coupling constants (a_p) between 17 and 25 Gauß, indicating a tetrahedral structure with the single electron delocalized mainly over the 2,2'-biphenylylene groups [94,95,98]. Not entirely unexpected, radical 78 is also formed in reactions of spirosalt 34-I with bulky lithium or sodium alkyls or substituted amides, whereas with sodium amide the unstable amino-bis-2,2'-biphenylylenephosphorane 84 is obtained [99].

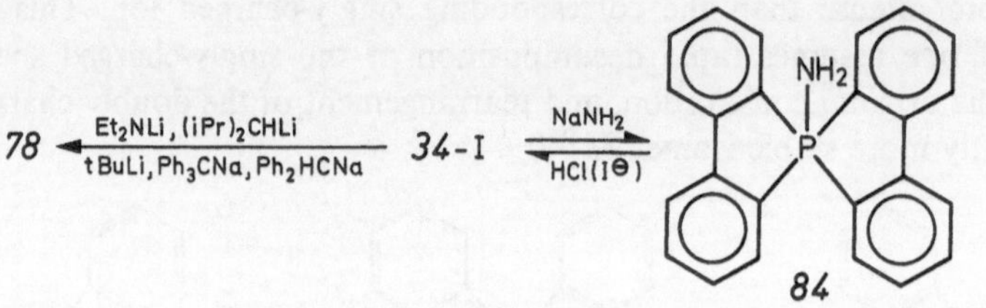

$$78 \xleftarrow[\text{t BuLi, Ph}_3\text{CNa, Ph}_2\text{HCNa}]{\text{Et}_2\text{NLi, (iPr)}_2\text{CHLi}} 34\text{-I} \xrightarrow[\text{HCl(I}^\ominus\text{)}]{\text{NaNH}_2} 84$$

In this connection it is most interesting that radical 78 is also generated when an etheral solution of 2,2'-dilithiobiphenyl is added to a similar solution of phosphorus trichloride at room temperature [100]. Even more surprising here was the fact that this reaction eventually led to the onium-ate complex 56, formerly only accessible from phosphorus compounds with oxidation level +5 [58,66,72,73].

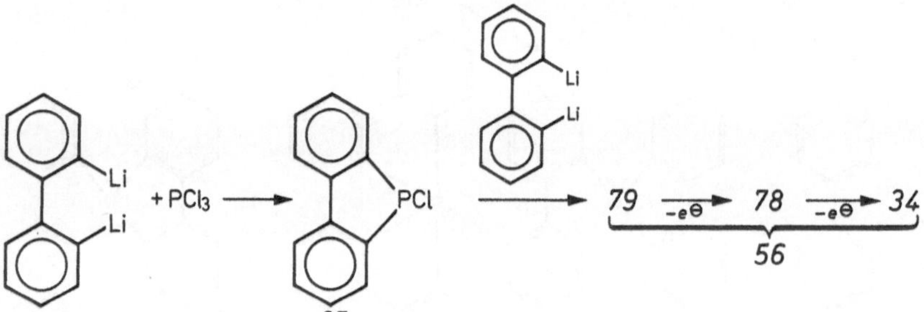

$$85 \quad + \text{PCl}_3 \longrightarrow \quad \text{PCl} \longrightarrow 79 \xrightarrow{-e^\ominus} 78 \xrightarrow{-e^\ominus} 34$$
$$\underbrace{\phantom{79 \xrightarrow{-e^\ominus} 78 \xrightarrow{-e^\ominus} 34}}_{56}$$

Although no explanation was given for the spontaneous, unexpected oxidation steps 79→78a and 78a→34, it is reasonable to assume that the initially present trivalent phosphorus compounds (phosphorus trichloride and 5-chloro-5H-dibenzo-phosphole 85, respectively), and/or the alkyl halide stemming from the halogen metal exchange reaction used for preparing 2,2'-dilithiobiphenyl, may play an active role here.

A second peculiarity of 77 expresses itself in the reaction with n-butyllithium in which the hydrogen atom at the phosphorus atom is displaced nucleophilically as hydride ion to give butyl-bis-2,2'-biphenylylenephosphorane (25b). Deprotonation to the phosphoranyl anion 79 occurs only to a minor extent here. For 79, an equilibrium with the ring-opened carbanionic structure 86 was proven [94,95,97].

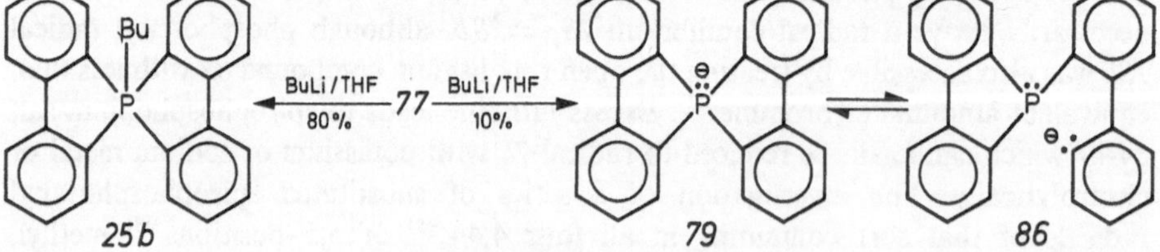

$$25b \xleftarrow[80\%]{\text{BuLi/THF}} 77 \xrightarrow[10\%]{\text{BuLi/THF}} 79 \rightleftharpoons 86$$

Whereas diethyl ether solutions of phosphoranyllithium *Li-79* are intensely blue-green at temperatures below −40 °C, above this temperature they become red [94]. This can be explained by transition from loose, solvent-separated ion pairs to more tightly-bound contact ion pairs with increasing temperature. In the better coordinating solvent tetrahydrofurane, only the blue-green color of the solvent-separated ion pairs is observable at all temperatures [101]. Similar tetraarylphosphoranyl anions have been formulated as unstable intermediates in nucleophilic exchange reactions on triphenylphosphine and its higher homologues with p-tolyllithium [102].

7 A Nitrogen Interlude

After it had thus been shown that with the help of the 2,2′-biphenylylene ligand system various types of hitherto unstable or even unknown phosphorus compounds could be obtained as remarkably persistent species, a resumption of the search for pentacoordinated nitrogen became unavoidable. This was all the more so, because in the meantime the restrictions implied by the octet principle had been somewhat mitigated in the light of newer bonding theories. Thus, for the description of the bonding situation in such "hypervalent" molecules [103] as discussed in this work, a more flexible bonding model, based on delocalized multicenter molecular orbitals and involving essentially only *s* and *p* base functions, became generally accepted [48, 75, 76, 103], and was in principle also applicable to pentacoordinate nitrogen compounds.

The crucial starting material for investigations in the nitrogen series, bis-2,2′-biphenylylenammonium iodide (*88-I*) (spiro[9,9′-bicarbazolium] iodide), was prepared analogously to the corresponding 9,9-diphenylcarbazolium salt [104, 105] by an intramolecular *N*-arylation via diazonium salt *87* [106]. As a byproduct, dibenz-[4,5:6,7]azepino[1,2,3-jk]carbazole (*89*) is formed here by attack of the phenylium ion formed in the decomposition of the diazonium ion at the π-system of the carbazole part of the molecule.

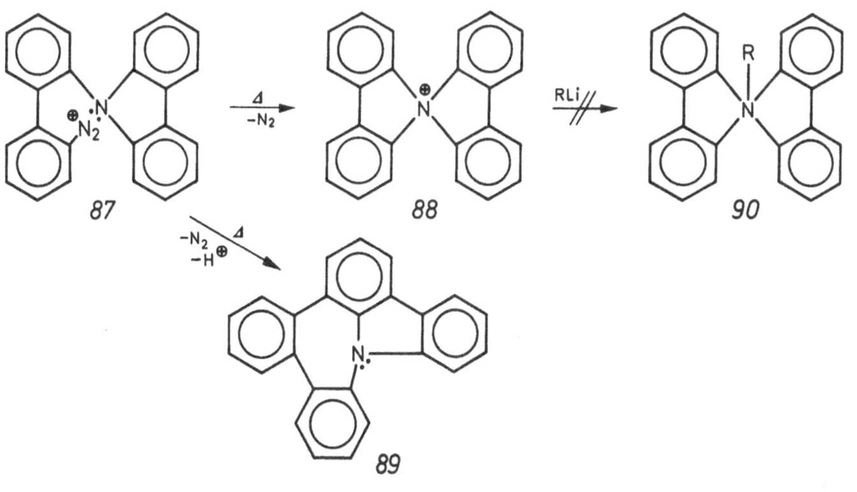

29

When *88-I* was treated with phenyl- or methyllithium, no compound of type *90* containing pentacoordinated nitrogen could be obtained. The isolated products, however, could reasonably be attributed to a dehydrobenzene reaction sequence initiated by nucleophilic attack at a hydrogen atom ortho to the ammonium center [107]. Elimination of triarylamine [108] from *91*, here in the form of an aryl-carbazole, led to the substituted aryne *92*, which added the nucleophiles present to give *93a*, *94a* and, after hydrolysis, *93b*, *94b*, respectively.

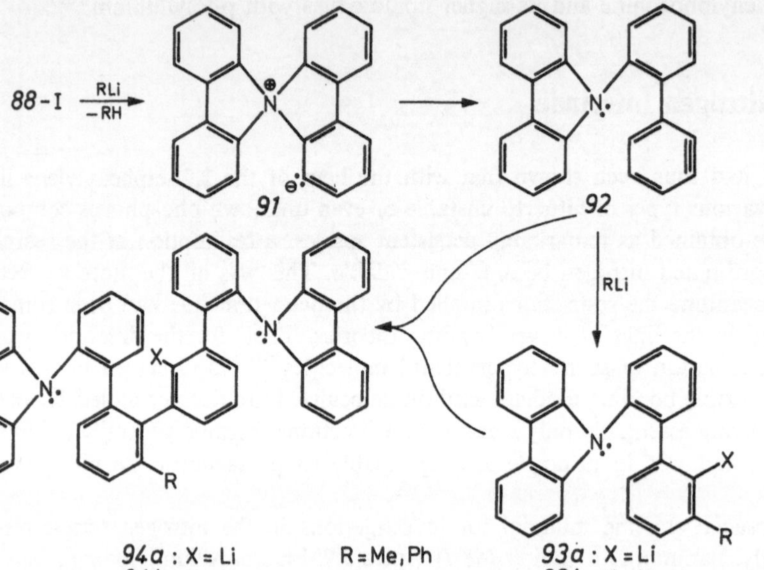

88-I $\xrightarrow[-RH]{RLi}$

91 *92*

94a : X = Li   R = Me, Ph   *93a* : X = Li
94b : X = H   *93b* : X = H

Once again, an initially very promising idea of constructing molecules containing a pentacoordinated nitrogen center had thus been disproved experimentally, the reason of this failure being essentially the same as in previous cases, namely, trapping of the incoming nucleophile by protons that were too acidic.

8 Alkylphosphoranes

Whereas the reaction of tetraphenylphosphonium salts with methyllithium had yielded solely degradation products [33,109], the analogous reaction of bis-2,2'-bi-phenylylenephosphonium ion *34* with methyl- and n-butyllithium produced the first stable phosphoranes *25a, b* containing an alkyl group [58].

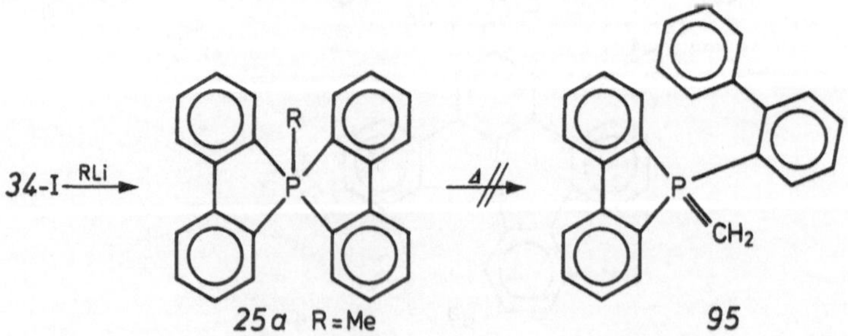

34-I $\xrightarrow{RLi}$

25a R = Me
25b R = Bu

95

Apparently, the spirocyclic skeleton which can very conveniently accomodate to the angle requirements of the trigonal-bipyramidal phosphorane structure [28,66) precludes here the potential decomposition to the tetrahedral ylide *95*, which would be much more strained.

Favourable strain factors of this sort were then explicitly utilized by Katz and Turnblom in the syntheses of a series of di- to pentaalkylphosphoranes *97* based on the phosphahomocubane system [110). The starting phosphonium salts *96* were

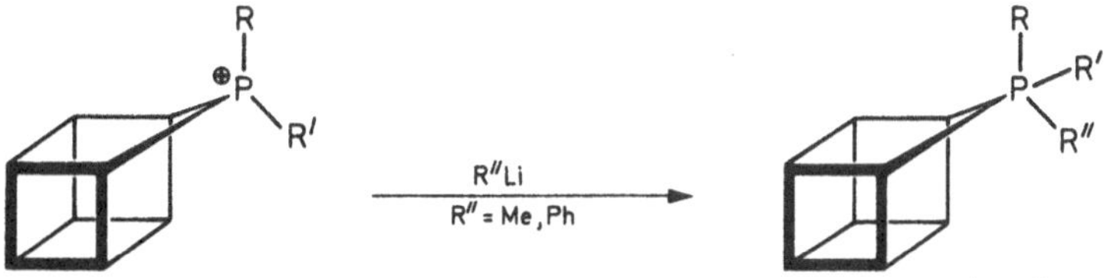

	R	R'
96a	Ph	Ph
b	Ph	Me
c	Me	Me

	R	R'	R"
97a	Ph	Ph	Me
b	Ph	Me	Me
c	Me	Me	Me
d	Ph	Ph	Ph

prepared in very elegant reaction sequences originating from cyclooctatetraene dianion.

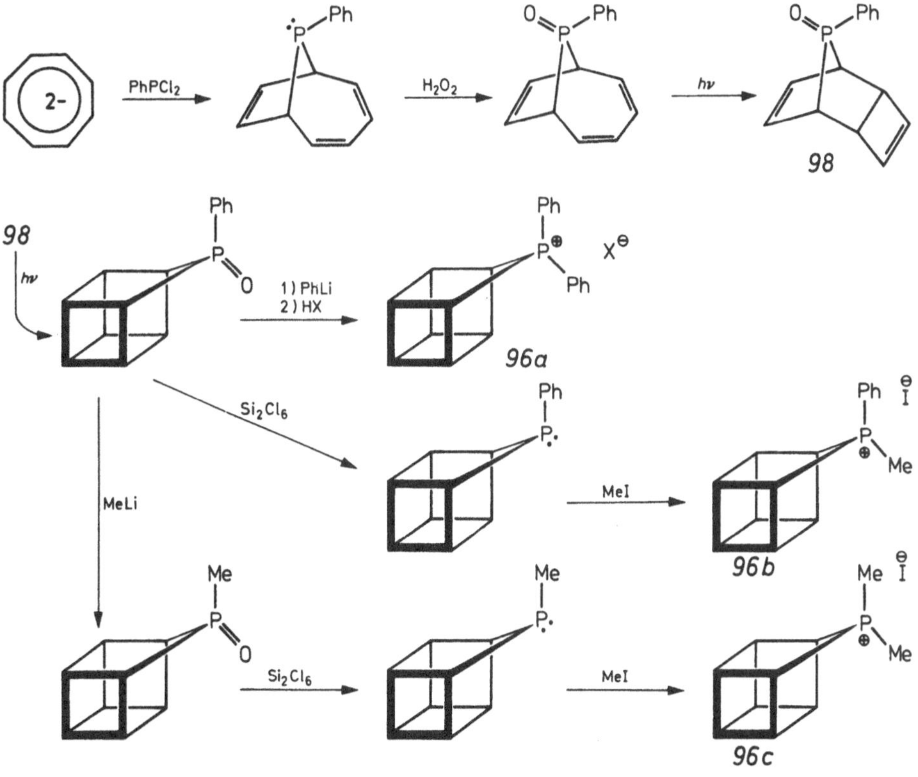

The alkylphosphoranes *97a–c* are remarkably stable but on heating can be cleaved to the corresponding phosphines *99* and cyclooctatetraene. Interestingly, the first known pentaalkylphosphorane *97c* is the most stable compound in this series, having a half life $t_{1/2}$ of 108 h at 75 °C. [110b, d]

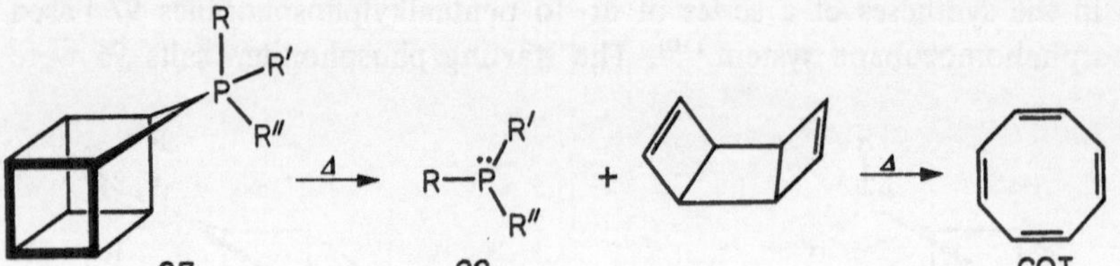

A comparable, bridged dialkylphosphorane *100*, derived from skeleton *98*, proved much less stable in that type of thermolysis, showing a $t_{1/2}$-value of only 0.6 h at 75 °C [110c, d].

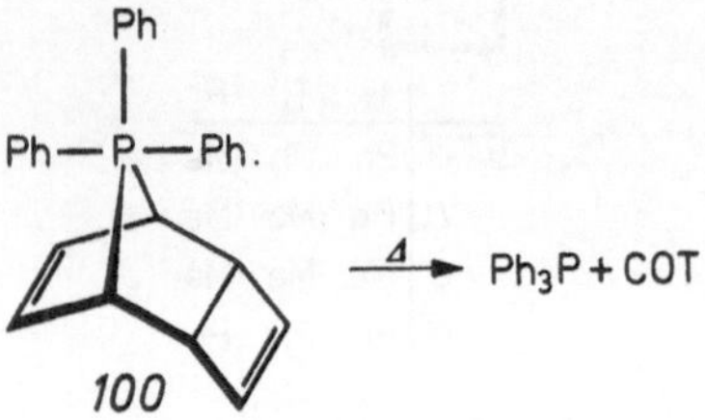

In the ¹H-NMR-spectra of the dimethyl derivative *97b* and the trimethyl derivative *97c*, significant methyl signal broadening is only observable at −184 °C. From this a ligand exchange barrier between two identical trigonal bipyramids of about 5 (21) kcal (kJ)/mol was derived [111].

In a similar attempt to transform the polycyclic phosphonium salt *101* into a phosphorane structure, only the decomposition products norbornadiene and tri-methylphosphine of the presumed intermediate pentaalkylphosphorane *102* could be isolated [112].

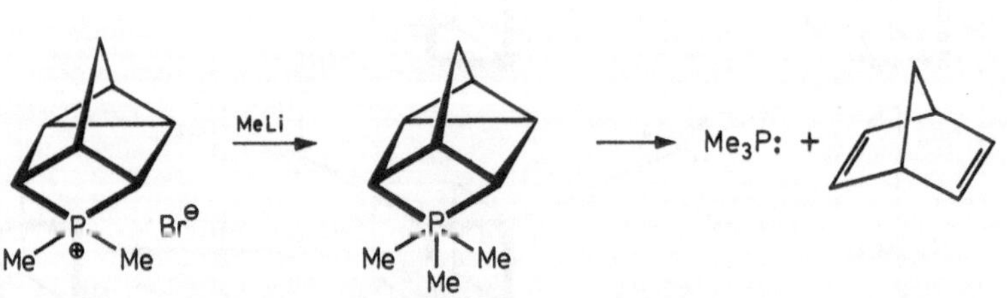

Combination of the stabilizing factors of a 2,2'-biphenylylene ligand with those of the phosphahomocubane system was achieved in the form of the mixed-skeleton phosphoranes *103*, synthesized according to the foregoing principles [113]. A hexa-coordinated structure, however, including the phosphahomocubane system could not be realized here [113].

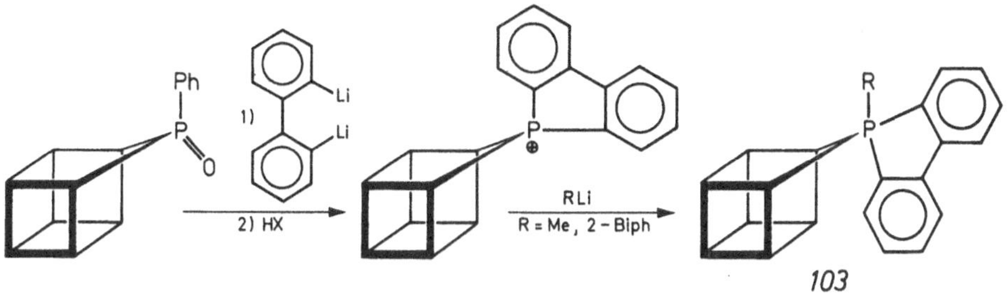

103

In their search for less complex pentaalkylphosphoranes [114], Schmidbaur and coworkers could provide convincing evidence for the intermediate occurence of monocyclic pentaalkylphosphoranes of type *104* in reactions of phosphorus ylids with silacyclobutanes. Isolation of such phosphoranes *104* was, however, never possible [115].

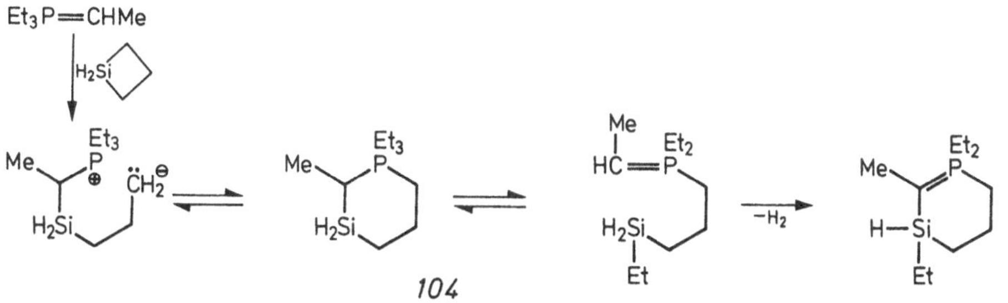

104

More successful was finally the reaction of 5-phosphoniaspiro[4,4]nonane chloride *105* with methyllithium at low temperatures which did yield, as hoped, 5-methyl-$5\lambda^5$-phosphaspiro[4.4]nonane *106* [116] which, in a way, can be considered as the fundamental skeleton of the alkyl-bis-2,2'-biphenylylenephosphoranes *25* investigated earlier [58]. Phosphorane *106* is a distillable liquid, which can be crystallized at −40 °C. Stereochemical assignments on the basis of NMR-data, though, were not yet possible.

Progress in the direction of still simpler, non-cyclic, pentaalkylphosphoranes could so far only be made in the case of some species bearing trifluoromethyl groups, e.g. dimethyl-tris-trifluoromethyl- and trimethyl-bis-trifluoromethylphosphorane *107*, *108*, for which, on the basis of detailed NMR-investigations, trigonal-bipyramidal structures with axial trifluoromethyl groups have been assigned [117]. For cyano-

(methyl)tris-trifluormethylphosphorane *109* only equivalent trifluoromethyl signals could be observed [117b)].

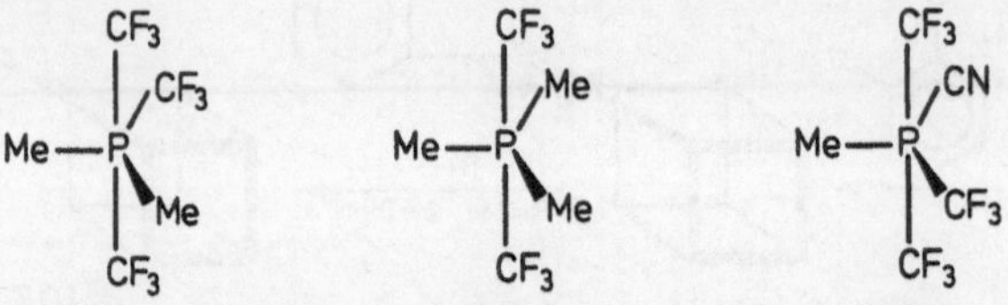

For other phosphoranes with strongly electron-withdrawing carbon groups, only vague structural descriptions are given. A dark red 1:1 adduct *110*, obtained from triphenylphosphine and tetracyanoethylene, has been described as being five co-ordinated with transfer of charge to the TCNE moieties [118)].

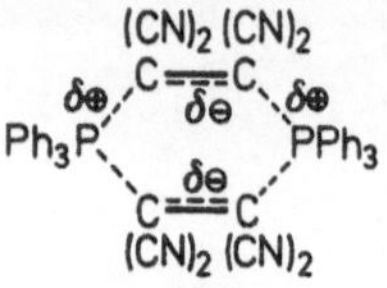

When methyl- and phenylphosphines *111* are treated with hexafluoro-1,3-butadiene in methylene chloride, solutions are obtained whose ^{31}P-NMR resonance signals (roughly between $\delta = -60$ and -80) indicate at least initial formation of the phosphorane mixtures *112, 113* [119)]. The only isolable products here were triorganyl-difluorophosphoranes.

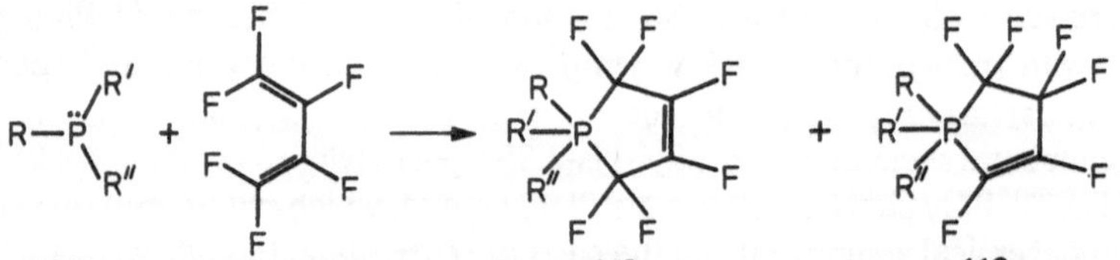

Before concluding the phosphorane chapters, a word has to be said concerning the ^{31}P-NMR spectra of pentaorganylphosphoranes which, at times, can be a very efficient diagnostic tool here. If the field of pentacoordinated phosphoranes as a whole is considered, ^{31}P-shifts are not very characteristic, reaching from com-paratively low phosphine-like or even phosphonium salt-like (de)shieldings to un-expected high shieldings as for bis-2,2'-biphenylylene-hydridophosphorane *77* (δ^{31}P $= -112$ [94)]) [120)]. For phosphoranes containing only P—C bonds, however, a quite characteristic and relatively narrow range for the δ^{31}P-values can be defined as lying around -88 ± 12 ppm [79,99,110,116,120)].

9 Newer Developments in the Area of Penta- and Hexaorganyl Arsenic Compounds

As in the case of the analogous phosphorus compounds, most of the systematic studies on pentaarylarsoranes were made with a variety of stable spirocyclic derivatives of the bis-2,2'-biphenylyleneorganylarsorane type. The original synthetic methods of Wittig and Clauß [25] were not applicable here, so that new preparative pathways had to be explored. Although triphenylarsine oxide and even phenylarsonic acid with excess phenyllithium gave fair yields of pentaphenylarsorane (13), the tosylimine method turned out to be superior because of its much more general applicability [44,121].

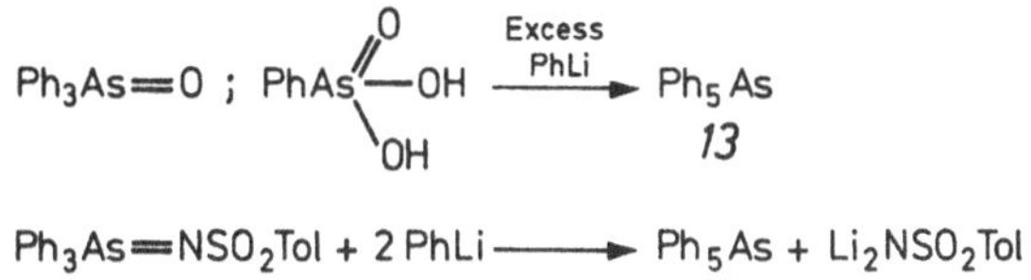

$$Ph_3As=O \ ; \ PhAs\overset{O}{\underset{OH}{\diagdown}}-OH \ \xrightarrow[\text{PhLi}]{\text{Excess}} \ Ph_5As$$
$$13$$

$$Ph_3As=NSO_2Tol + 2\,PhLi \longrightarrow Ph_5As + Li_2NSO_2Tol$$
$$13$$

Thus, 2,2'-biphenylylene-triphenylarsorane *115* could be prepared along two ways, either with 5-phenyldibenzarsole tosylimine *114* and phenyllithium or with triphenylarsine tosylimine *116* and 2,2'-dilithiobiphenyl [121,122]. Likewise, the spiro-arsoranes *117*, *118* were synthesized in high yields, *118* being rather unstable, probably because of unfavorable steric interactions at the reverse of the benzo-biphenylylene ligand [121].

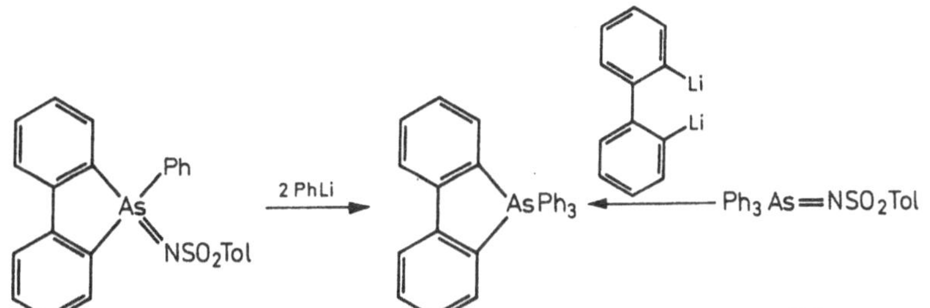

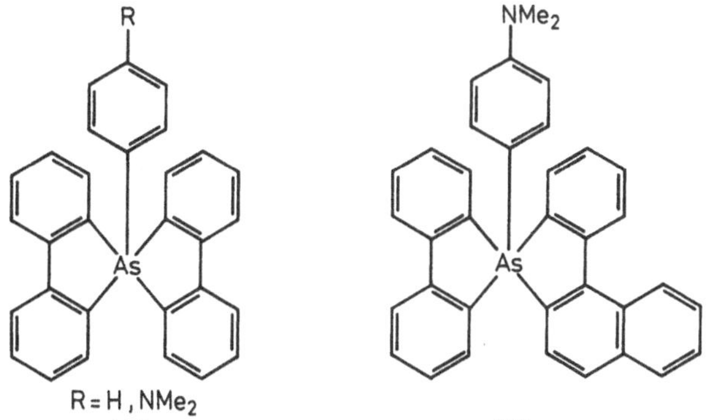

Again, for these compounds, names emphasizing the arsorane nature, e.g. bis-2,2'-biphenylylenearylarsorane for *117*, or names evaluating more the heterospirocyclic skeleton, e.g. 5'-(4-dimethylaminophenyl)spiro[11 H-benzo[b]naphth[2,1-d]arsole-11,5'-[5 H]dibenzarsole] for *118* are possible.

A still better synthesis for such spiroarsoranes was finally found in the reaction of the bis-2,2'-biphenylylenearsonium ion *120* (5,5'-spirobi[5 H-dibenzarsolium]) with a variety of organyllithium or Grignard reagents [121-123], *120* being readily accessible from appropriate chlorinated arsine precursors *119a, b* [121].

117, 121, 122	a	b	c	d	e	f	g	h	i	j
122										
R= Alkyl[122,123]	Me	Et	CD$_2$Me	i-Pr	n-Bu	t-Bu	Cyclopentyl	Benzyl	Neopentyl	PhMe$_2$CCH$_2$

In this way, not only aryl groups but also a broad selection of alkyl and alkenyl groups could be attached as the fifth ligand to the spiroarsorane skeleton. Systematic thermolysis studies revealed that all these compounds, at temperatures above their melting points, underwent diminution of the ligancy around arsenic from 5 to 3, but along quite different routes [123]. When R is a sterically inconspicuous group such as phenyl [121] or methyl [122], rearrangements to arsines incorporating the tetrabenzarsonin skeleton *123* occur. Stereochemically these compounds should closely resemble their phosphorus counterparts which have been discussed in an earlier chapter [47]. When the alkyl group R in a spiroarsorane *122* bears β-hydrogens, β-elimination is observed to give an olefin and 2-biphenylyl-2,2'-biphenylylenearsine (*124*), the ease of such reactions increasing with the bulkyness of the alkyl group [123]. Unsaturated groups R such as styryl in *121a, b* induce a stereospecific rearrangement to give (2'-alkenyl-2-biphenylyl)-2,2'-biphenylylenearsines *125*.

All attempts to combine sterically overcrowded alkyl Grignard reagents of the neopenthyl type with the spiroarsonium skeleton *120* led directly to the rearranged structures *126*. It is not clear whether these reactions involve the corresponding arsoranes *122i, j* as short lived intermediates or whether the nucleophilic attack is directed here from the beginning toward an arene ligand [123].

The inherent tendency of all such spiroarsoranes to thermally rearrange or decompose to arsine structures makes all assignments concerning their mass spectra

117a, *122a* → *123*; *122c* → *124*

121b,c → *125a* (*cis*), *b* (*trans*); *126*: R=Me, Ph

quite ambiguous. Thus, even for the molecular ions of phenyl- or methyl-bis-2,2′-biphenylylenearsorane (*117a*, *122a*), it is by no means clear whether they correspond to the original or to some rearranged constitution [29].

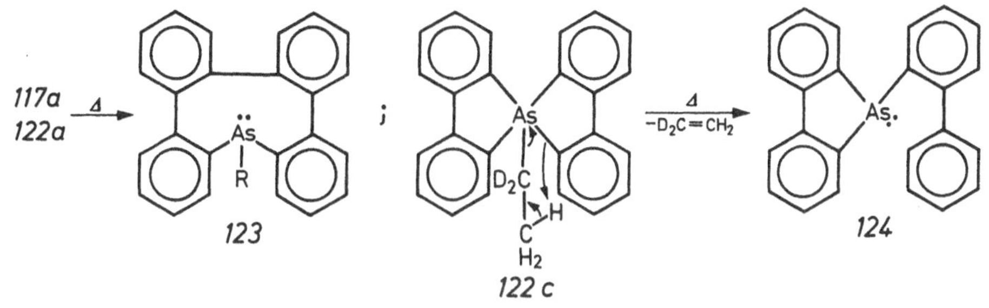

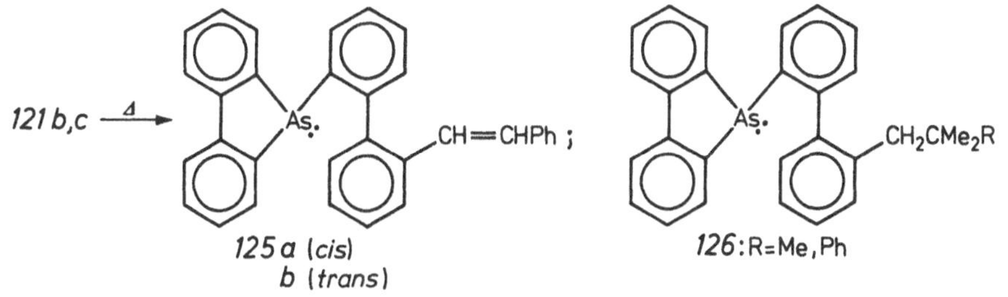

In electrophilic reactions bis-2,2′-biphenylylene-organylarsoranes react like the corresponding phosphoranes exclusively under cleavage of one arsafluorene system [121,123], thus reflecting the presence of considerable ring strain at least for the pentavalent state.

117, *122* → *127*

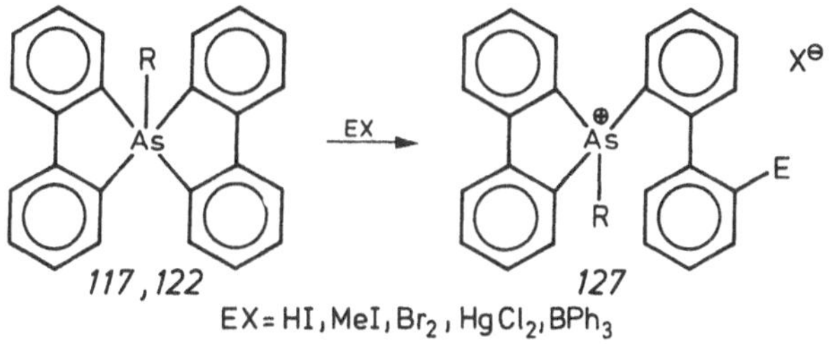

EX = HI, MeI, Br_2, $HgCl_2$, BPh_3

In certain cases even boiling alcohols are able to cleave spiroarsoranes of that type in a similar way to give more or less stable alkoxytetraarylarsoranes [121,123].

When spiroarsoranes of type *117*, *122* are subjected to strong nucleophiles such as lithium organyls, the spiro skeleton remains unaffected and only the single ligand R is exchanged [122,124]. A systematic study showed that the exchangability of R increases along the following sequence [122]:

$$n\text{-}C_4H_9 \ll p\text{-}Me_2NC_6H_4 \sim CH_3 \sim p\text{-}MeC_6H_4 < p\text{-}ClC_6H_4 < C_6H_5$$

For these reactions hexacoordinated intermediates *128* were proposed which can also be approached by a totally different route via the carbanionic precursor *129*.

It is thought, that ring closure from *129* to ate-complex *128* is essentially irreversible [122].

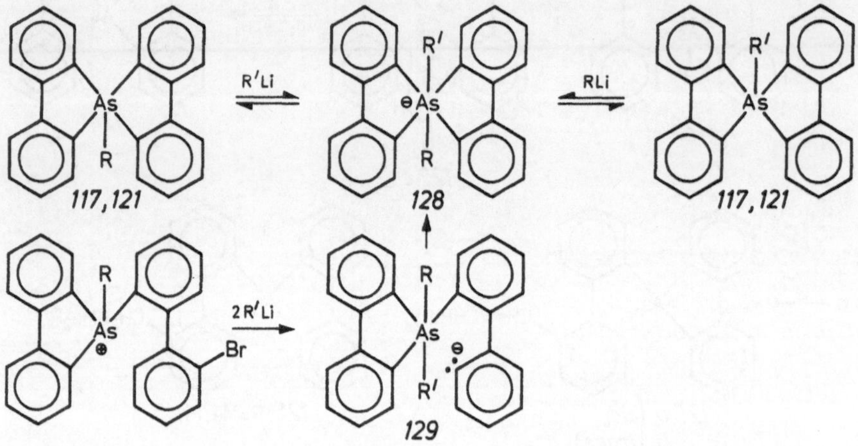

Similar reversibly formed ate-complexes, e.g. *132*, had been held responsible for the total disappearance of optical activity in the reaction of optically active spiroarsonium salt *130* with phenyllithium, where inactive spiroarsorane *131* was formed [121].

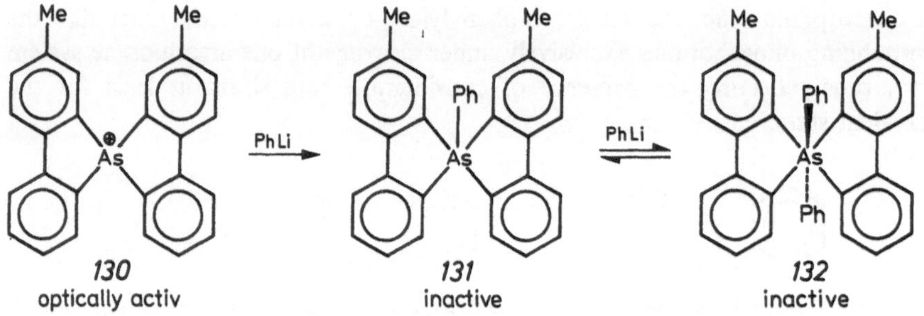

It should be remembered here that the reaction of pentaphenylarsorane with phenyllithium at low temperatures had not allowed a clear cut decision on the potential formation of a hexaphenylarsenate complex $(C_6H_5)_6As^{\ominus}Li^{\oplus}$ [30]. In this context it is also interesting to note that when pentaphenylarsorane is reacted with potassium amide in liquid ammonia at 120 °C all phenyl groups are displaced nucleophilically with the formation of benzene and $K_3[HN=As(NH)_3]$ [125].

A fairly stable hexaarylarsenate ion was finally obtained in the form of onium-ate complex *133* when spiroarsonium salt *120-I* was reacted with 2,2'-dilithiobiphenyl in ether. With excess 2,2'-dilithiobiphenyl in tetrahydrofurane the reaction proceeds to the lithium salt of the tris-2,2'-biphenylylenearsenate ion (*Li-134*) which is also obtained, besides arsoranes *117a*, *122a*, when *133* is treated with phenyl- or methyllithium. With sodium iodide in acetone onium-ate complex *133* is cleaved into its component salts *120-I* and *Na-134*. Both alkali metal salts *Li,Na-134* are moisture sensitive and react readily with water or alcohols so that the hexacoordinated structure is destroyed and the ring-opened arsorane *135* is produced [30].

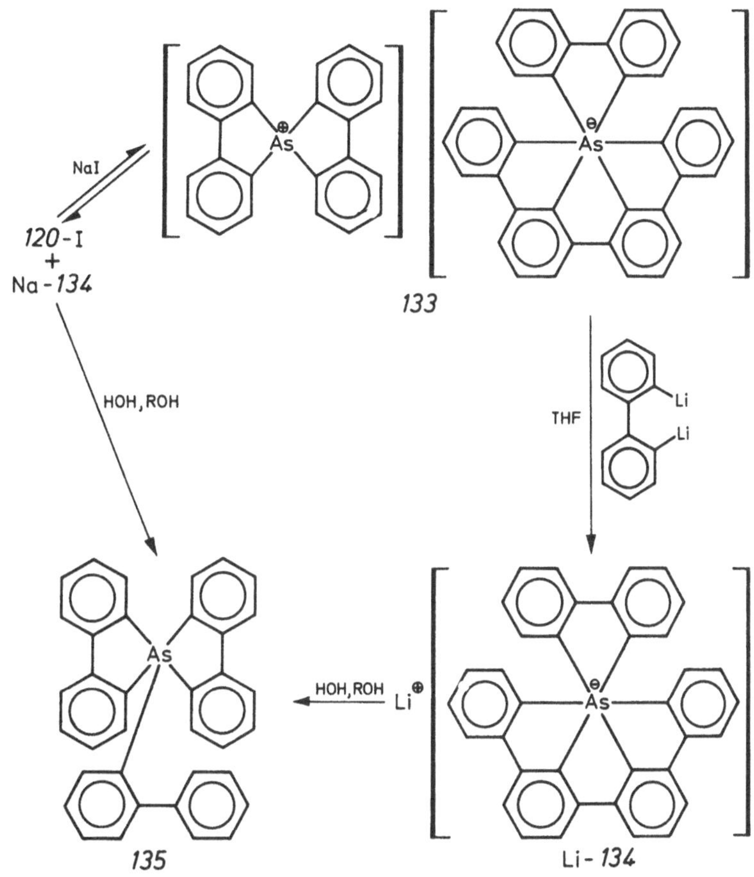

Thus, the arsenate ion *134* certainly has greater ring strain than the corresponding phosphate ion *57*, which is stable under these conditions.

The octahedral structure of the arsenate ion *134* was proven by its separation into optical antipodes via its stable methylbrucinium salts which showed the expected high molar rotations of $[M]_{578}^{24} = -10747$ and $+7742°$ respectively. When these salts were cleaved with aqueous hydrochloric acid in acetone, 2-biphenylyl-bis-2,2'-biphenylylenearsorane (*135*) was isolated, which showed not the slightest trace of optical activity. As with the analogous set of experiments in the phosphorus series this was one of the earliest indications of the non-rigid nature of the bis-2,2'-biphenyl-yleneorganylarsorane type whose racemization, as observed in the case of *135*, can easily be explained in terms of chiral trigonal-bipyramidal ground states, equilibrating rapidly over achiral tetragonal-pyramidal transition states [30]. For details on the stereochemical course of the above reaction sequences see the discussions pertaining to the corresponding phosphorus compounds.

Whereas earlier evidence for a trigonal-bipyramidal structure of pentaphenyl-arsorane and its derivatives could only be deduced indirectly on the basis of X-ray analogy considerations [62,63], IR-studies [126] and not very conclusive ¹H-NMR-experiments (only one methyl signal for penta-p-tolylarsorane [77]), a more recent crystal structure determination for a solvate of pentaphenylarsorane with one half molecule of cyclohexane has revealed a nearly undistorted trigonal-bipyramidal geometry with axial and equatorial bond lengths of 2.105 and 1.946 Å respectively [127].

An even clearer picture emerged from systematic ^{1}H-NMR-investigations of bis-2,2'-biphenylyleneorganylarsorane systems provided with appropriate monitors [79, 128]. Thus, in the case of spiroarsoranes *122*, *136*, temperature-dependent reversible coalescence phenomena for the various diastereotopic groups could be observed, exactly as in the case of analogous phosphorus compounds. The underlying modes of rearrangement were again analyzed in terms of the usual Berry pseudorotation process implying trigonal-bipyramidal ground states with the 2,2'-biphenylylene ligands in axial-equatorial arrangements and a tetragonal-pyramidal transition state with the bidentate groups at the basal positions. (For further details see chapter 4).

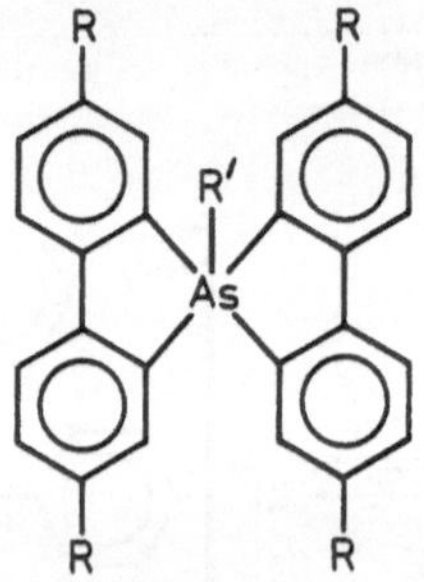

122 , 136

	R	R'	$\Delta G_c^{\pm}$ kcal (KJ)/mol
122 b	H	Et	———————
h	H	PhCH$_2$	———————
d	H	iPr	15.4 (64.5)
136 a	Me	PhCH$_2$	12.6 (52.7)
b	Me	2-Biph	15.2 (63.6)
c	Me	1-Np	15.2 (63.6)
d	Me	8-F-1-Np	15.4 (64.5)
e	Me	8-Cl-1-Np	19.1 (80.0)
f	Me	8-Me-1-Np	19.3 (80.8)
g	Me	8-Chinolyl	13.7 (57.5)

The calculated pseudorotation barriers were in general slightly lower than those for corresponding phosphorus compounds, pointing thus to a greater flexibility of the spiroarsorane skeleton which is probably due to the bigger average C—As bond distances as compared with corresponding P—C bond lengths [128].

When, as in spiroarsoranes of type *138*, an additional benzo-annelated across the 3,4-positions of one 2,2'-biphenylylene ligand renders the potential tetragonal-pyramidal transition state *137* highly improbable because of prohibitive steric overcrowding, inner flexibility of the molecule can express itself now only by virtue of a pseudorotation process passing eventually through the trigonal-bipyramidal transition state *139* with a diequatorially spanned 2,2′-biphenylylene group [82].

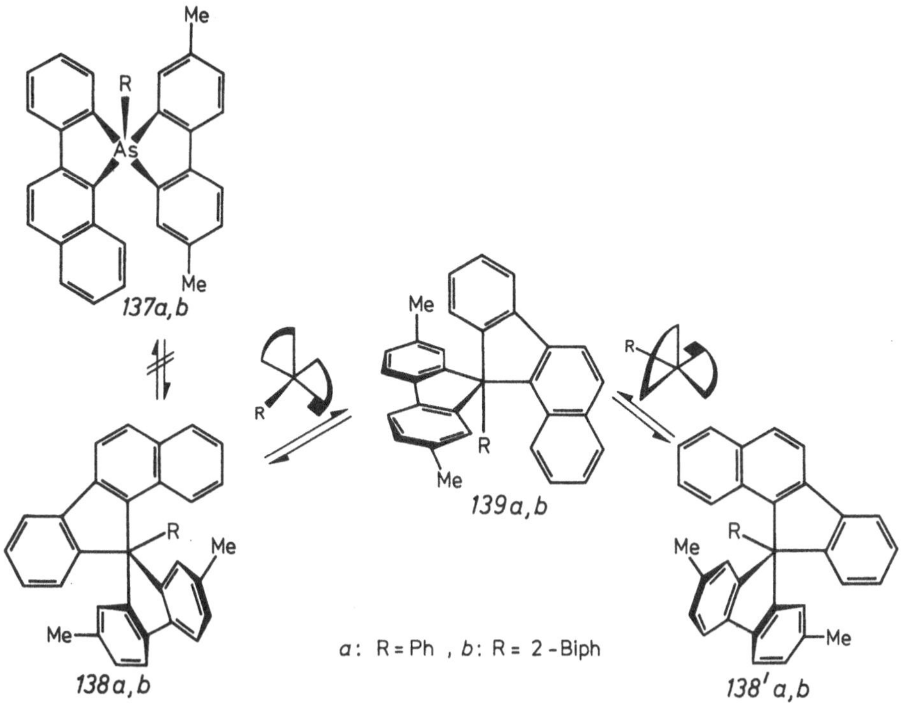

a: R = Ph , b: R = 2 - Biph

And indeed, on warming, the two well-resolved methyl signals of *138a* coalesce reversibly to one single line in a process for which an activation barrier of $\Delta G_c^{\ddagger} = 17.2$ (72.2) kcal (kJ)/mol is calculated, which again is only slightly lower than that for the corresponding phosphorus compound ($\Delta G_c^{\ddagger} = 18.2$ (76.0) kcal (kJ)/mol). When for R the spatially very clumsy 2-biphenylyl group is present, the process involving the now highly strained transitional conformation *139b* is so much slowed down that exchange broadening of the methyl signals begins only above 150 °C, from which a lower limit for the pseudorotation barrier of around 22.5 (94.2) kcal (kJ)/mol can be estimated [82].

Considering the high stability of the spiroarsoranes discussed so far, there was at least some hope that the prototype of this kind of compound, namely, bis-2,2′-biphenylylenehydridoarsorane (*140*), could be prepared in a similar way. All attempts, however, to generate this compound in the reaction of bis-2,2′-biphenylylenearsonium iodide (*120-I*) with lithium aluminum hydride gave at best only evidence of the very volatile existence of *140*. When the above reaction is performed in diethyl ether, hydrogen evolution is observed and variable amounts of dimer *143* and 2-biphenylyl-2,2′-biphenylylenearsine (*124*) are obtained. These results are best rationalized with a complex radical reaction sequence involving arsoranyl radical *141* and 2-biphenylyl radical *142* [129]. In no case, however, could these species or hydridoarsorane *140* be observed directly.

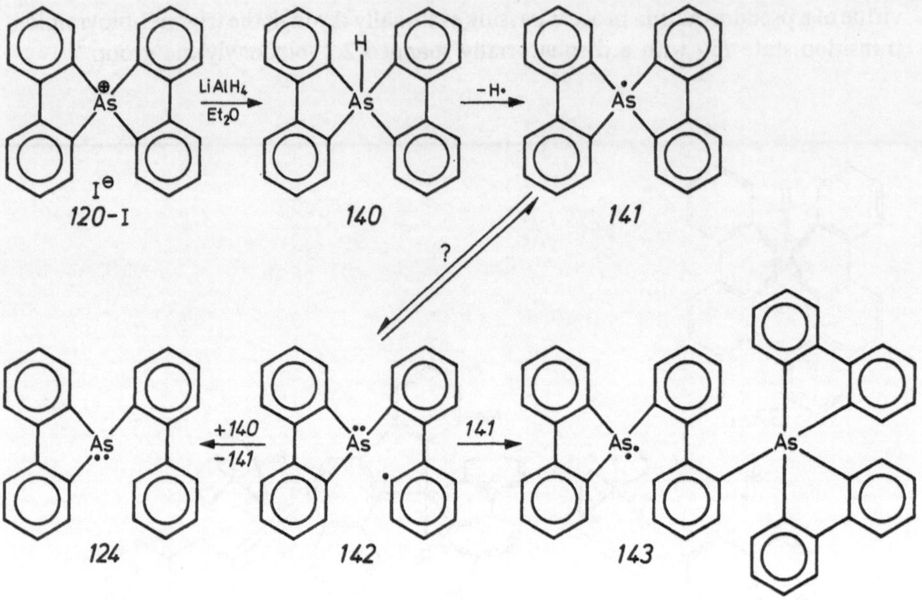

Similar results were obtained in corresponding reactions with the substituted spiro-arsonium salt 144-I [129].

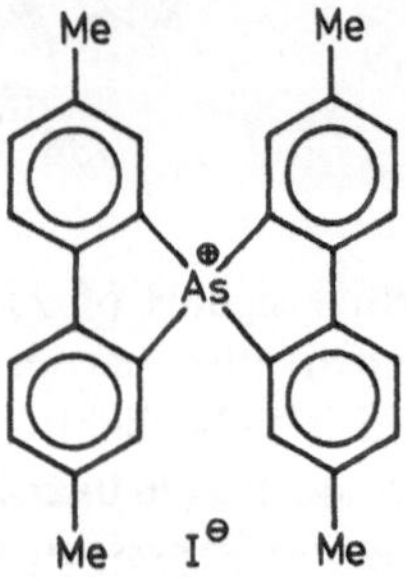

Dimer 143 is also produced in good yields when spiroarsonium salt 120-I is treated with t-butyl- or cyclohexylmagnesiumbromide [123], triphenylmethylpotassium or lithium amalgame [129], where again initial formation of arsorane 140 and/or radical 141 might play the key role. In the mass spectrum of dimer 143 the doubly charged parent ion is 4.2 times as intense as the corresponding singly-charged molecular ion which is explained, as in the case of the analogous phosphorus compound, by a preferred rearrangement of the dication to a stabilized species [96].

When arsonium salt 120-I is treated with lithium aluminum hydride in tetra-hydrofurane at —70 °C copious evolution of hydrogen occurs and a red solution is obtained, whose color fades on warming to room temperature. Hydrolysis yields only 2-biphenylyl-2,2'-biphenylylenearsine (124). The same deep red solution and colorless hydrolysis product 124 are obtained when 2'-bromo-2-biphenylyl-2,2'-biphenylylenearsine (147) is first reacted with t-butyllithium in tetrahydrofurane

42

and then hydrolyzed with ethanol/hydrochloric acid. When, on the other hand, the halogen metal exchange reaction on arsine *147* is performed with n-butyllithium, the initially obtained red solution turns yellowish within one minute and the only isolable product is butyl-bis-2,2'biphenylylenearsorane (*122e*). From these experiments it was concluded that in all three cases the same equilibrium mixture of the red bis-2,2'-biphenylylenearsoranyl anion (*145*) and the colorless ring-opened carbanion *146* is formed, most probably via the non-isolable hydridoarsorane *140* [129].

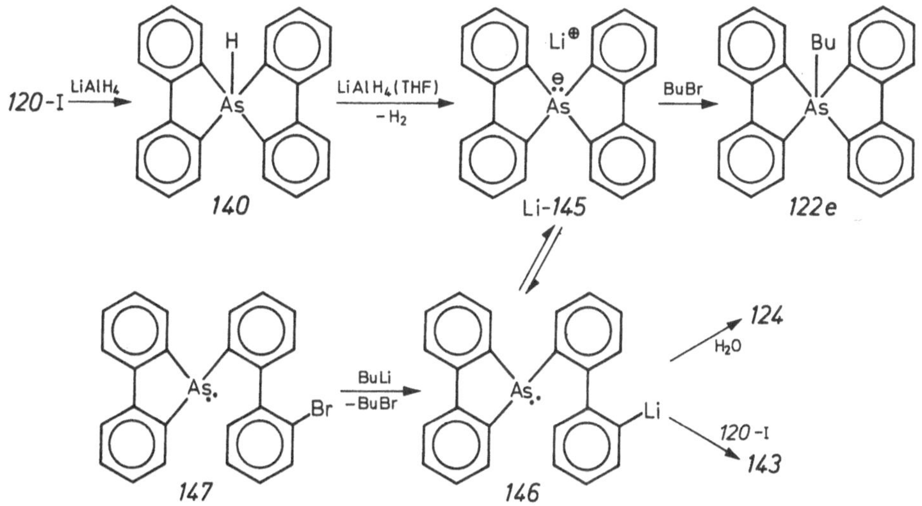

The equilibrium *145* ⇌ *146* seems to be very mobile since the hard acid proton traps exclusively the hard carbanion constituent *146*, whereas, the softer acid n-butylbromide (formed in the initial halogen metal exchange step *147* → *146*) combines solely with the soft arsoranyl component *145*. When the halogen metal exchange at *147* is performed in diethyl ether, a colorless precipitate of *Li-146* is obtained which again is hydrolyzed to arsine *124* and which, on the other hand, can be combined with spiroarsonium salt *120-I* to give high yields of dimer *143*, providing thus an independent proof of the latter's constitution [129].

With the exeption of the alkyl-bis-2,2'-biphenylylenearsoranes *122* whose chemistry and, especially, stability is determined by the spirocyclic bis-2,2'-biphenylylenearsoranyl part of the molecules, very little is known about alkylarsoranes. Reactions, which at least formally could have led to methyltetraphenylarsorane, $(C_6H_5)_4AsCH_3$, always produced arsenic ylides [130].

So far, pentamethylarsorane (*1*) is the only known arsorane with five alkyl ligands. First indirectly evidenced by Wittig and Torssell [31], this compound could finally be isolated and unambiguously characterized by Mitschke and Schmidbaur in the reaction of dichlorotrimethylarsorane and methyllithium at −60 °C [131]. *1* is a colorless liquid whose vibrational and NMR spectra are in accordance with a trigonal-bipyramidal structure. No broadening of the single methyl ^{1}H-NMR-signal is observed even at −95 °C which is indicative for the usual rapid pseudo-rotation of such molecules.

At 100 °C pentamethylarsorane (*1*) is quantitatively decomposed to trimethylarsine, methane and a little ethylene, thus supporting the assumption of intermediate formation of ylide *148*; ethane is only formed in traces [131]. With water and acids tetramethylarsonium salts are produced [131] whereas with alcohols, hydroxylamines and oximes covalent pentacoordinate arsoranes *149* are obtained in which one methyl group has been replaced by the respective electronegative group Y [132,133].

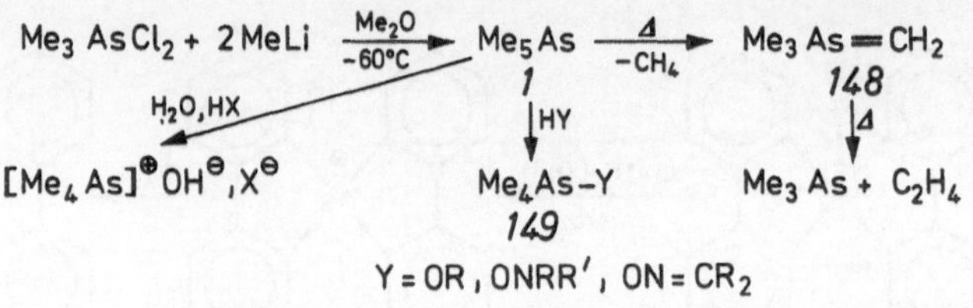
Me3 AsCl2 + 2MeLi
Me2O
-60°C
Me5As
1
Δ
-CH4
Me3 As=CH2
148
H2O,HX
HY
Δ
[Me4 As]⊕OH⊖,X⊖
Me4As-Y
149
Me3 As + C2H4
Y = OR, ONRR′, ON = CR2

10 Recent Developments in the Field of Pentaaryl Antimony Compounds

For the pentaarylphosphorane and pentaarylarsorane type the combination of each two phenyl groups to form a 2,2′-biphenylylene ligand had led to spirocyclic species which generally were of much higher stability then the simple pentaphenyl derivatives. In the antimony series, however, such a distinct stabilizing effect was no longer observable.

Although bis-2,2′-biphenylylenephenylstiborane (5-phenyl-5,5′-spirobi[5*H*-dibenzostibole]) (*150*) was first prepared by the tosylimine method [134], it was later shown that the more traditional procedure starting from appropriate stibine dihalides [25] was more convenient for the synthesis of such compound types [135].

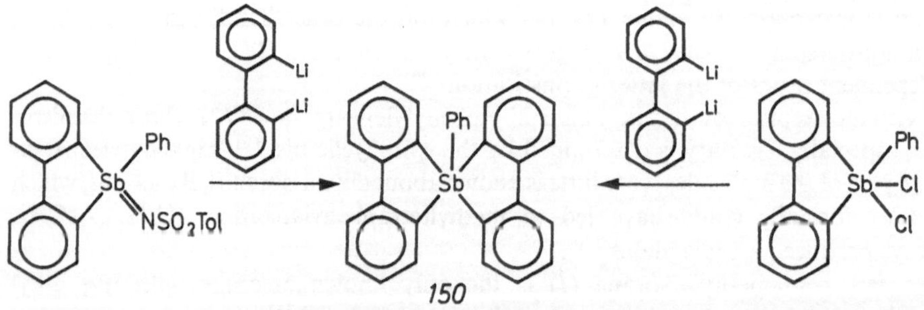
Li
Li
Ph
Sb
NSO2Tol
Ph
Sb
Li
Li
Ph
Sb—Cl
Cl

150

Thus, the mixed spirocyclic stiborane *151* as well as the monocyclic derivative *153* were obtainable by both possible ways as shown in the reaction scheme.

Attempts to synthesize by analogous procedures bis-2,2′-biphenylylenemethylstiborane (*157*) from 2,2′-biphenylylenemethylstibine dibromide *154* and 2,2′-dilithiobiphenyl failed, because of the obvious instability of the initially formed stibonium structure *155* which readily loses methyl bromide. The carbanionic stibine *156* so formed then attacks a second molecule of dibromostiborane *154* whereafter again methyl bromide is expelled under generation of the end product *158*. Similarly, reaction of *154* with phenyllithium gave 2,2′-biphenylylenephenylstibine as the only isolable product [135].

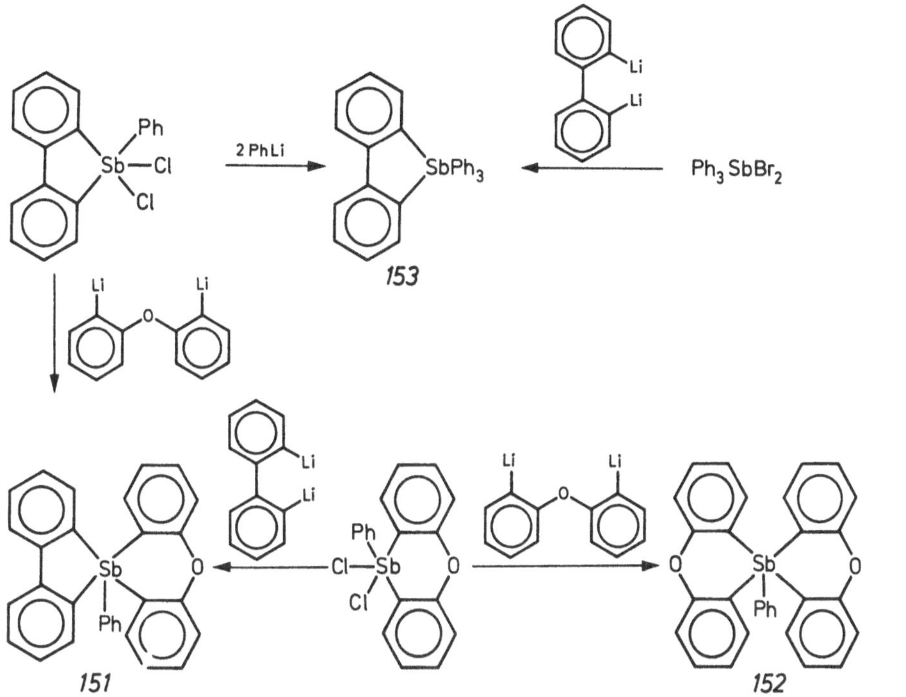

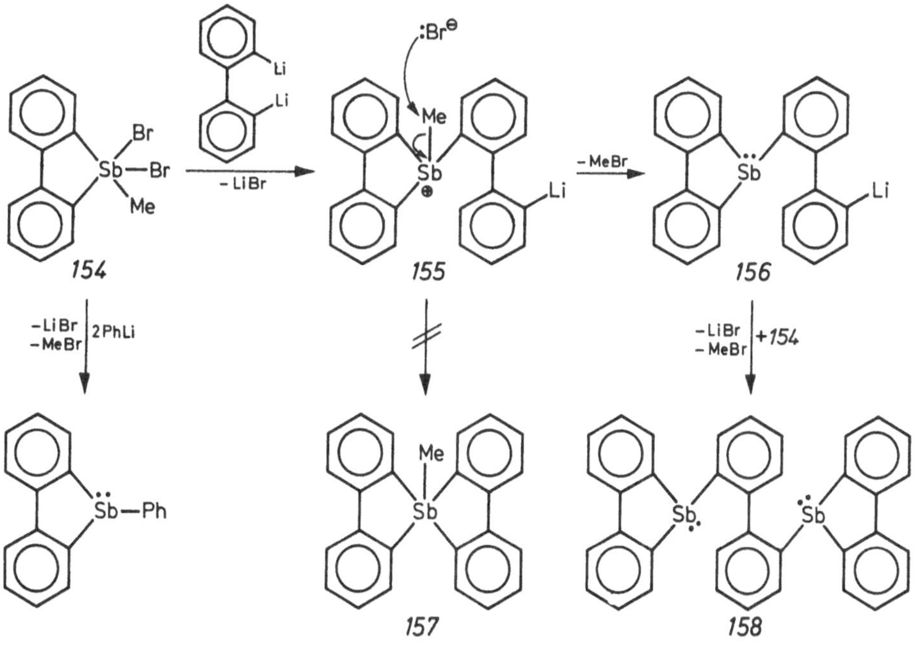

In contrast to these results simple alkyltetraarylstiboranes, namely, ethyl- and methyltetraphenylstiborane (*159a, b*) could be obtained in reactions of appropriate fluorostiboranes with alkyl- or aryl Grignard reagents [136].

$$Ph_2RSbF_2 \xrightarrow{PhMgBr} Ph_4SbR \xleftarrow{RMgBr} Ph_4SbF$$

$$159a: R = Me$$
$$b: R = Et$$

With electrophilic agents spirocyclic stiboranes of type *150* react even more readily than the corresponding phosphorus and arsenic derivatives, *150* for example being already decomposed by boiling ethanol to give an ethoxytetraarylstiborane *160*. With boiling aqueous hydrochloric acid even both stibafluorene systems are cleaved and dichlorostiborane *162* is obtained which can also be produced from *160* and hydrochloric acid [134].

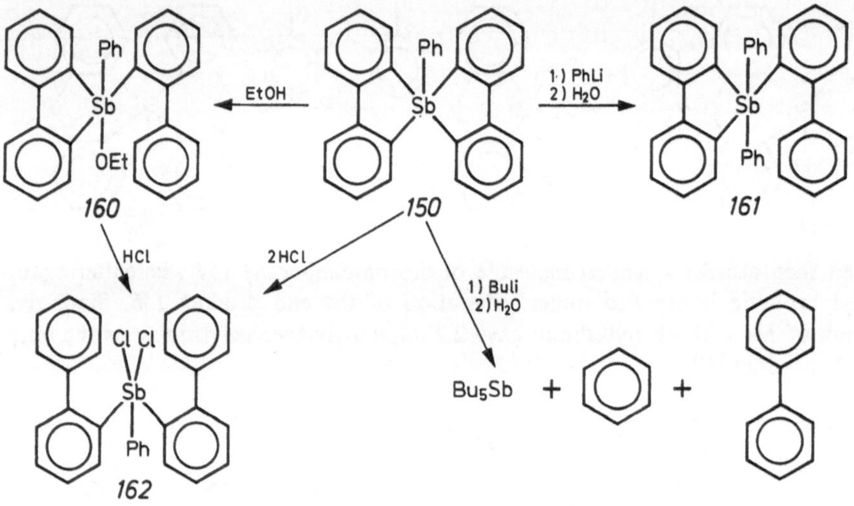

Under nucleophilic attack from excess n-butyllithium, *150* loses all its aryl ligands and pentabutylstiborane is produced [137]. Similar exhaustive aryl displacements had been observed in reactions of pentaphenylstiborane with butyl- and methyllithium [46]. With the less nucleophilic phenyllithium, for *150* the reaction stops essentially after the first substitution step and the new monocyclic pentaarylstiborane *161* is generated. Corresponding monocyclic pentaarylstiboranes *163, 164* are obtained from the oxydi-o-phenylene group-containing spirocyclic stiboranes *151* and *152* with phenyl-lithium; in the mixed-cyclic case (*151→163*) only the six-membered ring is cleaved for entropic reasons [137].

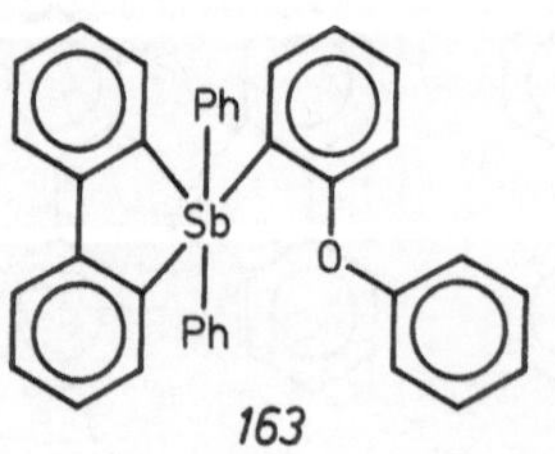

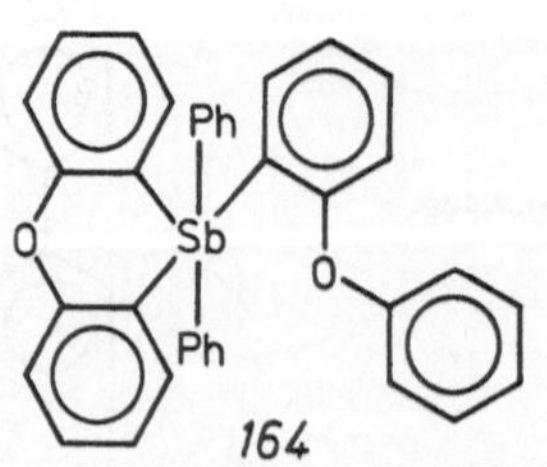

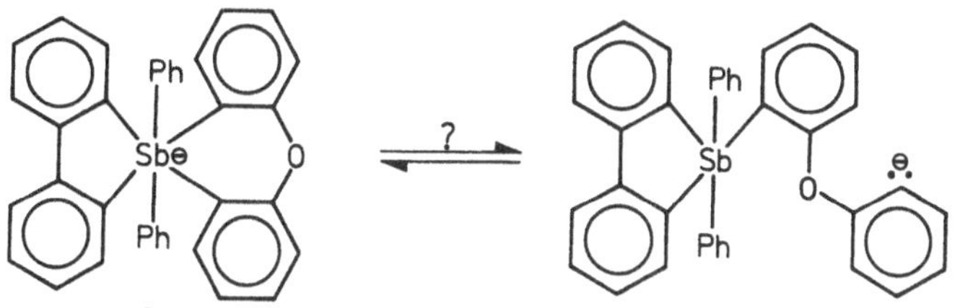

165   166

Initially, in all these nucleophilic substitution reactions thick white precipitates are formed for which it is by no means clear whether they correspond to ate-complexes of type *165* or to ring-opened structures such as *166*. From both intermediates, though, hydrolysis products such as *163* should be generated with comparable ease [137].

All attempts to prepare a spirocyclic hexacoordinate antimonate complex *168* and a tetracoordinate spirostibonium ion *167* or onium-ate combinations of these, according to structural and synthetic principles elaborated with corresponding phosphorus and arsenic compounds, met with failure [138]. When, for example,

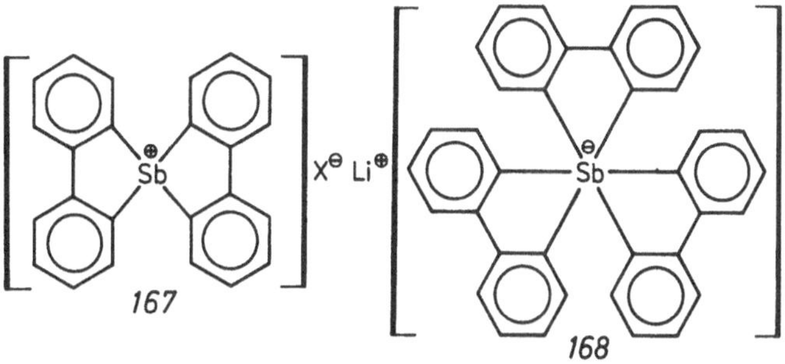

167    $X^{\ominus}$ $Li^{\oplus}$ 168

antimony pentachloride or the trichlorostiborane *169* were treated with 2,2'-dilithiobiphenyl, only poor yields of a "dimeric" product *171* could be isolated in which one antimony center had been reduced to the trivalent state. Several reaction sequences can be given for the generation of *171*, the simplest one, starting with a reduction process by the aryllithium, not even including as intermediates the desired species *167* and *168*.

The initial orange color, observed in these reactions at —70 °C, points to an at least peripheral occurence of the tetracoordinated antimonate anion *170*, which seems to be much less viable then the analogous and also intensely colored phosphate and arsenate complexes *79* and *145*. Equally, the inaccessibility of *167* and *168* in these reactions points to a considerably reduced stability with respect to analogous phosphorus and arsenic species, which again is explained by the higher ring strains caused by the larger antimony central atom [138].

When bis-2,2'-biphenylylenephenylstiborane (*150*) is treated with lithium aluminum hydride in diethyl ether or better in tetrahydrofurane, a deep red solution results, from which after addition of iodine good yields of 2,2'-biphenylyleneiodostibine (*175*) can be isolated [137]. Again a nucleophilic substitution, here by the hydride ion, is involved, leading first to the carbanionic hydridostiborane *172* (or the

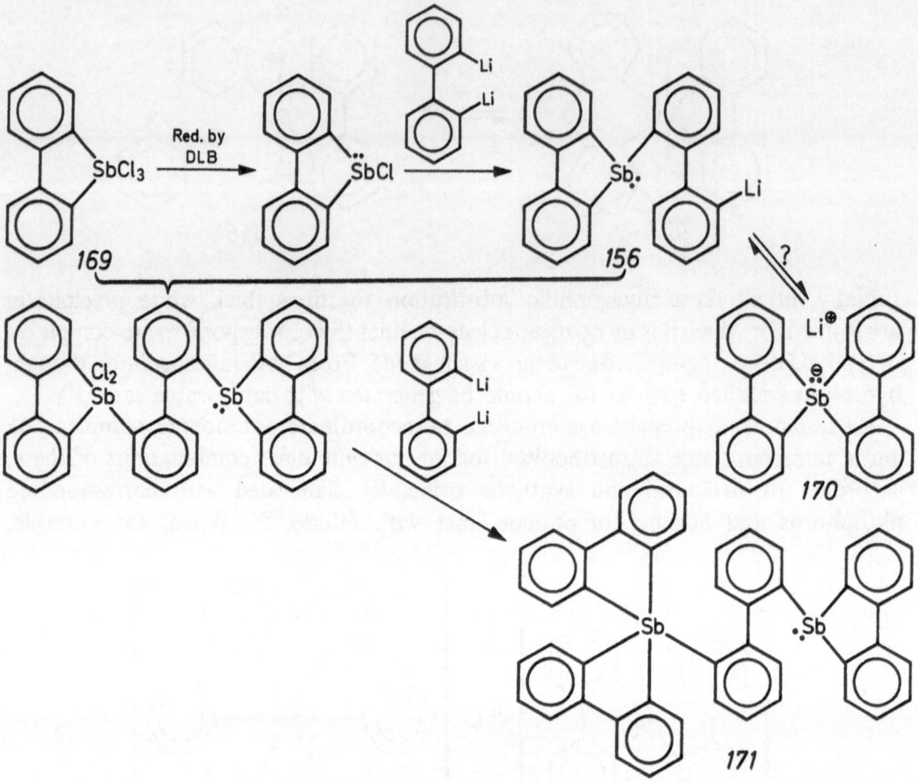
Red. by
DLB
SbCl3
SbCl
Li
Li
Sb
Li
169
156
?
Li⊕
Cl2
Sb
Sb
Li
Li
Sb
170
Sb
Sb
171

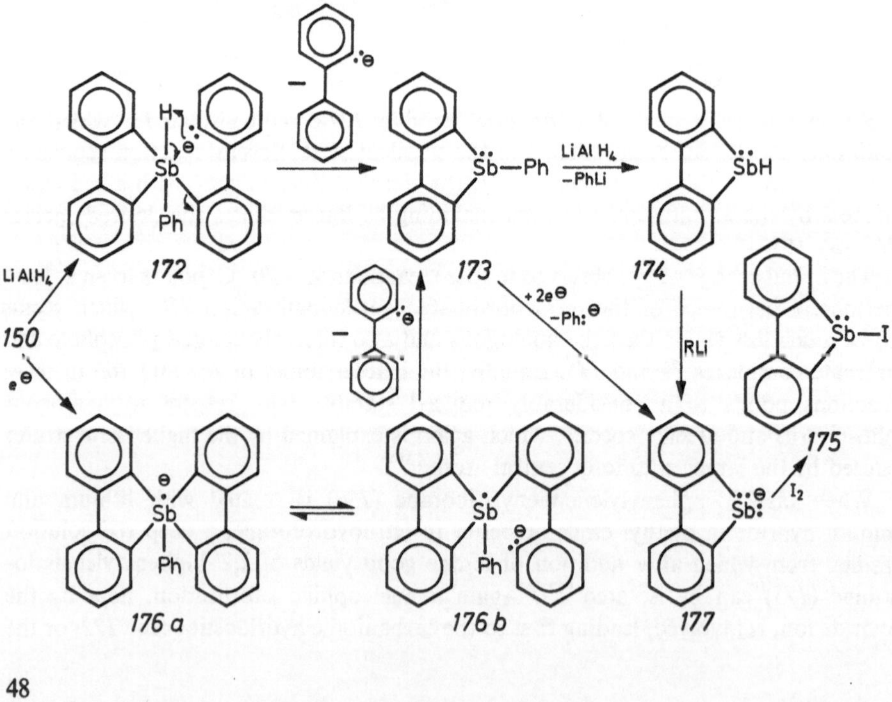
H
Sb
Ph
172
Sb—Ph
Li Al H4
−PhLi
SbH
173
174
Li AlH4
150
e⊖
+2e⊖
−Ph:⊖
RLi
Sb—I
175
I2
Sb
Ph
Sb
Ph
Sb
176 a
176 b
177

corresponding ate-complex), which decomposes rapidly to stibine *173* and 2-lithio-biphenyl. (A similar decomposition mode has been proposed for the also non-isolable tetraphenylhydridostiborane $(C_6H_5)_4SbH$ from which triphenylstibine and benzene are obtained [139]). A second nucleophilic attack by hydride displaces the phenyl anion from stibine *173* to give 5*H*-dibenzostibole (*174*), which then is metalated by the various lithium aryls present. Trapping of *177* with iodine finally leads to 5-iodo-5*H*-dibenzostibole (*175*).

The same final product *175* is obtained in the reaction of spirostiborane *150* with liquid sodium-potassium alloy, most probably via initial one-electron-transfer generating the anion radicals *176a* and/or *176b* [137].

Whereas the investigations discussed above were in the first place concerned with the preparation of new types of cyclic and spirocyclic pentaarylstiboranes, most of the basic chemistry of that class of compounds has been studied exemplarily with the prototype proper, pentaphenylstiborane [25]. In more recent investigations the thermal decomposition of pentaphenylstiborane to triphenylstibine and biphenyl [25] on the basis of ^{14}C-tracer studies was shown to proceed intramolecularly with no involvement of phenyl radicals [140]. Photolysis, on the other hand, in complex reaction sequences leads to triphenylstibine, biphenyl, p-quaterphenyl etc., where free radicals account for many of the products [140]. In this context it might be of interest that β-decay of $^{125}Sb(C_6H_5)_5$ is believed to occur under structure perser-vation to generate $[^{125}Te(C_6H_5)_5]^{\oplus}$ as an intermediate [141].

At elevated temperatures and in a sealed tube pentaphenylstiborane reacts with alkyl halides such as chloroform, tetrachloromethane [142] or iodomethane [143] to give tetraphenylstibonium halides, benzene, halobenzene and other products. Homo-lytic cleavage of Sb—C bonds is assumed to play a leading role here. In a more detailed study, a radical chain reaction was proposed for such decompositions, e.g.: [144]

$$Ph_5Sb \xrightarrow{\Delta} Ph\cdot + Ph_4Sb\cdot$$

$$Ph\cdot + CCl_4 \longrightarrow PhCl + \cdot CCl_3$$

$$Ph_5Sb + \cdot CCl_3 \longrightarrow [Cl_3C\cdot SbPh_5]$$

$$[Cl_3C\cdot SbPh_5] \longrightarrow Ph\cdot + Cl_3C-SbPh_4$$

Termination occurs by dimerization of phenyl and trichloromethyl radicals. An interesting subsequent reaction consists in the decomposition of (trichloromethyl)-tetraphenylstiborane to tetraphenylstibonium chloride and dichlorocarbene, which inserts into tetrachloromethane to yield hexachloroethane [144]. Recently a systematic study of the cleavage of parasubstituted pentaphenylstiboranes by antimony penta-chloride was published. In these reactions p-disubstituted benzenes are produced regioselectively together with diaryltrichlorostiboranes. Also, preferred cleavage of the axially-orientated aryl groups was observed, irrespective of their electronic nature [145].

$$\left(X-\!\!\left\langle\bigcirc\right\rangle\!\!- \right)_5 Sb + SbCl_5 \longrightarrow \left(X-\!\!\left\langle\bigcirc\right\rangle\!\!- \right)_2 SbCl_3 \quad + \quad 2\ X-\!\!\left\langle\bigcirc\right\rangle\!\!-Cl$$

$$X = Cl, CH_3$$

It might be mentioned in this connection that on the basis of IR-studies it has been concluded that the $M(C_6H_5)_4$ groups in pentaphenylstiborane, -arsorane and -phosphorane are probably weak electron acceptors [146].

A very different kind of electrophile, benzenediazonium tetrafluoroborate, has also been reacted with pentaphenylstiborane (and also with pentaphenylarsorane and -phosphorane) resulting again in the formation of tetraphenylstibonium species. [147] Furthermore, pentaphenylstiborane can be cleaved by formic acid [148] or boiling alcohols [142,149] to give benzene and formyl- or alkoxytetraphenylstiboranes, respectively. Similarly, reaction with water in dioxane leads to benzene and tetraphenylstibonium hydroxide [150]. In a systematic kinetic study of the decomposition of variously substituted pentaarylstiboranes with alcohols, McEwen and coworkers were able to conclude that these reactions are essentially of the $S_N1(Sb)$-type with nucleophilic or electrophilic participation by a solvent molecule, depending on the substituents. In the presence of alcoholate, however, an $S_N2(Sb-ate)$-mechanism involving intermediate hexacoordinate antimonate complexes is thought to be operating [151].

$$Ph_4Sb-Ar + ROH \longrightarrow \begin{cases} Ph_3Sb\begin{smallmatrix} Ar \\ OR \end{smallmatrix} + PhH \\ Ph_4Sb-OR + ArH \end{cases}$$

Ar = p-MePh, p-MeOPh, m-ClPh

Finally, pentaphenylstiborane has also been used as a catalyst in the reaction of epoxides with carbon dioxide to give dioxolanones [152].

Stereochemical studies with pentaarylstiboranes began essentially with a two-dimensional X-ray structure investigation of pentaphenylstiborane which most surprisingly revealed structural characteristics peculiar to a distorted tetragonal pyramid (see *178*) [62,153]. This was then confirmed by a three-dimensional crystal structure determination from which an apical C—Sb bond length of 2.115 and average basal bond lengths of 2.216 Å were derived [154]. The SbC_4-skeleton has only C_{2v}-symmetry since the two trans-basal C—Sb—C bond angles are different (149.3 and 163.5°). On the basis of detailed IR-studies in dichloro- and dibromomethane, it was suggested that pentaphenylstiborane remains tetragonal-pyramidal in solution too [126]. However, in the ^{1}H-NMR-spectrum of fully m,p-deuterated pentaphenylstiborane no corresponding signal splittings could be observed even at $-140\ °C$ [126].

Most contrastingly, the crystal structure of a solvate of pentaphenylstiborane with half a molecule of cyclohexane [155] revealed a nearly perfect trigonal-bipyramidal atomic arrangement [156]. A somewhat more distorted trigonal bipyramid was found for penta-p-tolylstiborane (see *179*) with axial and equatorial bond lengths of 2.26 and 2.16 Å respectively [157].

178   *179*

Again, in the ^{1}H-NMR-spectrum of this compound no splitting of the single methyl signal is observable [77,157] even at -130 °C in dichlorofluoromethane solution [158].

From these investigations it follows that simple pentaarylstiboranes have a rather flexible skeleton which can adapt easily to the stereochemical particularities imposed by packing forces upon the solid state structures. Recent detailed lattice energy calculations for pentaphenylstiborane, -arsorane and -phosphorane also point in this direction. It was thus derived that the most stable conformations for the solvates of these compounds grown from cyclohexane and containing one half equivalent of solvent should indeed be trigonal bipyramids, which should also prevail as equilibrium geometries in solution. On the other hand, some justification could be found for the fact that unsolvated pentaphenylstiborane is distorted strongly towards a tetragonal-pyramidal build. Because in this case the inherent energy difference between the two limiting conformations is rather small, external factors such as packing forces could easily gain predominance in determining the actual structure [159].

Also from X-ray investigations there were indications that the spirocyclic bis-2,2'-biphenylylenephenylstiborane (150) belonged to the trigonal-bipyramidal structure type [160]. When di- or tetramethyl substituted derivatives thereof, such as 180a–c, were cooled in carbon disulfide/pyridine solutions, significant broadening of the single methyl signal occured in each case. Only for 180c, however, incipient signal splitting into four lines was observable at -60 °C [135,161]. Again, as in the case of the corresponding phosphoranes 39 and arsoranes 136, this is consistent with a trigonal-bipyramidal ground state undergoing rapid intramolecular ligand exchange

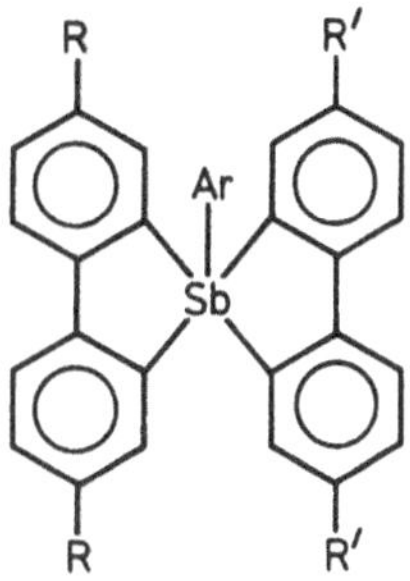

	Ar	R	R'
180 a	Ph	Me	H
b	2–Biph	Me	H
c	2–Biph	Me	Me

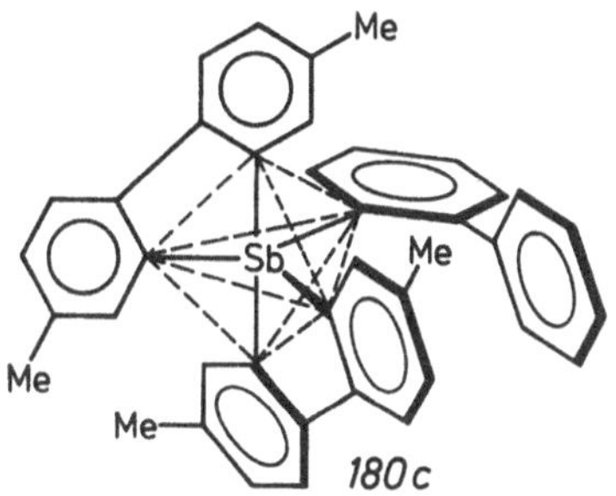

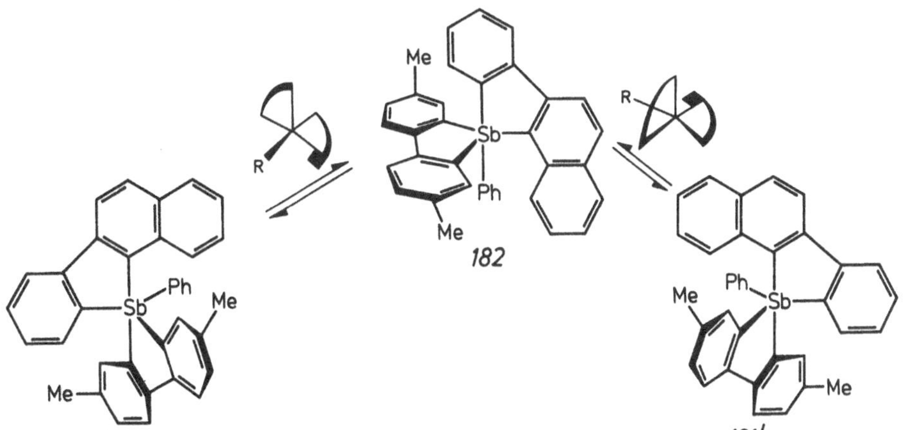

at room temperature via Berry pseudorotation and concomitant rotation of the 2-biphenylyl ligand. Here, tetragonal-pyramidal transition states with dibasal 2,2'-biphenylylene groups are decisive.

When such a transition state is sterically highly hindered, as for *181* (cf. *71, 137*), methyl signal equilibration again can only be explained by virtue of an achiral trigonal-bipyramidal transition state *182* containing a diequatorial 2,2'-biphenylylene group which is formed via tetragonal-pyramidal conformations with one apical-basal 2,2'-biphenylylene ligand as abstracted in the foregoing scheme. The calculated racemization barrier $\Delta G_c^{\neq}$ amounts here to 15.4 (64.5) kcal (kJ)/mol [82].

11 Pentaalkylstiboranes and Related Compounds

Simple acyclic pentaalkyl derivatives of antimony are in general quite stable, in contradistinction to analogous phosphorus and arsenic derivatives. Consequently, a fairly representative number of such compounds has been investigated in recent times. Their syntheses followed closely the original procedures as applied by Wittig and Torssell [31] in that trialkyldihalo- and dialkyltrihalostiboranes or tetraalkylstibonium salts were reacted with alkyl lithium or magnesium compounds.

Thus, pentaethyl- [162], pentacyclopropyl- [163] and pentabutylstiborane [164] have been obtained as yellowish liquids, which were distillable under reduced pressure. Pentabutylstiborane is also formed when pentaphenylstiborane [46] or spirocyclic pentaarylstiboranes [137] are subjected to nucleophilic displacement reactions with excess butyllithium. Similar nucleophilically-induced exchange phenomena interfered in reactions of triethyldichloro- and trimethyldichlorostiborane with methyllithium or diethylmagnesium, respectively, where the whole spectrum of mixed compounds Me_nSbEt_{5-n} could be observed [165]. Only under carefully controlled conditions at low temperatures can dimethyltriethylstiborane be prepared from triethyldichloro- and dimethyltrichlorostiborane with methyl- and ethyllithium, respectively [166]. By similar procedures a series of silyl group containing pentaalkylstiboranes $Me_nSb(CH_2SiMe_3)_{5-n}$ (n = 0–4) were recently prepared. The distilled pure compounds are quite stable and decompose only above 150 °C [167].

Because of the usually noncrystalline nature of pentaalkylstiboranes, structural investigations have so far been limited to vibrational spectroscopical studies and ^{1}H-NMR measurements. Thus, on the basis of IR and Raman spectra in the solid, liquid, vapor and solution states, a trigonal-bipyramidal structure was derived for pentamethylstiborane [168]. From force constant considerations, the energetic separation from a tetragonal-pyramidal conformation was estimated to be ~ 8.1 (34) kcal (kJ)/mol [169]. Quite on the contrary, IR and Laser Raman data for pentacyclopropylstiborane were concluded to be consistent with a tetragonal-pyramidal molecular framework [163].

In no case, however, not even at temperatures as low -100 to -120 °C, could any splitting of the alkyl NMR-signals be observed for pentamethylstiborane [170] and other pentaalkylstiboranes [166] which would have been indicative of different bonding sites for the alkyl groups.

The chemistry of pentaalkylstiboranes [171] is closely related to that of pentaphenyl- and other pentaarylstiboranes. With strong nucleophiles various exchange reactions take place, of the sort already discussed in connection with the synthetic procedures. With proton acids usually one alkyl group is removed and stibonium salts or new pentacoordinated stiboranes are formed. Thus alcohols [133,162,166,172], thiols [173], hydroxylamine and its derivatives [133], carboxylic acids [174] and phosphonic or phosphinic acids [175,176] yield covalent compounds R_4Sb-Y where Y is the corresponding acid residue. Surprisingly enough, fluorotetramethyl- and hydroxytetramethyl-stiborane which are obtainable from pentamethylstiborane with trimethyltin fluoride and hydroxide, respectively, also show no evidence of ionic character [172]. With other acids [166,176] and also with halogens [31] tetraalkylstibonium salts are formed as well as with Lewis acids of the triphenylborane [31], trialkylalane [162], tin dichloride and dicyclopentadienyltin type [177]. Finally, with two equivalents of carboxylic acid or an appropriate dicarboxylic acid, hexacoordinate chelate complexes such as *183* with strong intramolecular hydrogen bonds are obtained [178].

Decomposition of pentabutylstiborane in tetrachloromethane evidently follows a radical reaction sequence [164] analogous to that proposed for pentaphenyl-stiborane [144].

Not only pentaalkyl but also pentaalkenylstiboranes are so stable, that a greater variety of derivatives could be investigated. Synthetically, the usual methods starting from trialkenyldihalostiboranes or even antimony pentachloride and appropriate alkenyl magnesium or lithium compounds were applied. Thus, pentavinyl [179], cis- and trans-pentapropenyl and pentaisopropenylstiborane [180] were accessible without difficulties [181]. Similarly, diethyltrialkenyl- and diphenyltrialkenylstiboranes have been prepared [182]. The chemistry of all these compounds resembles closely that of the other pentaorganylstiboranes known. Thermally two Sb—C bonds are cleaved with the formation of trialkenylstibines; with bromine one or two alkenyl groups can be removed with the formation of tetraalkenylstibonium salts or trialkenyldibromostiboranes, respectively [179-182].

Very recently an alkynyl group containing stiborane, namely, trimethyldipropynylstiborane was synthesized according to the preparative principles outlined above. A crystal structure determination has shown that this compound again adheres to the trigonal-bipyramidal structure domain with the two propynyl ligands in axial positions [166].

12 Penta- and Hexaaryl Bismuth Compounds

Because of the low stability of pentaorganyl bismuth compounds it is not surprising that after the first report of Wittig and Clauß on the preparation of pentaphenyl-bismuthorane [26] only very few further investigations were communicated. In derivatization studies, it was shown that neither the violet triphenyl-bis-p-chlorophenyl-bismuthorane (184) nor the orange 2,2'-biphenylylenetriphenylbismuthorane (185) (5,5,5-triphenyl-5,5-dihydro-5H-dibenzobismuthole) were significantly more stable than pentaphenylbismuthorane itself [183]. With acids, the dibenzobismuthole ring of 185 is cleaved and 2-biphenylyltriphenylbismuthonium salts 186 are formed. With excess phenyllithium, 185 reacts, as does pentaphenylbismuthorane, reversibly to a yellow ate-complex 187 which is only observable at low temperatures [183].

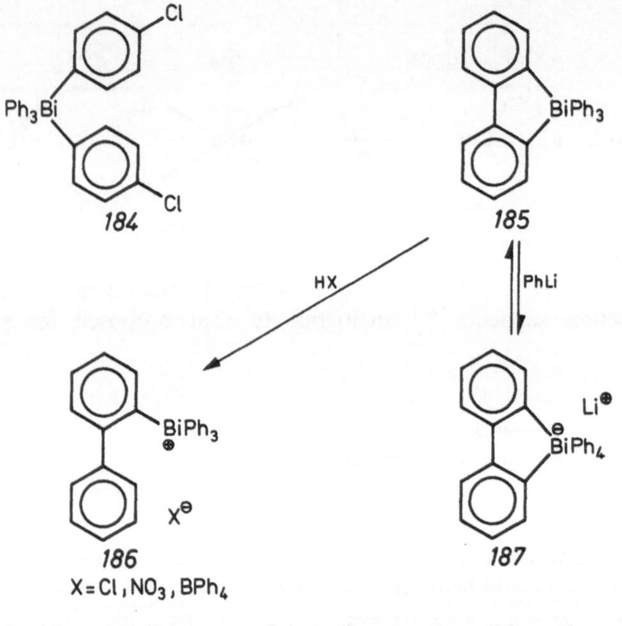

More recently, decomposition reactions of pentaphenylbismuthorane in organic solvents have been investigated by Russian workers. Thus with isopropanol and phenol, triphenylbismuthine, benzene and acetone or diphenylether, respectively, are formed [184]. The decomposition in pyridine, leading (among other products) again to triphenylbismuthine and benzene, was explained by means of a dehydrobenzene pathway [185]. Interesting phenylation reactions occur when pentaphenylbismuthorane is treated with phenols, naphthols or 2-nitropropane under mild conditions, ortho-phenylated phenols, naphthols and 2-phenyl-2-nitropropane being the isolable products [186]. With thiophenol a reduction/elimination reaction takes place, and with benzylalcohol a redox process leading to benzaldehyde is observed [186].

$$ArSH + Ph_5Bi \longrightarrow PhH + Ar-S-BiPh_4 \longrightarrow Ar-S-Ph + Ph_3Bi$$

Finally, in the β-decay of solid pentaphenylbismuthorane tagged with ^{210}Bi various phenyl polonium compounds have been detected [187].

The only other compound known containing five Bi—C bonds is dicyanotriphenyl-bismuthorane which is obtained from dichlorotriphenylbismuthorane and aqueous! potassium cyanide. The compound is described as a white solid for which on the basis of detailed IR and Raman studies a trigonal-bipyramidal structure with axial cyano groups is proposed [188].

13 Concluding Remarks

The incentives for the present review have been twofold in so far as a comprehensive treatment of derivatives of the nitrogen/phosphorus group elements comprising five covalent bonds to carbon groups should be combined with a specific evaluation of Georg Wittig's pioneering contributions to this field. Therefore, particularly in the first few chapters, an attempt has been made to outline as chronologically as possible the conflux of experiments and conceptions that ultimatly led Wittig to his systematic studies of pentaarylphosphoranes and their congeners. In that context, outstanding significance was attributed to the concomitant discovery of the Wittig reaction, which demonstrated very distinctly how a genuine fundamental research program can load quite incidentally and even accidentally to results of highest practical importance. In the present times of widely propagated misunderstandings concerning the role and relevance of basic research, the explicit demonstration of such fortuitous interrelationships between purpose and outcome in science is certainly of some necessity.

Now, after more than 30 years of more or less systematic research in this area, some general features have emerged which can be assembled as follows. When, in the pentaorganyl derivatives of phosphorus, arsenic, antimony and bismuth, aryl groups are prevailing, readily crystallizable substances are obtained. When, on the other hand, alkyl groups are in the majority, distillable liquids result for the more viable systems. The overall stabilities and reactivities of such compounds seem to be governed mainly by stereochemical factors — bulkyness of the groups when only monodentate ligands are present and angle strain when di- or multidentate subtituents prevail. Pentaorganylbismuth compounds are generally of low stability, which apparently reflects the low strength of Bi—C bonds [189]. Thermally all these compounds are cleaved or isomerized to give tricoordinated and/or tetracoordinated (ylidic) species, depending on the number and bonding situation of the aryl and alkyl groups present.

With electrophilic agents, in kinetically controlled reactions usually one ligand is removed with the formation of tetraorganyl-onium salts. With strong nucleophiles, such as organyllithium compounds, ligand exchange reactions can occur which involve hexacoordinated intermediates or transition states. In special cases, when certain basic stereochemical requirements are met, such hexacoordinate ate-complexes can attain considerable stability. Exceptional stabilizing effects have been brought about with the bidentate 2,2'-biphenylylene ligand type, particularly for penta- and hexacoordinated phosphorus and arsenic compounds. In passing it should be noted that this particular ligand system proved equally effective in stabilizing similar tri- and tetracovalent derivatives of iodine [190], selenium and tellurium [191], respectively.

Deceivingly, after all, analogous work aimed at stable pentacoordinated nitrogen derivatives was met with failure.

From the beginning, the stereochemical discussions pertaining to pentaorganylphosphoranes, -arsoranes, -stiboranes and -bismuthoranes were related to a trigonal-bipyramidal arrangement of the five carbon ligands around the hetero center. With very few exceptions this model has been substantiated in its essential features by a wealth of stereochemical investigations in solution, of neat liquids and of crystalline samples. Thus, the stereochemistry of pentaorganyl derivatives of the phosphorus group elements can be described in terms of more or less significantly distorted trigonal-bipyramidal ground states, which, as a rule, are subject to more or less rapid intramolecular ligand exchange processes of the Berry type running through tetragonal-pyramidal transition states.

Although very many pentacoordinated derivatives of phosphorus, arsenic, antimony and bismuth of all sorts have so far been synthesized, the fundamental hydrogenated skeletons PH_5, AsH_5, SbH_5 and BiH_5 are still unknown materially. Theoreticians, on the other hand, have been familiarizing themselves for a long time with those prototypic species by virtue of more or less elaborate calculations [48, 76, 192].

Over the years, partial aspects of the material discussed in this report have been treated under widely varying headings throughout the literature; therefore, a representative bibliography of pertinent review articles and monographs coming before the references to the original literature is thought appropriate.

14 Acknowledgements

The author wants to take this opportunity to express his most sincere thanks to Prof. Dr. G. Wittig for nearly a quarter century of close collaboration, fruitful interaction and manifold support. Parts of the author's own research in this field were generously supported by Deutsche Forschungsgemeinschaft, Fonds der Chemischen Industrie, BASF-Ludwigshafen/Rh. and a NATO-research grant; this is gratefully acknowledged.

15 Bibliography

Über metallorganische Komplexverbindungen. *G. Wittig*, Angew. Chem. *62*, 231 (1960).

Über Ylide und Ylid-Reaktionen. *G. Wittig*, Angew. Chem. *63*, 15 (1951).

Fortschritte auf dem Gebiet der organischen Aniono-Chemie. *G. Wittig*, Angew. Chem. *66*, 10 (1954).

Synthesen über Stickstoff-ylide und Phosphor-ylene. *G. Wittig*, Experientia *XII*, 41 (1956).

Ursprung und Entwicklung in der Chemie der Phosphin-alkylene. *G. Wittig*, Angew. Chem. *68*, 505 (1956).

Komplexbildung und Reaktivität in der metallorganischen Chemie. *G. Wittig*, Angew. Chem. *70*, 65 (1958).

Neuere präparative Methoden der organischen Chemie II. Carbonyl-Olefinierung mit Triphenyl-phosphin-methylenen. Wittig Reaktion. *U. Schöllkopf*, Angew. Chem. *71*, 260 (1959).

Absicht und Zufall in der Chemie der alkalimetallorganischen Verbindungen. *G. Wittig*, Verhandlungen der Schweiz. Naturforschenden Gesellschaft, Zürich *1964*, 19.

Variationen zu einem Thema von Staudinger; ein Beitrag zur Geschichte der phosphororganischen Carbonyl-Olefinierung. *G. Wittig*, Pure and Appl. Chem. *9*, 245 (1964).

The Role of Ate Complexes as Reaction-Determining Intermediates. *G. Wittig*, Quarterly Reviews, *20*, 191 (1966).

Pentacoordination. *E. L. Muetterties* and *R. A. Schunn*, Quarterly Reviews, *20*, 245 (1966).

De la Chimie du Phosphore Pentavalent. *G. Wittig*, Bull. Soc. Chim. France, 1966, 1162.

Pseudorotation in the Hydrolysis of Phosphate Esters. F. H. Westheimer, Acc. Chem. Res. *1*, 70 (1968).

Stereochemical Aspects of Phosphorus Chemistry. *M. J. Gallgher* and *I. D. Jenkins*, Top. Stereochem. *3*, 1 (1968).

Hypervalente Moleküle. *J. I. Musher*, Angew. Chem. *81*, 68 (1969); Angew. Chem. Int. Ed. Engl. *8*, 54 (1969).

Organometallic Compounds of Arsenic, Antimony and Bismuth. *G. O. Doak* and *L. D. Freedman*, Wiley-Interscience, 1970.

The Heterocyclic Derivatives of Phosphorus, Arsenic, Antimony and Bismuth. *F. G. Mann*, Wiley-Interscience, 1970.

Chimie et Stéréochimie de Nouveaux Dérivés du Phosphore Penta et Hexavalent. *D. Hellwinkel*, Colloques internationaux du Centre National de la Recherche Scientifique, *182*, 177 (1970).

Bewegliche Molekülgerüste — Pseudorotation und Turnstilerotation pentakoordinierter Phosphorverbindungen und verwandte Vorgänge. *P. Gillespie, P. Hoffmann, H. Klusacek, D. Marquarding, S. Pfohl, F. Ramirez, E. A. Tsolis* und *I. Ugi*, Angew. Chem. *83*, 691 (1971); Angew. Chem. Int. Ed. Engl. *10*, 687 (1971).

Penta- and Hexaorganophosphorus Compounds. *D. Hellwinkel*, in Organic Phosphorus Compounds, Ed. by G. M. Kosolapoff and L. Maier, Vol. 3, p. 185, Wiley, 1972.

Hypervalent Organic Derivatives of Tellurium and Selenium. *D. Hellwinkel*, Annals of the New York Academy of Sciences, *192*, 158 (1972).

Dynamic Stereochemistry of Pentacoordinated Phosphorus and Related Elements. *R. Luckenbach*, Georg Thieme Verlag, Stuttgart, 1973.

The structural Chemistry of Phosphorus. *D. E. C. Corbridge*, Elsevier, Amsterdam, 1974.

Organo-Phosphorus Stereochemistry, Part 2, P(V) Compounds. Edit. *W. E. McEwen, K. Darrell Berlin*, Hutchinson and Ross, Inc., Stroudsburg, Pennsylvania, 1975.

Pentaalkyls and Alkylidene Trialkyls of the Group V Elements. *H. Schmidbaur*, Advances in Organomet. Chem., *14*, 205 (1976).

Polycyclic Carbon-Phosphorus Heterocycles. *S. D. Venkataramu, G. D. MacDonell, W. R. Purdum, M. El-Deek* and *K. D. Berlin*, Chem. Rev. 77 121 (1977).

Phosphorus Heterocycles (Including some Arsenic and Antimony Heterocycles). *M. J. Gallgher* in W. L. F. Armarego, Stereochemistry of Heterocyclic Compounds, Part II, p. 339, Wiley, 1977.

Stereochemistry of Penta- and Hexacoordinate Phosphorus Derivatives. *W. S. Sheldrick*, Top. Curr. Chem. 73, 1 (1978).

Methoden zur Darstellung und Umwandlung von Organo-arsen-, antimon- und wismuth-Verbindungen. *S. Samaan* in Methoden der Organischen Chemie (Houben-Weyl) Band XIII/8, Thieme Verlag, 1978.

Arsenheterocyclen. *A. Tzschach* und *J. Heinicke*, VEB Deutscher Verlag für Grundstoffindustrie Leipzig, 1978.

A Model for Calculating Conformational Energies in Pentacoordinate Phosphorus Compounds. *R. R. Holmes*, J. Am. Chem. Soc. 100, 433 (1978).

Pentacoordinate Phosphoranes in Synthesis. *K. Burger*, in Organophosphorus Reagents in Organic Synthesis, Edit. J. I. G. Cadogan, p. 467, Academic Press, London, 1979.

Structure of Cyclic Pentacoordinated Molecules of Main Group Elements. R. R. Holmes, Accounts Chem. Res. 12, 257 (1979).

The Polytopal Rearrangement at Phosphorus. *F. H. Westheimer* in Rearrangements in Ground and Exited States, Vol. 2, p. 229, Ed. P. de Mayo, Academic Press, (1980).

Pentacoordinated Phosphorus Compounds, Spectroscopy and Structure. *R. R. Holmes*, ACS Monograph No 175, 1981.

16 References

1. Cahours, A.: Liebigs Ann. Chem. *122*, 329 (1862)
2. Lachman, A.: Am. Chem. J. *18*, 372 (1896)
3. For early discussions see: Willgerodt, C.: J. Prakt. Chem. *37*, 449 (1888); *41*, 291, 526 (1890)
   Behrend, R.: Ber. Dtsch. Chem. Ges. *23*, 454 (1890)
   Bischoff, C. A.: ibid. *23*, 1967 (1890)
4. Schlenk, W., Holtz, J.: Ber. Dtsch. Chem. Ges. *49*, 603 (1916)
5. Schlenk, W., Holtz, J.: ibid. *50*, 274 (1917)
6. see also: Werner, A.: Lehrbuch der Stereochemie, Fischer, Jena, 1904, p. 311 and
   Werner, A.: Neuere Anschauungen auf dem Gebiete der anorg. Chemie, Viehweg + Sohn, Braunschweig 1905, p. 96
7. Kossel, W.: Ann. Physik [4] *49*, 229 (1916)
8. Lewis, G. N.: J. Amer. Chem. Soc. *38*, 762 (1916)
9. Langmuir, T.: ibid. *41*, 868, 892 (1919); *42*, 274 (1920)
10. Staudinger, H., Meyer, J.: Helv. Chim. Acta 2, 608, 619 (1919)
11. Hager, F. D., Marvel, C. S.: J. Amer. Chem. Soc. *48*, 2689 (1926)
12. Coffman, D. D., Marvel, C. S.: ibid. *51*, 3496 (1929)
13. a) Staudinger, H., Meyer, J.: Helv. Chim. Acta 2, 612 (1919)
    b) Staudinger, H., Meyer, J.: ibid. *2*, 635 (1919)
14. Friedrich, M. E. P., Marvel, C. S.: J. Amer. Chem. Soc. *52*, 376 (1930)
15. Wittig, G.: Angew. Chem. *53*, 241 (1940)
16. Wittig, G.: Naturwissenschaften *30*, 696 (1942)
17. Wittig, G.: Angew. Chem. *69*, 245 (1957); see also: l. c. [16], p. 699. 700
    Wittig, G., Pieper, G., Fuhrmann, G.: Ber. Dtsch. Chem. Ges. *73*, 1193 (1940)
    Wittig, G.: Suomen Kemistilethi *A29*, 283 (1956)

18. Wittig, G., Heintzeler, M., Wetterling, M.-H.: Liebigs Ann. Chem. *557*, 201 (1947)
19. Wittig, G., Wetterling, M.-H.: ibid. *557*, 193 (1947)
20. Wittig, G., Felletschin, G.: ibid. *555*, 133, Footnote 2, (1944)
21. Brockway, L. O., Beach, J. Y.: J. Amer. Chem. Soc. *60*, 1836 (1938); Rouault, M.: C. R. Hebd. Séances Acad. Sci. *207*, 620 (1938)
22. Wittig, G., Rieber, M.: Liebigs Ann. Chem. *562*, 177 (1949); see also: Wittig, G.: Angew. Chem. *63*, 15 (1951)
23. Wittig, G., Rieber, M.: Naturwissenschaften *35*, 345 (1948); Liebigs Ann. Chem. *562*, 187 (1949)
24. a) Razuvaev, G. A., Osanova, N. A., Grigor'eva, I. K.: Izvest. Akad. Nauk SSSR, Ser. Khim. *1969*, 2234 (C.A. *72*, 31952c (1970))

    b) Razuvaev, G. A., Osanova, N. A.: Dokl. Akad. Nauk. SSSR *104*, 552 (1955), (C.A. *50*, 11268c (1956))

    Razuvaev, G. A., Osanova, N. A.: Zhur. Obshch. Khim. *26*, 2531 (1956), (C.A. *51*, 1875b (1957))

    Razuvaev, G. A., Osanova, N. A., Shlyapnikova, I. A.: Zhur. Obshch. Khim. *27*, 1466 (1957), (C.A. *52*, 3715b (1958))
25. Wittig, G., Clauß, K.: Liebigs Ann. Chem. *577*, 26 (1952)
26. Wittig, G., Clauß, K.: ibid. *578*, 136 (1953).
    See also: Gilman, H., Yablunky, H. L.: J. Amer. Chem. Soc. *63*, 839 (1941) for earlier unsuccessful attempts
27. Bach, M.: Dissertation, Heidelberg 1968
28. Hellwinkel, D.: Habilitationsschrift, Heidelberg 1966
29. Hellwinkel, D., Wünsche, C., Bach, M.: Phosphorus *2*, 167 (1973)
30. Hellwinkel, D., Kilthau, G.: Liebigs Ann. Chem. *705*, 66 (1967). This yellow color at low temperature had already been observed by Wittig and Clauß l.c. [26] p. 142
31. Wittig, G., Torssell, K.: Acta Chem. Scand. *7*, 1293 (1953)
32. An earlier report on the preparation of dialkyltriphenylphosphoranes by Grignard, V., Savard, J.: C. R. Hebd. Séances Acad. Sci. *192*, 592 (1931) could not be reproduced by later workers: Blount, B. K.: J. Chem. Soc. *1931*, 1891; *1932*, 337; Denney, D. B., Gross, F. J.: J. Org. Chem. *32*, 3710 (1967)
33. Wittig, G., Geissler, G.: Liebigs Ann. Chem. *580*, 44 (1953)
34. Wittig, G.: Pure and Appl. Chem. *9*, 245 (1964)
35. Vedejs, E., Snoble, K. A. J.: J. Amer. Chem. Soc. *95*, 5778 (1973)
    Schlosser, M. et al.: Chimia *29*, 341 (1975)
    Vedejs, E., Meier, G. P., Snoble, K. A. J.: J. Amer. Chem. Soc. *103*, 2823 (1981)
    See also: Schlosser, M., BaTuong, H.: Angew. Chem. *91*, 675 (1979); Angew. Chem. Internat. Ed. Engl. *18*, 633 (1979)
    Giese, B., Schoch, J., Rüchardt, C.: Chem. Ber. *111*, 1395 (1978)
    Bestmann, H. J.: Pure and Appl. Chem. *52*, 771 (1980)
36. Some selected reviews: Wittig, G.: Angew. Chem. *68*, 505 (1956)
    Schöllkopf, U.: Angew. Chem. *71*, 260 (1959)
    Maercker, A.: Org. Reactions *14*, 270 (1965)
    Johnson, A. W.: Ylid Chemistry, Academic Press, New York 1966
    Schlosser, M.: in Korte, F., Zimmer, H., Niedenzu, K.: Methodicum Chimicum, Band 7, p. 529, Thieme Verlag, Stuttgart 1976; Schlosser, M.: Topics in Stereochemistry *5*, 1 (1970)
    Pommer, H.: Angew. Chem. *89*, 437 (1977); Angew. Chem. Internat. Edit. Engl. *16*, 423 (1977)
    Gosney, I., Rowley, A. G.: in Cadogan, J. I. G.: Organophosphorus Reagents in Organic Synthesis, Academic Press, London 1979, chapter 2, p. 17
37. Nast, R., Käb, K.: Liebigs Ann. Chem. *706*, 75 (1967)
38. Seyferth, D., Fogel, J., Heeren, J. K.: J. Amer. Chem. Soc. *88*, 2207 (1966)
39. Razuvaev, G. A., Petukhov, G. G., Osanova, N. A.: Dokl. Akad. Nauk SSSR, *104*, 733 (1955); (C.A. *50*, 11268f (1956))
    Razuvaev, G. A., Petukhov, G. G., Osanova, N. A.: Zhur. Obshch. Khim. *29*, 2980 (1959); (C. A. *54*, 12030e (1960))
    ibid. *31*, 2350 (1961); (C. A. *56*, 3506i (1962)
40. Kochendoerfer, E.: Dissertation, Heidelberg 1959

41. Gielen, M.: in Chemical Applications of Graph Theory (ed. Balaban, A. T.) Academic Press, London 1976. Chapter 9, p. 261
42. Wittig, G., Kochendoerfer, E.: Angew. Chem. 70, 506 (1958); Chem. Ber. 97, 741 (1964)
43. Mann, F. G., Chaplin, E. J.: J. Chem. Soc. 1937, 527
44. Hellwinkel, D.: Diplomarbeit, Heidelberg 1960
    Wittig, G., Hellwinkel, D.: Angew. Chem. 74, 76 (1962)
45. Wittig, G., Maercker, A.: Chem. Ber. 97, 747 (1964)
46. Schlosser, M., Kadibelban, T., Steinhoff, G.: Liebigs Ann. Chem. 743, 25 (1971)
47. Hellwinkel, D., Lindner, W.: Chem. Ber. 109, 1497 (1976)
48. see Hoffmann, R., Howell, J. M., Muetterties, E. L.: J. Amer. Chem. Soc. 94, 3047 (1972)
    Hall, C. D., Bramblett, J. D., Lin, F. F. S.: J. Amer. Chem. Soc. 94, 9264 (1972)
    Howell, J. M.: J. Amer. Chem. Soc. 99, 7447 (1977)
49. see for example Dewar, M. J. S., Dougherty, R. C.: The PMO Theory of Organic Chemistry, Plenum Press, New York 1975
50. Hellwinkel, D., Krapp, W.: Chem. Ber. 110, 693 (1977)
51. Hellwinkel, D., Wilfinger, H. J.: ibid. 103, 1056 (1970)
52. John, K. P., Schmutzler, R., Sheldrick, W. S.: J. Chem. Soc. Dalton Transact. 1974, 1841, 2466
53. Munoz, A. et al.: C. R. Acad. Sci. Ser. C 280, 395 (1975)
54. On apicophilicity see Bone, S., Trippett, S., Whittle, P. J.: J. Chem. Soc. Perkin Transact. 1, 1974, 2125
    Buono, G., Llinas, J. R.: J. Amer. Chem. Soc. 103, 4532 (1981)
55. Hellwinkel, D., Krapp, W.: Chem. Ber. 111, 13 (1978)
56. For reviews on ate complexes see: Wittig, G.: Quarterly Reviews 20, 191 (1966)
    Tochtermann, W.: Angew. Chem. 78, 355 (1966); Angew. Chem. Internat. Ed. Engl. 5, 351 (1966)
57. Wittig, G., Angew. Chem. 62, 231 (1950)
58. Hellwinkel, D.: Chem. Ber. 98, 576 (1965)
59. Schlosser, M., Kadibelban, T., Steinhoff, G.: Angew. Chem. 78, 1018 (1966); Angew. Chem. Internat. Ed. Engl. 5, 969 (1966)
60. Daniel, H., Paetsch, J.: Chem. Ber. 101, 1451 (1968)
61. Ross, B. W., Reetz, K. P.: ibid. 107, 2720 (1974)
62. Wheatley, P. J., Wittig, G.: Proc. Chem. Soc. 1962, 251
63. Wheatley, P. J.: J. Chem. Soc. 1964, 2206
64. For a preliminary note see also: Hellwinkel, D.: Angew. Chem. 76, 756 (1964)
65. Richards, E. M., Tebby, J. C.: J. Chem. Soc. Ç 1970, 1425
66. Hellwinkel, D.: Chem. Ber. 99, 3628 (1966)
67. Hellwinkel, D., Mason, S. F.: J. Chem. Soc. B 1970, 640
68. Hellwinkel, D., Wilfinger, H. J.: Phosphorus 2, 87 (1972)
69. Berry, R. S.: J. Chem. Phys. 32, 933 (1960)
70. Gillespie, R. J.: J. Chem. Soc. 1963, 4672
71. Muetterties, E. L., Mahler, W., Schmutzler, R.: Inorg. Chem. 2, 613 (1963)
72. Hellwinkel, D.: Chem. Ber. 99, 3660 (1966)
73. Hellwinkel, D.: ibid. 99, 3642 (1966)
74. Hellwinkel, D., Krapp, W.: ibid. 112, 292 (1979)
75. Ugi, I. et al.: Angew. Chem. 82, 766 (1970); Angew. Chem. Internat. Ed. Engl. 9, 729 (1970)
    Gillespie, P. et al.: Angew. Chem. 83, 691 (1971); Angew. Chem. Internat. Ed. Engl. 10, 687 (1971)
76. Altmann, J. A., Yates, K., Csizmadia, I. G.: J. Amer. Chem. Soc. 98, 1450 (1976)
    Shih, S. K., Peyerimhoff, S. D., Buenker, R. J.: J. Chem. Soc. Faraday Transact. 2, 75, 379 (1979), and earlier lit. cited therein
76a. Finocchiaro, P., Gust, D., Mislow, K.: J. Amer. Chem. Soc. 95, 8172 (1973); 96, 3198, 3205 (1974)
76b. Hellwinkel, D. et al.: Chem. Ber. 108, 2219 (1975)
    Glaser, R., Blount, J. F., Mislow, K.: J. Amer. Chem. Soc. 102, 2777 (1980)
    For the first experimental studies see Hellwinkel, D., Melan, M., Degel, C. R.: Tetrahedron 29, 1895 (1973)

77. Hellwinkel, D.: Angew. Chem. *78*, 749 (1966); Angew. Chem. Internat. Ed. Engl. *5*, 725 (1966)
78. Hellwinkel, D.: Chimia *22*, 488 (1968)
79. Hellwinkel, D., Lindner, W., Wilfinger, H. J.: Chem. Ber. *107*, 1428 (1974)
    Hellwinkel, D., Wilfinger, H. J.: Tetrahedron Letters *1969*, 3423
80. For a critical evaluation of the coalescence temperature method see: Kost, D., Carlson, E. H., Raban, M.: J. Chem. Soc. D *1971*, 656
    Kost, D., Zeichner, A.: Tetrahedron Letters *1974*, 4533
81. Whitesides, G. M., Eisenhut, M., Bunting, W. M.: J. Amer. Chem. Soc. *96*, 5398 (1974)
    Whitesides, G. M., Bunting, M.: ibid. *89*, 6801 (1967)
82. Hellwinkel, D., Lindner, W., Schmidt, W.: Chem. Ber. *112*, 281 (1979)
83. Day, R. O., Husebye, S., Holmes, R. R.: Inorg. Chem. *19*, 3616 (1980)
84. Day, R. O., Holmes, R. R.: ibid. *19*, 3609 (1980)
85. Muetterties, E. L., Schunn, R. A.: Quarterly Reviews *20*, 245 (1966)
86. Holmes, R. R., Deiters, J. A.: J. Amer. Chem. Soc. *99*, 3318 (1977)
    Holmes, R. R.: Accounts Chem. Res. *12*, 257 (1979)
87. Sheldrick, W. S.: Top. Curr. Res. *73*, 1 (1978)
88. Holmes, R. R.: J. Amer. Chem. Soc. *100*, 433 (1978)
89. Luckenbach, R.: Dynamic Stereochemistry of Pentacoordinated Phosphorus and Related Elements, Thieme, Stuttgart 1973
90. Hamilton, W. C. et al.: J. Amer. Chem. Soc. *95*, 6335 (1973)
    Carrell, H. L. et al.: J. Amer. Chem. Soc. *97*, 38 (1975)
91. Hellwinkel, D., Schenk, W.: Angew. Chem. *81*, 1049 (1969); Angew. Chem. Internat. Ed. Engl. *8*, 987 (1969)
    Hellwinkel, D., Schenk, W., Blaicher, W.: Chem. Ber. *111*, 1798 (1978)
92. Hellwinkel, D. et al.: Chem. Ber. *113*, 1406 (1980)
93. Heal, H. G.: J. Inorg. Nucl. Chem. *16*, 208 (1961)
94. Hellwinkel, D.: Chem. Ber. *102*, 528 (1969)
95. Hellwinkel, D., Wilfinger, H. J.: Phosphorus *6*, 151 (1976)
96. Hellwinkel, D., Wünsche, C.: J. Chem. Soc. Chem. Commun. *1969*, 1412
97. Hellwinkel, D.: Chem. Ber. *102*, 548 (1969)
98. Rothuis, R., FontFreide, J. J. H. M., Buck, H. M.: Rec. Trav. Chim. Pays Bas *92*, 1308 (1973)
    see also Rothuis, R. et al.: Rec. Trac. Chim. Pays Bas *93*, 128 (1974) and
    van Dijk, J. M. F., Pennings, J. F. M., Buck, H. M.: J. Amer. Chem. Soc. *97*, 4836 (1975)
99. Hellwinkel, D., Wilfinger, H. J.: Liebigs Ann. Chem. *742*, 163 (1972)
100. Rothuis, R., Luderer, T. K. J., Buck, H. M.: Rec. Trav. Chim. Pays Bas *91*, 836 (1972)
101. For a recent review on the role of ion pairs in carbanion chemistry see Hogen-Esch, T. E.: Adv. Phys. Org. Chem. *15*, 154 (1977)
102. Wittig, G., Maercker, A.: J. Organomet. Chem. *8*, 491 (1967)
     see also Kyba, E. P.: J. Amer. Chem. Soc. *98*, 4805 (1976)
103. Musher, J. I.: Angew. Chem. *81*, 68 (1969; Angew. Chem. Internat. Ed. Engl. *8*, 54 (1969)
104. Nesmeyanov, A. N., Tolstaya, T. P., Grib, A. V.: Dokl. Akad. Nauk SSSR, *153*, 608 (1963), (C. A. *60*, 6812b (1964))
105. Nesmeyanov, A. N. et al.: Dokl. Akad. Nauk SSSR *199*, 610 (1971), (C. A. *75*, 140574p (1971))
     Nesmeyanov, A. N. et al.: Izvest. Akad. Nauk SSSR, Ser. Khim. *1973*, 2632, (C. A. *81*, 49503c (1974))
106. Hellwinkel, D., Seifert, H.: Chem. Ber. *105*, 880 (1972)
107. Hellwinkel, D., Seifert, H.: Liebigs Ann. Chem. *762*, 29 (1972)
108. See Wittig, G., Benz, E.: Chem. Ber. *92*, 1999 (1959)
     Hoffmann, R. W.: Dehydrobenzene and Cycloalkynes, Verlag Chemie, Winheim/Bergstraße 1967, p. 173
109. Seyferth, D., Hughes, W. B., Heeren, J. K.: J. Amer. Chem. Soc. *87*, 2847, 3467 (1965)
110. a) Katz, T. J., Turnblom, E. W.: J. Amer. Chem. Soc. *92*, 6701 (1970)
     b) Turnblom, E. W., Katz, T. J.: ibid. *93*, 4065 (1971)
     c) Turnblom, E. W., Katz, T. J.: J. C. S. Chem. Commun. *1972*, 1270
     d) Turnblom, E. W., Katz, T. J.: J. Amer. Chem. Soc. *95*, 4292 (1973)
111. Bushweller, C. H. et al.: Tetrahedron Letters *1972*, 2401
112. Cremer, S. E. et al.: J. C. S. Chem. Commun. *1975*, 374

113. Turnblom, E. W., Hellwinkel, D.: ibid. *1972*, 404
114. For a recent review see Schmidbaur, H.: Adv. Organomet. Chem. *14*, 205 (1976)
115. Schmidbaur, H., Wolf, W.: Chem. Ber. *108*, 2834, 2842, 2851 (1975)
116. Schmidbaur, H., Holl, P., Köhler, F. H.: Angew. Chem. *89*, 748 (1977); Angew. Chem. Internat. Ed. Engl. *16*, 722 (1977)
    Schmidbaur, H., Holl, P.: Z. Anorg. Allg. Chem. *458*, 249 (1979)
117. a) The, K. I., Cavell, R. G.: J. C. S. Chem. Commun. *1975*, 716
    b) Cavell, R. G., Gibson, J. A., The, K. I.: J. Amer. Chem. Soc. *99*, 7841 (1977)
    c) Cavell, R. G., Gibson, J. A., The, K. I.: Inorg. Chem. *17*, 2880 (1978)
118. Douglas, J. E.: Inorg. Chem. *11*, 654 (1972)
    See also Reddy, G. S., Weis, C. D.: J. Org. Chem. *28*, 1822 (1963)
119. Denney, D. B., Denney, D. Z., Hsu, Y. F.: Phosphorus *4*, 217 (1974)
120. See for example Hellwinkel, D. in: Organic Phosphorus Compounds, Vol. 3, (ed. Kosolapoff, G. M., Maier, L.) Wiley, New York 1972, p. 185
121. Wittig, G., Hellwinkel, D.: Chem. Ber. *97*, 769 (1964)
122. Hellwinkel, D., Kilthau, G.: ibid. *101*, 121 (1968)
123. Hellwinkel, D., Knabe, B.: ibid. *104*, 1761 (1971)
124. Hellwinkel, D., Kilthau, G.: Angew. Chem. *78*, 1018 (1966), Angew. Chem. Internat. Ed. Engl. *5*, 969 (1966)
125. Ross, B. W., Marzi, W. B.: Chem. Ber. *108*, 1518 (1975)
126. Beattie, I. R. et al.: J. C. S. Dalton Transactions *1972*, 784
    See also McKay, K. M., Sowerby, D. B., Young, W. C.: Spectrochim. Acta *A24*, 611 (1968)
127. Brock, C. P., Webster, D. F.: Acta Crystall. *B 32*, 2089 (1976)
128. Hellwinkel, D., Knabe, B.: Phosphorus *2*, 129 (1972)
129. Hellwinkel, D., Knabe, B., Kilthau, G.: J. Organometal. Chem. *24*, 165 (1970)
130. Grim, S. O., Seyferth, D.: Chem. Ind. *1959*, 849
    Wittig, G., Henry, M. C.: J. Amer. Chem. Soc. *82*, 563 (1960)
    Seyferth, D., Cohen, H. M.: J. Inorg. Nucl. Chem. *20*, 73 (1961)
131. Mitschke, K. H., Schmidbaur, H.: Chem. Ber. *106*, 3645 (1973)
132. Ott, R., Weidlein, J., Mitschke, K. H.: Chimia *29*, 262 (1975)
133. Eberwein, B., Ott, R., Weidlein, J.: Z. Anorg. Allg. Chem. *431*, 95 (1977)
134. Wittig, G., Hellwinkel, D.: Chem. Ber. *97*, 789 (1964)
135. Hellwinkel, D., Bach, M.: J. Organometal. Chem. *17*, 389 (1969)
136. Doleshall, G., Nesmeyanov, N. A., Reutov, O. A.: ibid. *30*, 369 (1971)
137. Hellwinkel, D., Bach, M.: ibid. *28*, 349 (1971)
138. Hellwinkel, D., Bach, M.: ibid. *20*, 273 (1969)
139. Sauers, R. R.: Chem. Ind. *1960*, 717
140. Shen, K. W., McEwen, W. E., Wolf, A. P.: J. Amer. Chem. Soc. *91*, 1283 (1969)
141. Nefedov, V. D., Kirin, I. S., Zaitsev, V. M.: Radiokhimiya *6*, 78 (1964), (C. A. *61*, 3876g, 3877d (1964))
142. Razuvaev, G. A. et al.: Zhur. Obshch. Khim. *30*, 3234 (1960), (C. A. *55*, 21010c (1961))
143. Razuvaev, G. A., Osanova, N. A., Sangalov, Y. A.: Zhur. Obshch. Khim. *37*, 216 (1967), (C. A. *66*, 95139b (1967))
144. McEwen, W. E., Lin, C. T.: Phosphorus *4*, 91 (1974)
145. Raynier, B. et al.: Nouv. J. Chem. *3*, 393 (1979)
146. Angelelli, J. M. et al.: J. Amer. Chem. Soc. *91*, 4500 (1969)
    See also Taft, R. W. et al.: ibid. *85*, 3146 (1963)
147. Nesmeyanov, A. N., Mikul'shina, V. V., Reutov, O. A.: Izv. Akad. Nauk SSSR, Ser. Khim. *1976*, 2397), (C. A. *86*, 72788m (1977)); Dokl. Akad. Nauk SSSR *237*, 1111 (1977), (C. A. *88*, 120704w (1978))
148. Sowerby, D. B.: J. Chem. Res. (S) *1979*, 80
149. Briles, G. H., McEwen, W. E.: Tetrahedron Letters *1966*, 5191
150. Chupka jr., F. L., Knapczyk, J. W., McEwen, W. E.: J. Org. Chem. *42*, 1399 (1977)
151. Lanneau, G. F. et al.: J. Organometal. Chem. *85*, 179 (1975)
    McEwen, W. E., Lin, C. T.: Phosphorus *3*, 229 (1974)
152. Matsuda, H., Ninagawa, A., Nomura, R.: Chem. Letters *1979*, 1261
153. Wheatly, P. J.: J. Chem. Soc. *1964*, 3718

154. Beauchamp, A. L., Bennett, M. J., Cotton, F. A.: J. Amer. Chem. Soc. *90*, 6675 (1968)
155. Such solvates had already been identified by Wittig and Clauß [25]
156. Brabant, C., Blanck, B., Beauchamp, A. L.: J. Organometal. Chem. *82*, 231 (1974)
157. Brabant, C., Hubert, J., Beauchamp, A. L.: Canad. J. Chem. *51*, 2952 (1973)
158. Kuykendall, G. L., Mills, J. L.: J. Organometal. Chem. *118*, 123 (1976)
159. Brock, C. P.: Acta Crystall. *A 33*, 193, 898 (1977)
    Brock, C. P., Ibers, J. A.: ibid. *A32*, 38 (1976)
160. Weiss, J.: Heidelberg, 1964, unpublished
161. Hellwinkel, D., Bach, M.: Naturwissenschaften *54*, 214 (1969)
162. Takashi, Y.: J. Organometal. Chem. *8*, 225 (1967)
163. Cowley, A. H. et al.: J. Amer. Chem. Soc. *93*, 2150 (1971)
164. Nesmeyanov, A. N., Borisov, A. E., Kizim, N. G.: Izv. Acad. Nauk SSSR, Ser. Khim. *1974*, 1672 (C. A. *81*, 105652q (1974))
165. Meinema, H. A., Noltes, J. G.: J. Organometal. Chem. *22*, 653 (1970)
166. Tempel, N., Schwarz, W., Weidlein, J.: ibid. *154*, 21 (1978)
167. Schmidbaur, H., Haßlberger, G.: Chem. Ber. *111*, 2702 (1978)
168. Downs, A. J., Schmutzler, R., Steer, I. A.: J. C. S. Chem. Commun *1966*, 221
169. Holmes, R. R., Deiters, R. M., Golen, J. A.: Inorg. Chem. *8*, 2612 (1969)
170. Muetterties, E. L. et al.: ibid. *3*, 1298 (1964)
171. For a review see Schmidbaur, H.: Adv. Organomet. Chem. *14*, 232 (1976)
172. Schmidbaur, H., Weidlein, J., Mitschke, K. H.: Chem. Ber. *102*, 4136 (1969)
173. Schmidbaur, H., Mitschke, K. H.: ibid. *104*, 1837, 1842 (1971)
174. Schmidbaur, H., Mitschke, K. H., Weidlein, J.: Z. Anorg. Allg. Chem. *386*, 147 (1971)
175. Graves, G. E., Van Wazer, J. R.: J. Organometal. Chem. *150*, 233 (1978)
176. Eberwein, B., Weidlein, J.: Z. Anorg. Allg. Chem. *420*, 229 (1976)
177. Bos, K. D. et al.: J. Organometal. Chem. *168*, 159 (1979)
178. Schmidbaur, H., Mitschke, K. H.: Angew. Chem. *83*, 149 (1971), Angew. Chem. Internat. Ed. Engl. *10*, 136 (1971)
179. Nesmeyanov, A. N., Borisov, A. E., Novikova, N. V.: Izv. Akad. Nauk SSSR Otdel. Khim. Nauk *1960*, 952, (C. A. *54*, 24351a (1960))
180. Nesmeyanov, A. N., Borisov, A. E., Novikova, N. V.: ibid. *1960*, 147; *1961*, 612 (C. A. *54*, 20853d (1960), *55*, 22100c (1961)); Tetrahedron Letters *1960* (8), 23
181. Nesmeyanov, A. N., Borisov, A. E., Novikova, N. V.: Izv. Akad. Nauk SSSR Otdel. Khim. Nauk *1961*, 1578; *1964*, 1202 (C. A. *56*, 4792d (1962); *61*, 12032f (1964))
182. Nesmeyanov, A. N., Borisov, A. E., Novikova, N. V.: ibid. *1961*, 730; *1964*, 1197 (C. A. *55*, 22101e (1961); *61*, 12032a (1964))
183. Hellwinkel, D., Bach, M.: Liebigs Ann. Chem. *720*, 198 (1968)
184. Razuvaev, G. A., Osanova, N. A., Sharutin, V. V.: Dokl. Akad. Nauk SSSR *225*, 581 (1975) (C. A. *84*, 105704v (1976))
185. Razuvaev, G. A. et al.: ibid. *238*, 361 (1978) (C. A. *88*, 105499q (1978))
186. Barton, D. H. et al.: J. C. S. Chem. Commun. *1980*, 827
187. Nefedov, V. D. et al.: Radiokhimiya *6*, 632 (1964) (C. A. *62*, 3600d (1965))
188. Goel, R. G., Prasad, H. S.: J. Organometal. Chem. *50*, 129 (1973)
189. Tel'noi, V. I., Rabinovich, I. B.: Usp. Khim. *49*, 1137 (1980)
190. Clauß, K.: Chem. Ber. *88*, 268 (1955). See also Hellwinkel, D., Haltmeier, M., Reiff, G.: Liebigs Ann. Chem. *1975*, 249
191. Wittig, G., Fritz, H.: ibid. *577*, 39 (1952)
    Hellwinkel, D., Fahrbach, G.: ibid. *712*, 1 (1968); *715*, 68 (1968); Chem. Ber. *101*, 574 (1968)
    Hellwinkel, D.: Annals of the New York Academy of Sciences *192*, 158 (1972)
192. For some key references see Kutzelnigg, W., Wallmeier, H., Wasilewski, J.: Theor. Chim. Acta *51*, 261 (1979)
    Strich, A., Veillard, A.: J. Amer. Chem. Soc. *95*, 5574 (1973)
    Florey, J. B., Cusachs, L. C.: ibid. *94*, 3040 (1972)
    Rauk, A., Allen, L. C., Mislow, K.: ibid. *94*, 3035 (1972)

Enantioselective Synthesis
of Nonproteinogenic Amino Acids

Ulrich Schöllkopf

Institut für Organische Chemie der Universität, Tammannstraße 2, 3400 Göttingen, FRG

Table of Contents

1 Introduction 66

2 Concept 67

3 Enantioselective Synthesis of α-Methyl Amino Acids
  from the Bis-lactim Ether (6) of Cyclo(L-Ala-L-Ala) (5) 69

4 Enantioselective Synthesis of α-Methyl Amino Acids
  from the Mixed Bis-lactim Ether (20a) of Cyclo(L-Val-Ala) 73
  4.1 Synthesis of the Bis-lactim Ether (20a) of Cyclo(L-Val-Ala)
      and other Mixed Bis-lactim Ethers (20) 74
  4.2 Alkylation of the Lithiated Bis-lactim Ether 22 of Cyclo(L-Val-Ala) . . 74
  4.3 Hydrolysis of the Alkylation Products 23; (R)-α-Methyl
      Amino Acid Esters 11 . . . . : . . . 76
  4.4 Reactions of the Lithiated Bis-lactim Ether 22
      with Carbonyl Compounds and Benzoylchloride 77

5 Enantioselective Synthesis of α-Unsubstituted Amino Acids
  from the "mixed" Bis-lactim Ethers 20b of Cyclo(L-Val-Gly)
  and 20d of Cyclo(L-tLeu-Gly) 78
  5.1 α-Unsubstituted Amino Acids by Alkylation
      of the Lithiated Bis-lactim Ether 31 of Cyclo(L-Val-Gly) 78
  5.2 α-Unsubstituted Amino Acids by Alkylation
      of the Lithiated Bis-lactim Ether (35) of Cyclo(L-tLeu-Gly) 79
  5.3 Reactions of the Lithiated Bis-lactim Ether (22) of Cyclo(L-Val-Gly)
      with Carbonyl Compounds 80

6 Enantioselective Synthesis of α-Unsubstituted Amino Acids
  from the Bis-lactim Ethers of Cyclo[(S)-α-Methyl(3,4-dimethyl)Phe-Gly] . . . 81

7 References 83

Ulrich Schöllkopf

1 Introduction

Optically active, non-proteinogenic (uncommon) amino acids deserve attention because of their documented or potential biological activity. Some are valuable pharmaceuticals, such as L-Dopa (*1*), (S)-α-Methyldopa (*2*), D-Penicillamine (*3*) or D-Cycloserine (*4*). Others are components of pharmaceuticals, for instance D-phenylglycine or D-thyrosine in the semisynthetic penicillines Ampicillin or Amoxycillin.

In biochemisty, uncommon amino acids are valuable tools to investigate the mechanism of enzyme reactions [1,2]. Some 250 uncommon amino acids have been isolated so far from microorganisms [2,3]. They are antibiotically active, i.e. they prevent cell growth of certain microorganisms. Some of them contain double bonds — either in β, γ- or in γ,δ-position — others contain leaving groups or possess the unusual D-configuration [2,3]. As much as they differ in structure, they have one thing in common: They are all enantiomerically pure; only one enantiomer displays the specific activity. This activity discloses how they work. They are antagonists of the physiological (proteinogenic) amino acids [1,2]. Interacting in the amino acid metabolism of the living cell they either inhibit or deceive an enzyme, preventing thus the proper biosynthesis or biodegradation of a proteinogenic amino acid or the correct biosynthesis of a protein [2]. In fact, enzyme inhibition studies with such uncommon amino acids furnished valuable information about the mode of action of certain enzymes [1,2]. This lead to a biochemist's recent remark [2]: "Selective alteration of some amino acids, or total synthesis of such compounds, could produce further biochemical tools for gaining more information about the mechanisms of enzymatic recognition and reaction. Such compounds could be of great value in aiding our understanding of the role played by these enzymes in cell physiology". Obviously, there is a demand for optically active — if possible optically pure — amino acids both for pure and applied organic chemistry or bioorganic chemistry.

It is, therefore, a challenge for the synthetic organic chemist to develop asymmetric syntheses of non-proteinogenic amino acids [4]. Asymmetric synthesis is the shortest and most economic way to optically active compounds [4,5]. However, a modern asymmetric synthesis has to fulfil several requirements which are not easily met in practice:

1. It should be easily performed and must give good chemical yields.
2. It must proceed with high optical yields, if possible with de or ee[1], respectively, close to 100%.
3. In a stoichiometric asymmetric synthesis the chiral auxiliary must be recoverable for recyclization.
4. The chiral auxiliary should be readily available, if possible from nature's chiral pool and in both enantiomers.
5. The configuration of the newly created asymmetric center should be predictable.

2 Concept

Over the past five years Schöllkopf et al. have tried to develop asymmetric syntheses of amino acids, that fulfil the above mentioned criteria [6]. Some of the results, achieved so far, are reported here. Their approach is based on heterocyclic chemistry and on the following strategy.

1. From a readily available racemic lower amino acid and a chiral auxiliary an heterocycle is built up, that is CH-acidic adjacent to the potential amino group and that contains two sites susceptible to hydrolysis.
2. An electrophile is introduced diastereoselectively via the anion of the heterocycle.
3. Subsequently the heterocycle is cleaved by hydrolysis to liberate the chiral auxiliary and the new optically active amino acid.

In this strategy, the heterocycle is of no interest by itself. It merely serves as a vehicle to construct an acyclic molecule with the proper structure and proper configuration. It makes use of the obvious fact, that an heterocyclic intermediate is necessaryly more rigid than its openchain analog so that a higher degree of asymmetric induction is to be expected.

Item 2 — the diastereoselectivity of the bond-forming step — is the crucial point in the concept. The underlying principle is shown schematically in Fig. 1. The circle symbolizes the heterocyclic ring with its anionic center. The diastereotopic faces

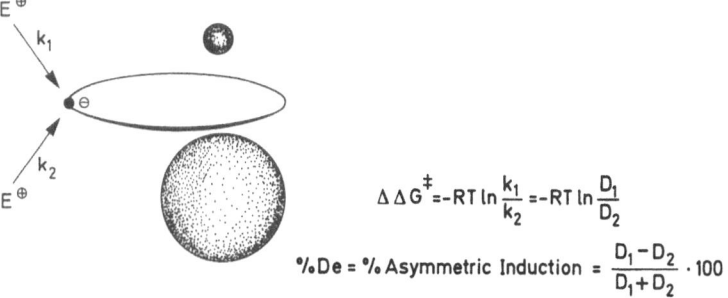

$$\Delta\Delta G^{\ddagger} = -RT \ln \frac{k_1}{k_2} = -RT \ln \frac{D_1}{D_2}$$

$$\%\,De = \%\,\text{Asymmetric Induction} = \frac{D_1 - D_2}{D_1 + D_2} \cdot 100$$

Fig. 1. Underlying Principle of the Asymmetric Induction, schematically (cf. text)

---

[1] de = diastereomeric excess (= asymmetric induction); ee = enantiomeric excess.

are shielded differently as indicated by the small and large ball, respectively. Provided kinetic control of product formation, the ratio of diastereomers $\dfrac{D_1}{D_2}$ is equal to $\dfrac{k_1}{k_2}$. This ratio is governed by $\Delta\Delta G^{\ast}$ — the difference of Free Enthalpy of activation of the two competing pathways — according to $\Delta\Delta G^{\ast} = -RT \ln \dfrac{D_1}{D_2} \cdot \%$. De is defined as $\dfrac{D_1 - D_2}{D_1 + D_2} \cdot 100$. The problem is to find a suitable system in which $\dfrac{D_1}{D_2}$ is at least 40, corresponding to de $> 95\%$.

So far, three types of heterocycles — I, II and III — turned out to be suitable. The bold arrow points to the acidic hydrogen, the thin ones to the two sites susceptible to hydrolysis, where the molecule is cleaved after introduction of the electrophile at the CH-acidic site. Type I and II possess an endocyclic inducing center, rigidly fixed by covalent bonds. *Cis/trans*-isomers are formed here in the course of asymmetric induction. Type III contains an exocyclic inducing center. It turned out-unexpectedly — that also here — at the stage of the anion — the inducing center is fairly rigidly frozen by non-bonding interactions [6].

I   II   III

$L^{\ast} = (S)$-Phenylethyl

⟹ CH - acide site   → Sites susceptible to Hydrolysis

This article deals with results achieved with the 2,5-dimethoxy-3,6-dihydropyrazines, the heterocycles of type I. Results obtained with the imidazolinones III are discussed elsewhere [6]. At first glance the heterocycles I look rather esoteric. However, the yare nothing but the bis-lactim ethers of the well known 2,5-diketopiperazines, the cyclic dipeptides. — At first, experiments with the "symmetrical" bis-lactim-ether (6) of cyclo(L-Ala-L-Ala) (5) are described and then results with several mixed bis-lactimethers. "Symmetrical" bis-lactimethers — i.e. those, build up from two identical amino acids — do have one disadvantage, inherent in the system, namely, only one half of the chiral auxiliary is recovered, the other half is incorporated in the product. But they are easily prepared and, hence, are good models to commence a study.

All studies were carried out with L-amino acids as chiral auxiliaries. With these, the newly created asymmetric center has the (R)-configuration (see below). Starting with D-amino acids the (S)-configuration would be induced. Moreover, bis-*methoxy*-lactim ethers were used throughout in this study, since the ^{1}H- and ^{13}C-nmr-spectra of methoxycompounds are simpler and more easily analyzed than the ones of ethoxycompounds. For application of the method the bis-*ethoxy*lactim ethers can be used as well; they give almost identical chemical and stereochemical results.

De-values were determined either ^{1}H- or ^{13}C-nmr-spectroscopically or by capillary gas chromatography. Ee-values were determined mainly by ^{1}H-nmr-spectroscopy using chiral shift reagents. >95% de or ee was assumed, if only one stereoisomer was detectable in the spectrum.

3 Enantioselective Synthesis of α-Methyl Amino Acids from the Bis-lactim Ether (6) of Cyclo(L-Ala-L-Ala) (5)

Active α-methyl amino acids are of interest in medicinal chemistry. In biochemistry, they are believed to be inhibitors of those enzymes that metabolize the corresponding α-unsubstituted proteinogenic amino acids [7] and they have been found to be constituents in antibiotically active peptides [8]. Moreover, if incorporated in a peptide instead of the corresponding proteinogenic amino acid they alter the conformation of the molecule in a certain — sometimes predictable — way. Therefore, they might be useful tools to study the mode of action of a peptide [9, 10].

On heating of L-alanine methyl ester, two molecules condense to give the diketopiperazine 5 [cyclo(L-Ala-L-Ala)]. In this reaction the optical purity is reduced. Recrystallization from water gives 5 with about 93% ee [9]. Compound 5 is not a suitable starting material; its CH-acidity is not high enough and, furthermore, any base would remove preferentially the protons from nitrogen. However, a suitable starting compound is obtained by converting 5 into its bis-lactim ether 6 [(3S, 6S)-2,5-dimethoxy-3,6-dimethyl-3,6-dihydropyrazine] with trimethyl oxoniumtetrafluoroborate (Meerwein's salt). Hydrolysis of the intermediate heterocyclic bis-tetrafluoroborate must be carried out with phosphate buffer to avoid racemization [10]. With butyllithium or LDA^2 (in THF or glyme, −78 °C) 6 reacts smoothly to the lithium derivative 7 which contains a stable diazapentadienyl anion and is probably best described as an ion pair. A second metallation at C-6 is very unlikely, because it would give an antiaromatic 8π-electron system [11]. With benzylbromide, 7 gives in ≈88% chemical yield 8a. The benzyl group enters trans to the methyl group at the inducing center C-6, i.e. (R)-configuration is induced at C-3 with de ≈ 93% [9]

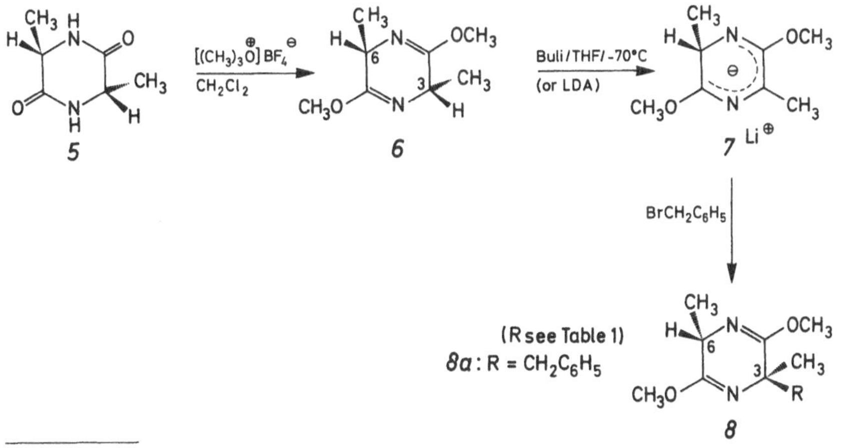

^{2} LDA = Lithiumdiisopropylamide.

69

The (3R)-configuration of the major epimer follows unambigously from the ^{1}H-nmr-spectrum. Obviously, the heterocycle adopts a boat shape and the benzyl group has the "folded" conformation, i.e. the aryl ring is turned inside [conformation 9 for the major, and conformation 10 for the minor isomer]. In 10 the C-6-methyl group is located within the shielding cone of the aryl ring and its signal suffers an upfield shift relativ to the one of 9 ($\Delta\delta \approx 0.75$ ppm).

9 (major isomer)   10 (minor isomer)

The nmr-assignment is confirmed by hydrolysis. With two equivalents of 0,25 N hydrochloric acid 8a is readily hydrolyzed at room temperature to liberate L-Ala-OCH$_3$ and (R)-α-methyl-phenylalanine methyl ester (11a). The two esters are separated by distillation. Remarkably, hydrolysis under these conditions does not proceed via the diketopiperazine; otherwise the amino acids would be formed and not the esters, and hydrolysis would require much more forcing conditions. (R)-11a is a configurationally known compound [6, 12]. Its enantiomeric purity is readily determined by nmr-spectroscopy using chiral shift reagents such as Eu(hfc)$_3$. The α-methyl group and the ester methyl group can both be used for the analysis.

8a

11 (R cf. Table 1)
a : R = CH$_2$C$_6$H$_5$

Table 1 contains the alkylation products of the lithium compound 7 with a variety of alkylating agents [9]. As for alkylations with benzyl- or heterobenzyl-type halides the (3R)-configuration in 8 can be derived unambigously from the nmr-spectrum because of the mentioned anisotropic shift phenomena. As for 8h, after hydrolysis (R)-isovaline was isolated, a configurationally known compound. For the remaining cases, (3R)-configuration can be assumed by analogy.

The asymmetric induction in the alkylation 7→8 may be explained by the following transition state model. The anion of 7 is planar; its two diastereotopic faces are shielded differently (hydrogen versus methyl). The alkyl halide enters predominantely by S$_N$2 from the "top side" where the relatively small hydrogen is located. 12 depicts the favored transition state, the one leading to the (3R)-diastereomer. Furthermore, the transition state is supposed to have the "folded" conformation as far as R' is concerned. In this conformation R' comes close to the

Table 1. Dihydropyrazines *8* by Alkylation of *6* with butyl lithium

(*8*)	R	Yield %	≈De %	Config. C-3
(*a*)	$C_6H_5CH_2$	88	93[a]	R
(*b*)	2-naphthyl-CH_2	88	92[a], 95[c]	R
(*c*)	2-chinolyl-CH_2	78	95[a]	R
(*d*)	3-pyridyl-CH_2	91	95[a]	R
(*e*)	3,4-$(CH_3O)_2C_6H_3CH_2$	93	95[a]	R
(*f*)	$CH_2=CHCH_2$	88	92[b]	R
(*g*)	$(CH_3)_2CH$	80	92[b]	R
(*h*)	C_2H_5	81	95[b]	R
(*i*)	n-C_8H_{17}	83	92[b]	R
(*j*)	$C_6H_5CH=CHCH_2$	85	91[b]	R

[a] By ^1H-nmr-spectroscopy.
[b] Indirectly, via ee of *11*;
[c] With 0,0-diethyl bis-lactim ether instead of *6*

inducing center, rendering $\Delta\Delta G^{\ne}$ for the two competing pathways relatively large. For R′ = benzyl or heterobenzyl the "folded" conformation has been proved for the product (see above), hence, it is very likely that already the transition state leading to this product adopts the "folded" conformation. As discussed below, also 3- or 6-alkyl sidechains of 2,5-dialkoxy-3,6-dihydropyrazines seem to prefer the "folded" arrangement. For R′ = aryl (π-system!) this conformation may be stabilized by an (HOMO)anion-(LUMO)aryl interaction, for R′ = alkyl by some sort of Van-der-Waals attraction — According to MO-calculations N-1 is the atom with the highest electron density. Hence, the lithium cation will probably be located above N-1. However, it is unclear, what role it plays. It is also unclear, whether the anion is really planar; it might be (slightly) bent.

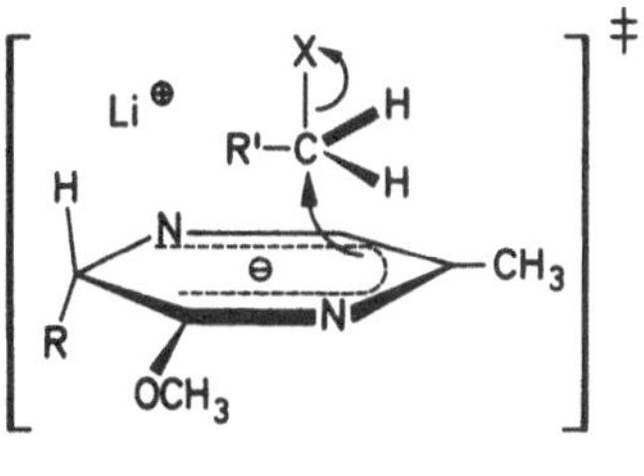

12	a	b
R	CH_3	iPr

On hydrolysis the alkylated bis-lactim ethers *8* yield (besides L-Ala-OCH_3) the corresponding (R)-α-methylamino acid methyl esters *11*. Their ee is ≈85% corresponding to the de of table 1 and taking into account ≈93% ee of *6* [9].

Also ketones and aldehydes react with the lithiated bis-lactim ether *7* with rather high asymmetric induction to give the aldol-type adduct *13* (Table 2). Like alkyl halides, the carbonyl compounds enter *trans* to the methyl group at C-6; i.e. (R)-configuration is induced at C-3 [13].

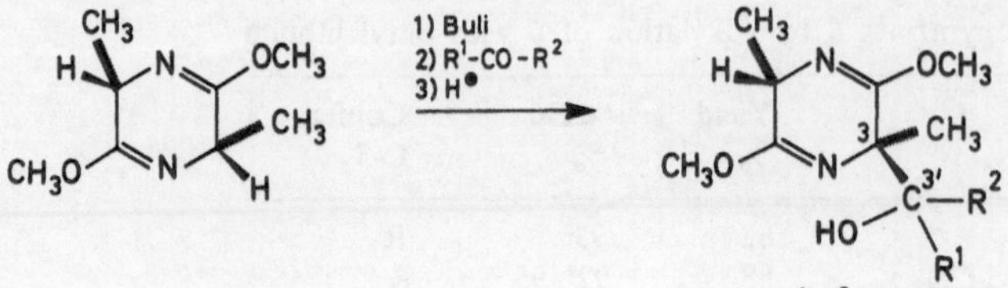

6 → 13 (R¹,R² cf. Table 2)

Table 2. Aldol-type addition products *13*

13	R¹	R²	C-3 ≈de (%)	C-3 Config.	C-3'[a] ≈de (%)	C-3'[a] Config.
a	H	H	81	R		
b	CH₃	CH₃	85[b]	77[c]		
c	C₆H₅	C₆H₅	>95[c]			
d	C₆H₅	H	88[b]	80[c]	52[b]	R
e	C₆H₄(4-OCH₃)	H	89[b]	71[c]	74[b]	R
f	C₆H₅	CH₃	>95[b]	86[c]	41[b]	R

[a] For the (6S, 3R)-isomers; [b] In Glyme; [c] In THF

For aromatic aldehydes or ketones, the (3R)-configuration can be derived from the ¹H-nmr-spectrum. Like the benzyl-substituted dihydropyrazines *8a–e* the hydroxy-benzyl-substituted compounds *13c–f* adopt the "folded" conformation. Hence, in the minor isomer with (3S)-configuration and conformation *14* the C-6-methyl signal suffers an upfield shift.

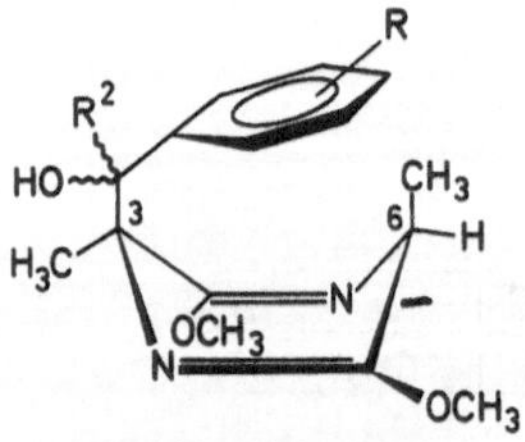

14 (6S,3S–isomer)

The formaldehyde adduct (*13a*) was hydrolyzed and R-(—)-α-methylserine isolated, a configurationally known compound. For the acetone-adduct (*13b*) (3R)-configuration was assumed by analogy. Its de was derived via ee of methyl α-methyl-β-hydroxyvalinate, (type *16*), obtained by hydrolysis of (*13b*)[13].

With aldehydes or unsymmetrical ketones, also C-3' becomes a chiral center. The asymmetric induction at this center — i.e. the enantioface selection at the carbonyl group — is also listed in Table 2. Details of assignment of configurations are described elseqhere[13].

Remarkably, the extent of induction at C-3 in (13) does not depend very much on the size of R^1 and R^2 (Table 2). Hence, interaction of these groups with the heterocyclic anion cannot be a decisive factor in the transition state of the addition. Since oxygen is the functional atom common to all carbonyl compounds, the transition state has probably the "oxygen inside" conformation as indicated in (15a) for the most favored transition state, the one leading to the (3R)-diastereomer. This arrangement may be stabilized by an (HOMO)anion-(LUMO)carbonyl attraction. In order to get bonding overlap the oxygen atom has to be located above N-1, the atom with the largest coefficient in the HOMO. Alternatively — or additionally —, the transition state (15) may be stabilized by complex formation between the lithium cation and the negatively charged oxygen [14]. The lithium cation is probably located above N-1 in an ideal position for complex formation.

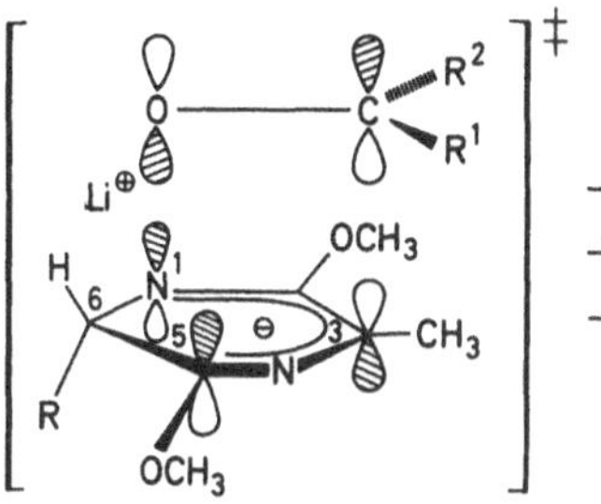

15	a	b
R	CH_3	iPr

Hydrolysis of the adducts 13 to liberate L-Ala-OCH$_3$ and α-methyl serine methyl esters 16 occurs under mild conditions (0.25 N HCl, r.t.) [13].

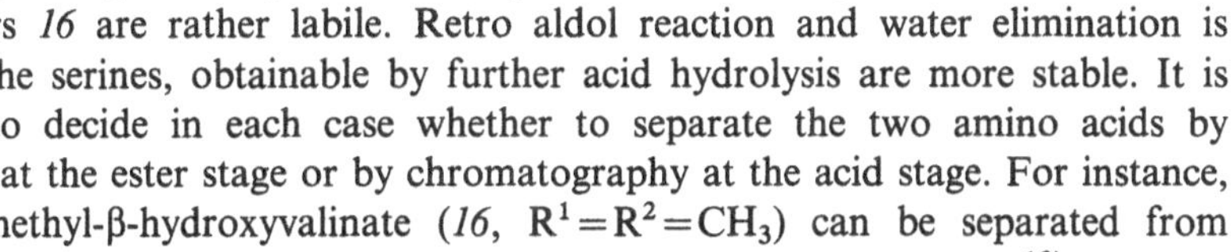

16

However, it might be a problem to separate L-Ala-OCH$_3$ because the α-methyl serine esters 16 are rather labile. Retro aldol reaction and water elimination is possible. The serines, obtainable by further acid hydrolysis are more stable. It is necessary to decide in each case whether to separate the two amino acids by distillation at the ester stage or by chromatography at the acid stage. For instance, methyl α-methyl-β-hydroxyvalinate (16, $R^1 = R^2 = CH_3$) can be separated from L-Ala-OCH$_3$ by distillation, although some decomposition takes place [13].

4 Enantioselective Synthesis of α-Methyl Amino Acids from the Mixed Bis-lactim Ether (20a) of Cyclo(L-Val-Ala)

As mentioned above, "symmetrical" bis-lactim ethers of type 6 have the inherent disadvantage, that only one half of the chiral auxiliary is recovered, the other half is incorporated in the product. This disadvantage can be overcome by using mixed

73

bis-lactim ethers, i.e. those built up from two different amino acids. Two types suggest themselves. Type I consists of an optically active amino acid (such as L- or D-valine, -leucine, etc.) which serves as chiral auxiliary and contributes C-6 and of a racemic amino acid which contributes C-3. Of course, one has to choose R^1 and R^2 in such a way, that metallation takes place regiospecifically at C-3.

As for the induction, type I looks promising, one of the diastereotopic faces of the corresponding anion is shielded by the tiny hydrogen, the other by the much bulkier R^1 which can be chosen rather freely. — Type II consists of an optically active α-methylamino acid which is the chiral auxiliary and contributes C-3 and of a racemic amino acid that contributes C-6.

With type II one has not to worry about the site of metallation, but one has to worry about the extent of asymmetric induction. Instead of hydrogen the chiral center contains a relatively big methyl group. Nevertheless, it is worthwhile to elaborate also type II, especially with R^1 = benzyl. Obviously in this case the anion has the "folded" conformation in which one diastereotopic face is effectively shielded by the aryl ring (see below).

Type I   Type II

The following discussion deals firstly with results achieved so far with mixed bis-lactim ethers of type I and secondly with studies on mixed bis-lactim ethers of type II with R^1 = 3,4-dimethoxybenzyl and R^2 = H.

4.1 Synthesis of the Bis-lactim Ether (20a) of Cyclo(L-Val-Ala) and other "Mixed" Bis-lactim Ethers (20)

"Mixed" bis-lactim ethers of type (20) are best prepared by the following route, outlined for the bis-lactimether (20a) of Cyclo(L-Val-Ala). L-Val, the chiral auxiliary, is converted with phosgene into its N-carboxyanhydride (L-Val-NCA, Leuchs anhydride) (17) [15]. This gives with D,L-Ala-OCH$_3$ the dipeptide (18) which on heating in toluene cyclizes to the diketopiperazine (19). This is converted into the bis-lactim ether (20a) [(3RS, 6S)-2,5-dimethoxy-6-isopropyl-3-methyl-3,6-dihydropyrazine] with methyl Meerwein's salt.

By this practical — and resonably inexpensive — method the bis-lactim ether (20b) of cyclo(L-Val-Gly) [16], (20c) of cyclo(L-tLeu-Ala) [17], (20d) of cyclo(L-tLeu-Gly) [18], (21a) of cyclo(L-α-Methyldimethoxyphe-Gly) [19] and (21b) of cyclo(L-α-Methyl-phe-Gly) [20] were also prepared advantageously.

4.2 Alkylation of the Lithiated Bis-lactim Ether 22 of Cyclo(L-Val-Ala)

The bis-lactim ether 20a reacts with butyllithium (THF, —78 °C) regiospecifically in the alanine part of the molecule to give the lithium compound 22. This gives with

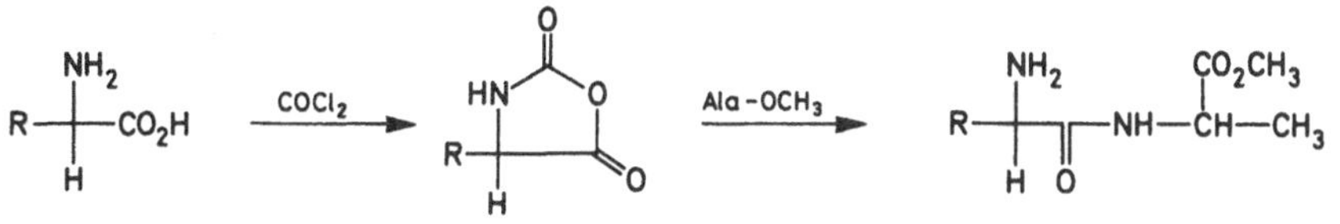

17

18

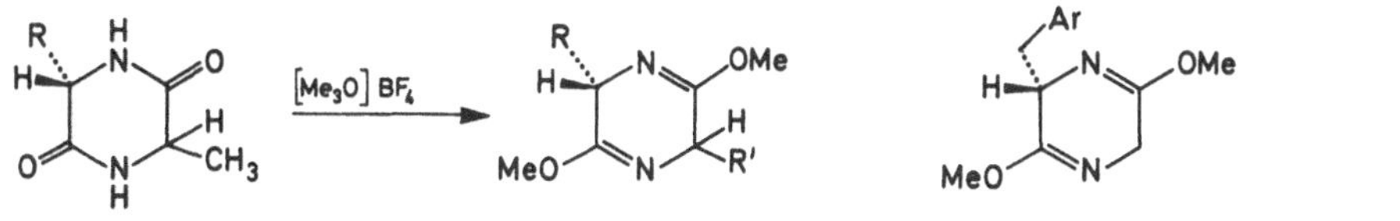

19

20	a	b	c	d
R	iPr	iPr	iBu	tBu
R'	CH₃	H	CH₃	H

21	a	b
Ar	C₆H₃(3,4-OCH₃)₂	C₆H₅

alkyl halides in good chemical yields and with >95% de the addition products *23* (table 3) [15]. The alkyl group enters *trans* to the isopropyl group at the inducing center, i.e. the (3R)-configuration is induced in *23*. For benzyl type halides this can be derived unambiguously from the ¹H-nmr-spectrum. In these cases the alkylation products adopt the "folded" conformation analogous to (*9*) with the C-6-H within the shielding cone of the aryl ring.

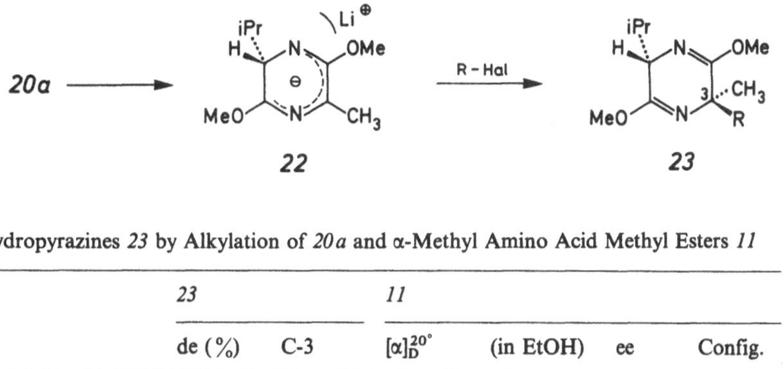

20a ⟶ **22** $\xrightarrow{\text{R - Hal}}$ **23**

Table 3. Dihydropyrazines *23* by Alkylation of *20a* and α-Methyl Amino Acid Methyl Esters *11*

		23		*11*			
		de (%)	C-3	[α]$_D^{20°}$	(in EtOH)	ee	Config.
a	C₆H₅CH₂	>95	R	− 2.8	(c = 1.0)	>95	R
b	(3,4-OCH₃)₂C₆H₃CH₂	>95	R	− 0.7	(c = 1.1)	>95	R
c	C₆H₅CH=CHCH₂	>95	R	−13.2	(c = 1.1)	>95	R
d	n-C₇H₁₅	>95	R	−12.9	(c = 0.6)	>95	R
e	CH₂=CHCH₂	>95	R	+ 2.33	(c = 0.4)	>95	R
f	CH≡CCH₂	>95	R	+ 2.08	(c = 0.8)	>95	R
g	(CH₃)₂C=CHCH₂	>95	R				
h	BzSCH₂CH₂	>95	R				
i	tBuO₂CCH₂	>95	R				
j	BzOCH₂CH₂	>95	R				

The asymmetric induction can be explained on the basis of the transition state model *12b*. Exchange of methyl in *12a* for isopropyl at the inducing center renders $\Delta\Delta G^{\ne}$ so large that practically only one diastereomer is formed. There is good reason to assume that one methyl group of the isopropyl moiety is turned inside and faces the heterocyclic anion (see below).

The (3S)-epimers *25* of *23* are prepared by reversing the sequence of group introduction at C-3. For instance, starting with the mixed bis-lactim ethers of type *24* the (3S)-compounds *25* are obtained with de ≈ 85–87% by metallation and subsequent addition of methyl iodide. The ^{1}H-nmr-spectra of *23* and *25* are distinctly different.

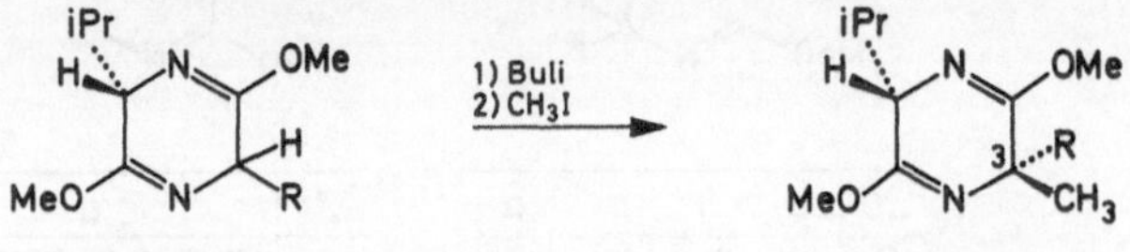

$$R = CH_2C_6H_5, \; CH_2C_6H_3 (3,4-OCH_3)_2, \; CH_2CH = CHC_6H_5$$

The mechanism of these reactions, which lead to the thermodynamically less stable *cis*-products, is instructive. Obviously, the alkylation step has an "early" transition state and product formation is governed mainly by "steric approach control" [21] — a fact that renders the induced configuration predictable.

Of course, the (3S)-compounds would also be formed if D-valine would be employed as chiral auxiliary. Hence, this method with valine as chiral auxiliary reagent solves the problem of enantioselective synthesis of α-methyl amino acids satisfactorily. Probably it can also be used — mutatis mutandis — for the asymmetric synthesis of a variety of α-alkyl amino acids, provided, the corresponding bis-lactim ether (type I) with valine as C-6 is regiospecifically metallated by butyllithium. This, for instance, is not be case with the "mixed" bis-lactim ether (*20c*) of cyclo(L-Leu-D,L-Ala) [17].

4.3 Hydrolysis of the Alkylation Products 23; (R)-α-Methyl Amino Acid Esters 11

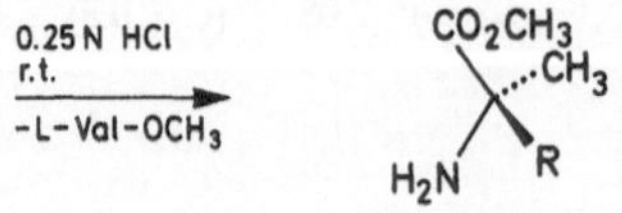

Hydrolysis of the alkylation products *23* with two equivalents of 0.25 N hydrochloric acid at room temperature occurs rather cleanly — although in some cases slowly — to give L-Val-OCH$_3$ and the (R)-α-methyl amino acid methyl esters *11*. The two esters can be separated either by distillation or — if boiling points are similar — by chromatography, eventually after further hydrolysis to the amino acids. The ee of the methyl esters are readily determined by ^{1}H-nmr-spectroscopy using chiral shift reagents such as Eu(hfc)$_3$. In all cases only one enantiomer is detectable, corresponding to ee $>95\%$ [15,22] (Table 3).

4.4 Reactions of the Lithiated Bis-lactim Ether 22 with Carbonyl Compounds and Benzoylchloride

The lithiated bis-lactim ether 22 reacts with acetophenone and acetone with $>95\%$ de at C-3 to give the (3R)-aldol-type adducts 26. With acetone, only one diastereomer 26a can be detected both in the ^{1}H-nmr- and ^{13}C-nmr-spectrum, with acetophenone a ≈ 1:1-mixture of the two diastereomers 26b with (R)- and (S)-configuration at C-3'. They show both in the ^{1}H-nmr-spectrum the expected high-field shift of the C-6-H, indicating the "folded" conformation 27 [17].

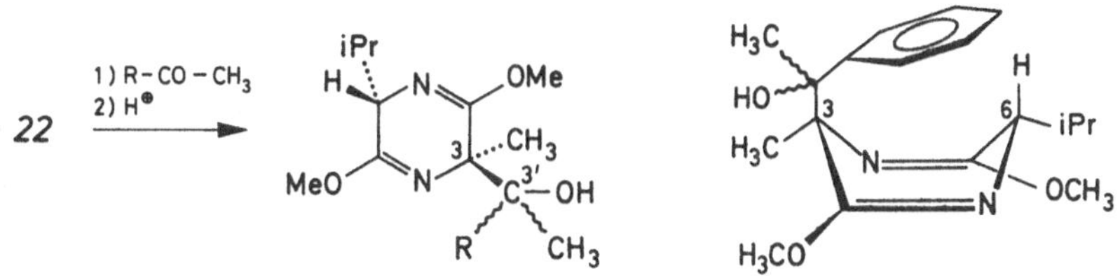

26	a	b
R	CH$_3$	C$_6$H$_5$

The high asymmetric induction can be explained on the basis of the transition state 15b with the bulky isopropyl group at the inducing center.

On treatment with thionylchloride/pyridine the compounds 26 yield the olefines 28 which on hydrolysis (2 equivalents 0.25 HCl, r.t.) are cleaved to L-Val-OCH$_3$ and (R)-α-Alkenylalanine methyl esters 29 with ee $>95\%$ [17,23].

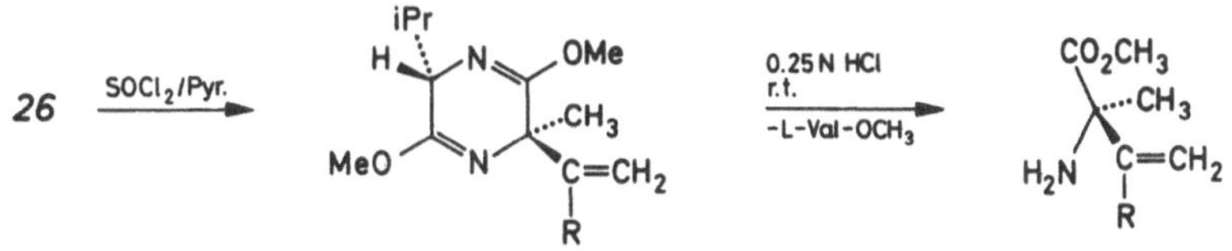

	a	b
R	CH$_3$	C$_6$H$_5$

Hydrolysis of the aldol-type adducts (26) proceeds to form a mixture of products due to retro aldol reactions.

Benzoylchloride reacts with 22 selectively at C-3 to give the (3R)-benzoyl-compound 30a (de $>95\%$) [24].

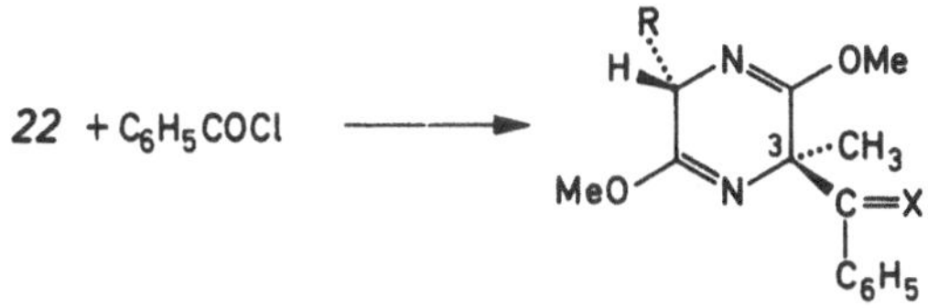

30	a	b
X	O	CH$_2$

A Wittig reaction, for instance, transforms *30a* in *30b* [24]. Thus, acylation of *22* followed by Wittig reactions and hydrolysis offers the possibility to prepare a variety of optically active α-alkenyl alanines of type *28*.

5 Enantioselective Synthesis of α-Unsubstituted Amino Acids from the "mixed" Bis-lactim Ethers 20b of Cyclo(L-Val-Gly) and 20d of Cyclo(L-tLeu-Gly)

In general, enantioselective hydrogenation of dehydroamino acids — elaborated mainly by Kagan et al. [4] and Knowles et al. [4] — seems to be a promising and elegant route to α-unsubstituted amino acids [4], especially for industrial chemistry. However, not all dehydroamino acids react properly with hydrogen and for each case the suitable catalyst and conditions must be found in preliminary studies. Furthermore, the method is limited to those dehydroamino acids carrying no additional functional groups susceptible to hydrogenation, such as double bonds, triple bonds, carbonyl groups, nitro groups, etc. Hence, also in the field of α-unsubstituted amino acids efficient stoichiometric asymmetric syntheses would be useful.

5.1 α-Unsubstituted Amino Acids by Alkylation of the Lithiated Bis-lactim Ether 31 of Cyclo(L-Val-Gly)

The bis-lactim ether *20b* of cyclo(L-Val-Gly) reacts regiospecifically with butyllithium to afford the lithium compound *31*. The addition products (*32*) are formed with alkyl halides in good chemical yields and — depending on R — with 65–>95% de [16]. They have the (3R)-configuration, as derived either from the ^{1}H-nmr-spectrum of *32* (Table 4) or, indirectly, from the sign of rotation of the (R)-amino acid methyl ester (*34*) obtained by hydrolysis.

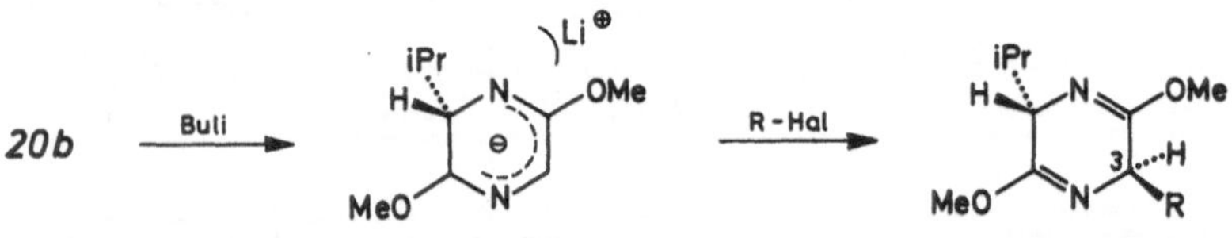

31

32 (R cf. table 4)

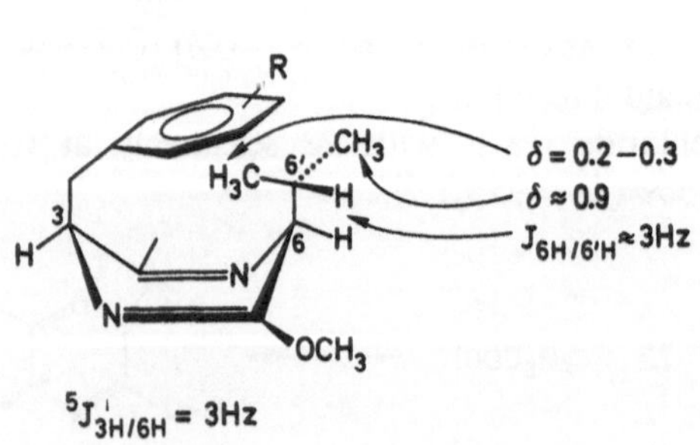

33

The ^{1}H-nmr-spectrum of the minor isomer of *32a* (isopropyl and benzyl *cis*) are remarkable. One of the methyl groups of the isopropyl moiety absorbs at much higher field than the other ($\delta = 0.25$ versus 0.9 ppm). $J_{6H/6'H} \approx 3$ Hz indicates a gauche relation of these hydrogens and $^5J_{6H/3H} \approx 3$ Hz a boat shape of the heterocycle. Obviously, in contrast to expectations the crowded arrangement *33* — stabilized by Van-der-Waals attraction — seems to be the preferred conformation.

Probably, also in the anion *31* and in the alkylation transition state *12b* one methyl group is turned inside and shields the bottom side of the molecule effectively. Since *these* alkyl halides give the highest inductions that contain an extended π-system (benzyl, cinnamyl, phenylpropargyl, etc.) it is plausible to assume also for the incoming alkyl group the "folded" conformation (cf. *12b*).

On hydrolysis of *32* (2 equivalents 0.25 N HCl, r.t.) L-Val-OCH$_3$ and the (R)-amino acid methyl esters *34* are liberated. Their ee can be determined ^{1}H-nmr-spectroscopically using chiral shift reagents (Table 4) [16].

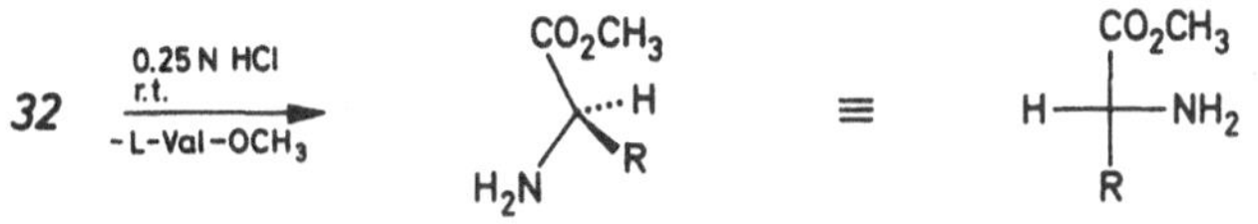

Table 4. (R)-Amino Acid Methyl Esters *34*

34	R	$\approx$ ee[a] (%)
a	C$_6$H$_5$CH$_2$	92–95
b	(3,4-OCH$_3$)$_2$C$_6$H$_3$CH$_2$	92
c	2-naphthyl-CH$_2$	92
d	C$_6$H$_5$CH=CHCH$_2$	>95
e	C$_6$H$_5$C≡CCH$_2$	>95
f	n-C$_7$H$_{15}$	80
g	CH≡CHCH$_2$	65
f	CH$_2$OBz	92

[a] ^{1}H-nmr-spectroscopically with Eu(hfc)$_3$ as chiral shift reagent; >95% is assumed if only one enantiomer is detectable

5.2 α-Unsubstituted Amino Acids by Alkylation of the Lithiated Bis-lactim Ether (35) of Cyclo(L-*t*Leu-Gly)

Apart from methyl iodide, all alkyl halides react with de >95% with the lithiated bis-lactim ether *35* of cyclo(L-*t*Leu-Gly) *20d* (Table 5). De can be determined either at the stage of the adducts *36* or, indirectly, via ee of the (R)-amino acid methyl esters (type *34*) liberated on hydrolysis (2 equivalents 0.1 N HCl, r.t.) [18]. Although this system works exceedingly well — in fact, it could be the solution to the problem as far as this approach is concerned —, the method has one disadvantage at the time being: *tert*-leucine — as *tert*-butylglycine is commonly called — is not available in nature's chiral pool. There is a good synthesis of the racemic mixture, but resolution to gain the optically pure amino acid is still a problem. An efficient synthesis of (R)- and (S) *tert*-leucine — perhaps an enzymatic one — is highly desirable [26].

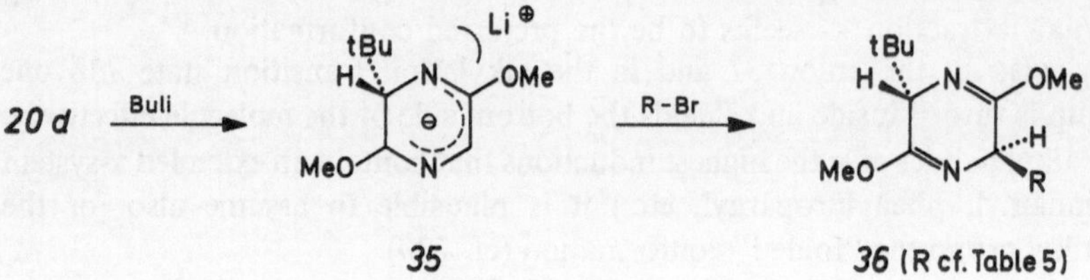

Table 5. De-values of the reaction $(35) \rightarrow (36)^a$

R	CH$_3$	H$_2$C=CHCH$_2$	HC≡CCH$_2$	n-C$_7$H$_{15}$	CH$_2$CO$_2$tBu
De (%)	≈80	>95	>95	>95	>95

^{a} Determined indirectly via ee of (34) obtained after hydrolysis of (36)

5.3 Reactions of the Lithiated Bis-lactim Ether (22) of Cyclo(L-Val-Gly) with Carbonyl Compounds

As studied so far [17], ketones seem to react with (22) with de >95% (Table 6). In the ^{1}H- and ^{13}C-nmr-spectra of (37a) only one diastereomer can be detected and in the spectra of (37b) only the (3R, 3'S)- and the (3R, 3'R)-diastereomers. Aldehydes

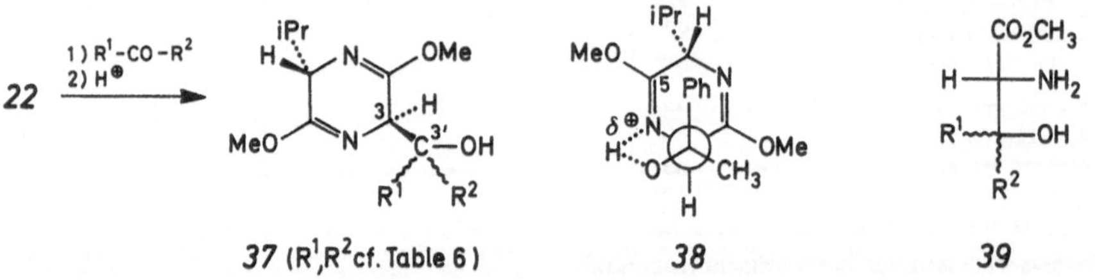

Table 6. Addition Products 37 from 22 and Carbonyl Compounds [17]

37	R^1	R^2	C-3		C-3'	
			≈de (%)	Config.	≈de (%)	Config.
a	CH$_3$	CH$_3$	>95	R		
b	C$_6$H$_5$	CH$_3$	>95	R	30^a	S
c	(CH$_3$)$_2$CH	H	94	R	85^a	S
d	C$_6$H$_5$	H	86	R	8–10^a	S
e	CH$_3$	H	84	R	60^a	S

^{a} For the (3R)-diastereomers

react with lower but still very high diastereoface differentiation (cf. Table 6). The enantioface selection at the carbonyl group (with unsymmetrical ketones and aldehydes) is in the range of 85–10% for the (3R)-diastereomers (cf. Table 6). Contrary to the compounds *13*, the (3'S)-configuration is favored here [17].

The (3R, 3'R)-configuration of the minor epimer of (*37b*) can be derived from the ¹H-nmr-spectrum. This epimer contains an hydrogen bond between C-3'-OH and N-4 (Newman projection (*38*)) which renders the C-5-OCH₃ somewhat positive. Hence, the two OCH₃-groups of this epimer differ more strongly in chemical shifts than the ones of the (3R, 3'S)-isomer [27]. — Hydrolysis of the aldol-type adducts (*37*) liberates the (R)-serine methyl esters of type (*39*) [17].

On treatment with thionylchloride/2,6-lutidine (*37b*) is converted into a 80:20-mixture of the "Hofmann-olefin" (*40*) and the "Saytzeff-olefin" (*41*). By hydrolysis of the mixture, followed by proper workup enantiomerically pure (R)-β-methylene phenylalanine methyl ester *42* can be obtained [28]. This compound was recently prepared in racemic form in a lengthy synthesis [29].

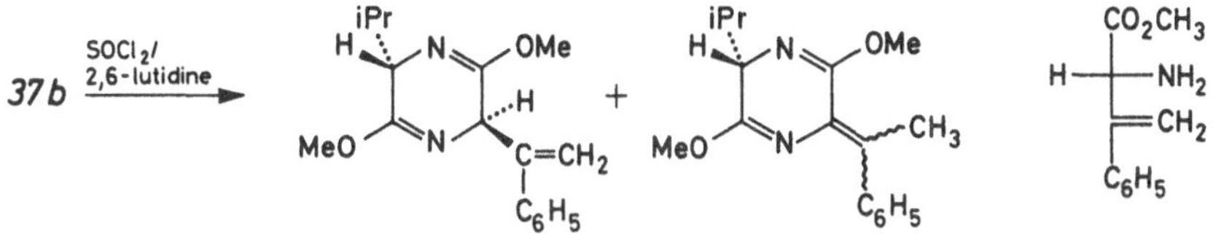

6 Enantioselective Synthesis of α-Unsubstituted Amino Acids from the Bis-lactim Ether of Cyclo[(S)-α-Methyl(3,4-dimethoxy) Phe-Gly]

The anions derived from mixed bis-lactim ethers of type II with R=benzyl seem to adopt a "folded" conformation of type *43* — stabilized possibly by an (HOMO)-anion-(LUMO)aryl attraction —, in which one of the diastereotopic faces is most effectively shielded. Consequently, electrophiles enter predominantely from the "bottom side" and show a much stronger face differentiation than they would do if the more size of the benzyl group would be the decisive factor. With (S)-α-methyl-(3,4-dimethoxy)-phenylalanine *44* — the precursor of α-methyldopa (*2*) — a suitable commercial chiral auxiliary is available.

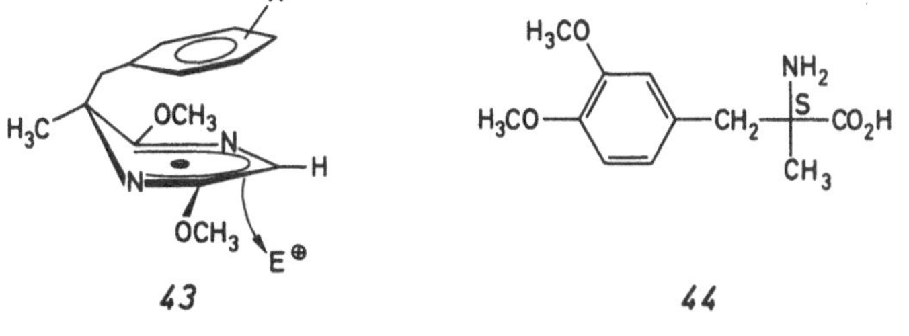

The bis-lactim ether *21a* of *44* and glycine reacts with butyllithium to the lithium compound *45* which affords with alkyl halides in good chemical yields the (3R)-alkylation products *46* with 70–95% de (Table 7) [19,25]. The de values were determined either from the nmr-spectrum of *46* or, indirectly, from ee of the (R)-amino acid esters (type *34*) obtained from *46* by hydrolysis (2 equivalents 0.25 N HCl, r.t.).

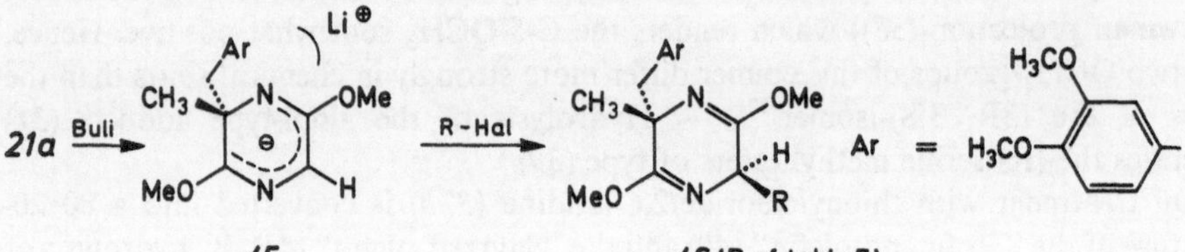

Table 7. De-values of the Reaction *45* → *46* [19,25]

R	H$_2$C=CHCH$_2$	CH$_3$CH=CHCH$_2$	CH$_2$CH=CHC$_6$H$_5$	(CH$_3$)$_2$C=CHCH$_2$
≈ de (%)	80	80	70	80
R	CH$_3$(CH=CH)$_2$CH$_2$	HC≡CCH$_2$ n-C$_7$H$_{15}$	BzOCH$_2$	H$_3$CO–C with C$_3$H$_7$-n and C$_3$H$_7$-n
≈ de (%)	80	85 85	90	95

Also ketones react with *45* with unusual high diastereoselectivity to give the (3R)-aldol-type adducts *47*.

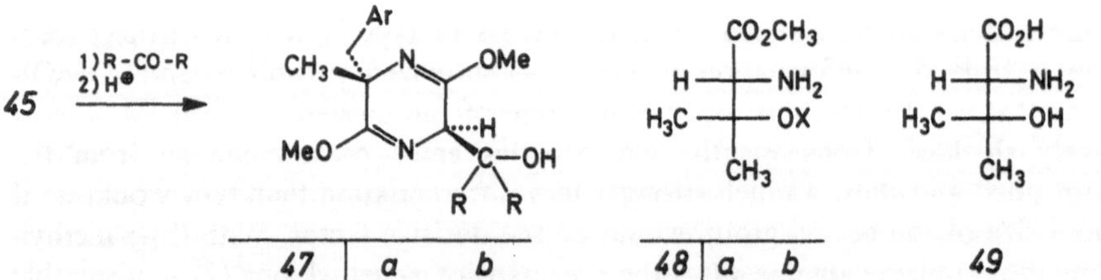

47	*a*	*b*
R	CH$_3$	C$_6$H$_5$
≈ de (%)	03	>05

48	*a*	*b*
X	H	CH$_3$

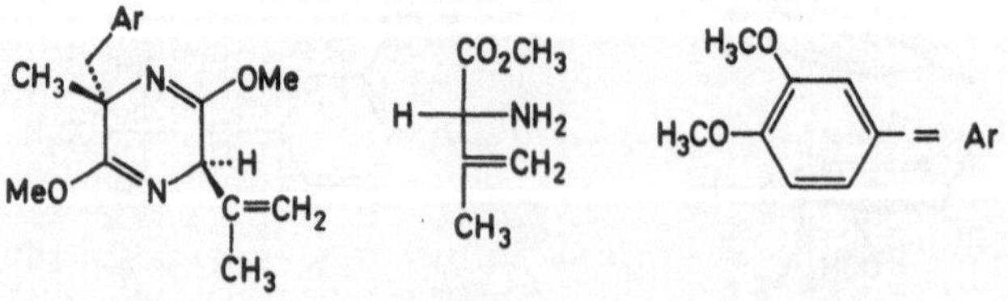

On hydrolysis of (*47a*) (2 equivalents 0.1 N HCl, r.t.) the methyl ester of (*44*) and methyl (R)-β-hydroxyvalinate (*48a*) are liberated. Separation of the two esters in difficult since (*48a*) is rather thermolabile. However, after 0-methylation of (*48a*) the thermostable methyl (R)-β-methoxyvalinate (*48b*) (ee ≈93%) is formed, which can be isolated by fractional distillation and subsequently converted into (R)-β-hydroxyvaline (*49*) by known methods[30]. With these experiments the absolute configuration of (*49*), which was previously assigned tentatively[31], could be established beyond doubt[30]. On treatment with thionyl chloride/2,6-lutidine (*47a*) gives the "Hofmann-olefin" (*50*) as the major elimination product. From this, methyl (R)-isopropenylglycinate (*51*) with ee ≈88% could be obtained[17].

The lithiated bis-lactim ether (*52*) in which benzyl at the inducing center is replaced by phenyl, reacts with electrophiles considerably less stereospecifically (Table 8)[20]. Since phenyl is commonly regarded as bigger than benzyl — although the apparent size of a group depends very much on the reacting system — this finding seems to be consistent with the "folded" conformation *43* of the anion and the transition state of it's reactions with electrophiles.

Table 8. De-Values of the Reaction *52* → *53*

$E^\oplus$	CH_3I	$C_6H_5CH_2Br$	$CH_2=CHCH_2Br$	$(CH_3)_2CO$
≈de (%)	14	29	58	72

Thanks for enthusiastic cooperation are due to W. Hartwig, who carried out the initial crucial experiments, but also to U. Groth, H. Kehne, R. Lonsky, H.-J. Neubauer, J. Nozulak, K.-H. Pospischil, K.-O. Westphalen, C. Deng and Y. Chiang.

7 References

1. Rando, R. R.: Acc. Chem. Res. 8, 281 (1975); Abeles, R. H., Maycock, A. L.: ibid. 9, 313 (1976); Abeles, R. H.: Pure & Appl. Chem. 53, 149 (1980); Trowitzsch, W., Sahm, H.: Z. Naturforsch., Teil C 32, 78 (1977)
2. Nass, G., Poralla, K., Zähner, H.: Naturwissenschaften 58, 603 (1971)
3. Bell, E. A.: Endeavour 4, 102 (1980)
4. Recent reviews on asymmetric synthesis, including amino acids ApSimon, J. W., Seguin, R. P.: Tetrahedron 35, 2797 (1979); Weinges, K., Stemmle, B.: Recent Develop. Chem. Nat. Carbon Compd. 7, 91 (1976); Valentine jr., D., Scott, J. W.: Synthesis 1978, 329; Kagan, H. B.: Pure & Appl. Chem. 43, 401 (1975)
5. Morrison, J. D., Mosher, H. S.: Asymmetric Organic Reactions, 1. Aufl., Prentice Hall, Englewood Cliffs 1971

6. Schöllkopf, U. et al.: Angew. Chem. *90*, 136 (1978); Angew. Chem. Int. Ed. Engl. *17*, 117 (1978); Schöllkopf, U. et al.: Liebigs Ann. Chem. 1981, 439
7. Cf. Pankaskie, M., Abdel-Monem, M.: J. Med. Chem. *23*, 121 (1980)
8. Jung, G., Brückner, H., Schmitt, H.: in (Voelter, W., Weitzel, G., Eds.) Structure and Activity of Natural Peptides, 1981, 75, Walter de Gruyter, Berlin 1981
9. Schöllkopf, U. et al.: Liebigs Ann. Chem. 1981, 696
10. Cf. Blake, K. W., Porter, A. E. A., Sammes, P. G.: J. Chem. Soc. Perkin I 1972, 2494
11. Cf. Schmidt, R. R.: Angew. Chem. *87*, 603 (1975); Angew. Chem. Int. Ed. Engl. *14*, 581 (1975)
12. Weinges, K. et al.: Chem. Ber. *104*, 3594 (1971). See this paper for alternative asymmetric syntheses of α-methyl amino acids as well as Kolb, M., Barth, J.: Tetrahedron Lett. 1979, 2999; Suzuki, M. et al.: Chem. Ind. (London) 1972, 687
13. Schöllkopf, U., Groth, U., Hartwig, W.: Liebigs Ann. Chem. 1981, 2407
14. This alternative model is analogous to the one proposed for the aldol addition. Cf. Kleschick, W. A., Buse, C. T., Heathcock, C. H.: J. Am. Chem. Soc. *99*, 247 (1977); Buse, C. T., Heathcock, C. H.: ibid. *99*, 8109 (1977); Fellmann, P., Dubois, J.-E.: Tetrahedron *34*, 1349 (1978); House, H. O., Crumrine, D. S., Teranishi, A. Y., Olmstead, H. D.: J. Am. Chem. Soc. *95*, 3310 (1973). Very likely, in the "real" transition state both factors, (HOMO)-LUMO) attraction and lithium complexation play a role.
15. Schöllkopf, U. et al.: Synthesis 1981, 969. For an alternative synthesis of amino acids NCAs with trichlormethyl chloroformate instead of phosgene cf. Oya, M., Katakai, R., Nakai, H.: Chem. Lett. 1973, 1143
16. Schöllkopf, U., Groth, U., Deng, C.: Angew. Chem. *93*, 793 (1981), Angew. Chem. Int. Ed. Engl. *20*, 798 (1981)
17. Groth, U.: Dissertation, Univ. of Göttingen 1981
18. Schöllkopf, U., Neubauer, J.: Synthesis 1982, 861
19. Schöllkopf, U. et al.: Synthesis 1981, 966
20. Kehne, H.: Dissertation, Univ. of Göttingen 1980
21. Dauben, W. G., Fonken, G. J., Noyce, D. S.: J. Am. Chem. Soc. *78*, 2579 (1956)
22. Lonsky, R.: Diplomarbeit, Univ. of Göttingen 1980
23. Groth, U., Schöllkopf, U., Chiang, Y.: Synthesis 1982, 864
24. Westphalen, K.-O.: Dissertation, Univ. of Göttingen, presumably 1983
25. Nozulak, J.: Dissertation, Univ. of Göttingen, presumably 1983
26. We recently learned, that tert-leucine can in fact be obtained in almost optically pure form by enzymatic resolution. Sauber, K.: Hoechst AG, Pharmaforschung, private communication
27. Cf. argumentation in ref. [13]
28. Schöllkopf, U., Groth, U.: Angew. Chem. *93*, 1022 (1981); Angew. Chem. Int. Ed. Engl. *20*, 977 (1981)
29. Chari, R. V. J., Wemple, J.: Tetrahedron Lett. 1979, 111
30. Schöllkopf, U., Nozulak, J., Groth, U.: Synthesis 1982, 868; this paper also describes the synthesis of 49 from 37a

Selected Topics of the Wittig Reaction in the Synthesis of Natural Products

Dedicated to Prof. Dr. Georg Wittig's 85th birthday

Hans Jürgen Bestmann and Otto Vostrowsky

Organic Chemistry Institute of the University Erlangen—Nürnberg, Henkestraße 42, D-8520 Erlangen, FRG

Table of Contents

1 The Wittig Carbonyl Olefination 86

2 Mechanism and Stereochemistry of the Wittig Reaction 86

3 Synthesis of Fatty Acids and Fatty Acid Esters 92

4 Prostaglandins 100

5 Synthesis of Insect Juvenile Hormones 111

6 Aliphatic Fragrance and Aroma Substances 117

7 Insect Pheromones 120
   7.1 Butterflies — Lepidoptera 120
   7.2 Beetles — Coleoptera 131
   7.3 Flies — Diptera 134
   7.4 Hymenoptera 135
   7.5 Orthoptera 136
   7.6 Kairomones by Wittig Synthesis 137

8 Terpenoids 138

9 Retenoids, Carotenoids, Isoprenoids, and Polyenes 146

10 Concluding Remarks 157

11 References 157

1 The Wittig Carbonyl Olefination

In a paper edited in 1953, concerned with the preparation of the stereoisomeric forms of pentaphenylphosphorus, Wittig and Geißler [1] described the reaction of methylene-triphenylphosphorane *1* and benzophenone *2*, forming 1,1-diphenylethylene *3* and triphenylphosphine oxide *4* (Scheme 1). Soon afterwards, it could be demonstrated that alkylidenephosphoranes (phosphine alkylenes, phosphorus ylides) generally react with carbonyl compounds such as aldehydes and ketones to give alkenes with the formation of phosphine oxide [1, 2].

This olefin synthesis, later generally called Wittig reaction or Wittig olefination, allows an alkylene group to be linked to a carbonyl compound generating a double bond exclusively at the carbon atom of the carbonyl group. Thus, this reaction often offers decisive advantages over comparable metalorganic alkene syntheses like the Grignard or Reformatzky reaction which frequently lead to the formation of structural isomers.

$$(C_6H_5)_3P=CH_2 + O=C(C_6H_5)_2 \rightarrow CH_2=C(C_6H_5)_2 + (C_6H_5)_3PO$$

$$\textit{(1)} \qquad \textit{(2)} \qquad \textit{(3)} \qquad \textit{(4)}$$

Scheme 1

Because of the preparative potential of this reaction the Wittig carbonyl olefination and its manifold synthetic variants gave rise to a number of new olefin syntheses or improvements of common preparation methods in the following 30 years, and led to investigations of its mechanistic aspects and the stereochemical course. The Wittig reaction nowadays represents one of the key reactions in the preparation of natural compounds and in polyene chemistry [3–5], and has found new areas of application in industrial practice [6].

The special potential for constructing double bonds stereoselectively, often necessary in natural material syntheses, makes the Wittig reaction a valuable alternative compared to partial hydrogenation of acetylenes. It is used in the synthesis of carotenoids, fragrance and aroma compounds, terpenes, steroides, hormones, prostaglandins, pheromones, fatty acid derivatives, plant substances, and a variety of other olefinic naturally occurring compounds. Because of the considerable volume of this topic we would like to consider only selected paths of the synthesis of natural compounds in the following sections and to restrict it to reactions of phosphoranes (ylides) only.

2 Mechanism and Stereochemistry of the Wittig Reaction

A mechanistic interpretation of the Wittig reaction, allowing the various experimental data collected during many years to bring into accord with theoretical demands, was reported only 25 years after its discovery [4, 7–9]. Wittig already formulated a zwitterionic adduct of the form of a P—O-betain *5* [2] (Scheme 2). This is better described by formula *6*, since at that time almost exclusively lithium alkyls have been used for the deprotonation of the phosphonium salts. The reversibilitiy of the formation of *5* and the different (*Z/E*)-isomer ratios of the olefinic reaction products led to the assumption of an equilibrium between the diastereomeric betains *5*. Thus, by kinetic control, the *erythro*-betains should preferently be formed from instable alkylidenephosphoranes, and from them the (*Z*)-alkenes generated. Retardation of the last reaction step (e.g. formation of associates with lithium salts) or the use of resonance-stabilized ylides causes the formation of the *threo*-form of *5* and thus the formation of (*E*)-olefins [10,11].

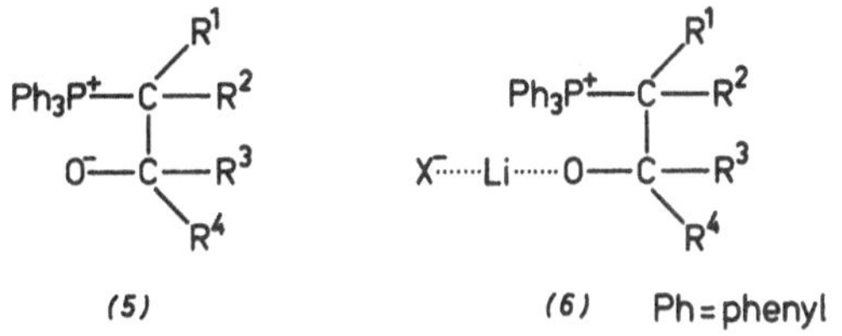

Scheme 2

Reports dealing with the synthesis of stable 1,2-oxaphosphetanes and the proof of four-membered ring compounds of the kind, respectively, as intermediates [12], necessarily led to a revision of the theory of the reaction mechanism. Investigations of the course of carbonyl olefination by means of low-temperature ^{31}P-nuclear magnetic resonance demonstrated that only four-membered phosphorus ring compounds (*9* and *10*) such as oxaphosphetanes occur as intermediates [13-17]. According to the rules for the occupation of the apical positions [18,19], for the formation of pentacovalent, bipyramidal phosphorus compounds and for the conversion and collapse of the latter [20], during ring formation of *9*, the oxygen atom must enter the apical ligand position of the trigonal bipyramid formed. This could be proved by X-ray analysis [8]. Cleavage of the original ylide P—C bond necessary for olefin formation now requires a ligand rearrangement process (pseudorotation) [18] which brings this bond into the apical position (Scheme 3).

Ab initio MO-calculations of model oxaphosphetanes ($R^1 = R^2 = R^3 = H$ in *9* and *10*, respectively) using (4 + 31G + d) basis sets (consideration of d-orbitals) gave pseudorotation energy differences of only 20 kJ/mol [21,22], i.e. the apicophilicity of OR groups is astonishingly small compared to e.g. fluorosubstituted phosphoranes. The four-membered ring causes a further lowering of the energy differences of the pseudorotameric forms [21]. *9* as well as *10*, the latter formed in about 1% yield of the sum of [*9* + *10*], could be detected in the NMR spectrum; the structure of *9* was deduced from coupling constants of the proton resonance spectrum [9,23].

During or after the conversion of *9* into the trigonal-bipyramidal structure *10* cleavage of the C—P bond occurs to give betain *11*. The electronic nature of the substituents R^1 and R^2 of *11* determines the lifetime of this zwitterionic species and thus the stereochemistry of the olefinic products *13* and *15*, respectively. If R^1 is phenyl and R^2 an electron-donating group, rapid phosphine oxide elimination occurs with the formation of (*Z*)-olefins *13*. Electron-withdrawing groups R^2 (when R^1 = phenyl) increase the lifetime of betain *11* which now isomerizes to the thermodynamically more stable conformer *14* yielding (*E*)-alkenes *15* by phosphine oxide

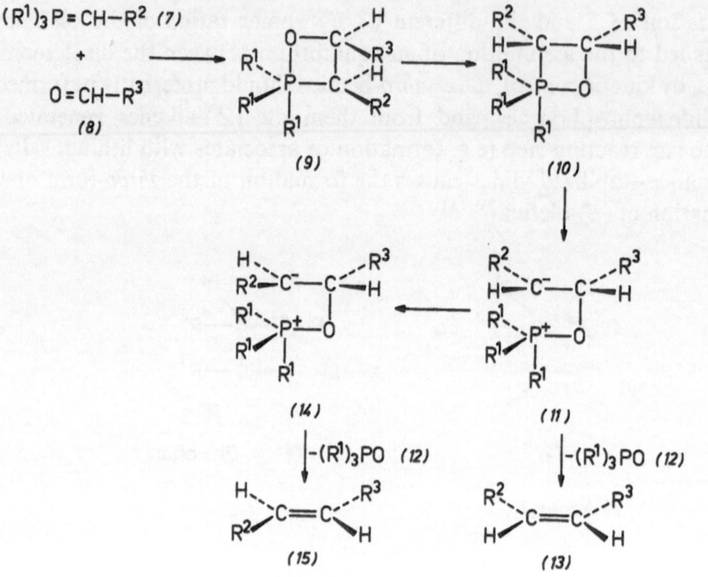

$(R^1)_3 P = CH - R^2$ (7)

+

$O = CH - R^3$

(8)

Scheme 3

elimination. When the phosphorus ligands R^1 are electron donors, e.g. alkyl or cyclohexyl groups, phosphine oxide cleavage from *11* also proceeds less rapidly, i.e. the lifetime of the species is increased. As a consequence, the conversion of *11* into *14* causes a shift of the (Z/E)-ratio to the (E)-form.

These considerations show that the (Z/E)-ratio of the final products of the Wittig reaction is influenced by the electronic character of the groups R^1 and R^2. On the other hand, however, the stereochemistry of the whole reaction sequence is substantially determined by the geometry of the (Z)-configurated oxaphosphetane *9* formed initially. The mechanism and the stereochemistry of the formation of the thermodynamically less stable (Z)-oxaphosphetane *9* still remain unknown. Vedejs and Snoble [12] proposed for this first reaction step a concerted, thermodynamically allowed [$\pi2_s + \pi2_a$]-cycloaddition. The sterically most favored transition state should be the one in which the bulky groups R^2 and R^3 point away from the reaction partner; the flattening of this transition state leads to the (Z)-arrangement of groups R^2 and R^3 [9, 12].

The necessary steric requirements for a concerted formation of *9* and the question of the applicability of symmetry rules to pericyclic reactions for the case of two π-electron systems which are strongly polarized by inclusion of d-orbitals, recently led to a new proposal for the generation of the oxaphosphetane *9* [4, 7, 9]. Thus, it is assumed that the ylide is approached by the carbonyl compound according to "Bürgi-Dunitz-Lehn-Wipf" [24]. MNDO-calculations of the reaction coordinates [25] of the model ylide $R_3P=CHCH_3$ ($R = H$, CH_3) and acetaldehyde CH_3CHO have been started from four different starting conformations and a C—C distance of 400 pm. The two reactants approach in 20 pm-steps and all parameters are always optimized. Thus, it could be demonstrated that the reaction runs through a quasi-betain transition state the calculated geometry of which is shown in Scheme 4.

The transition state *16*, however, does not give a real betain intermediate but directly an oxaphosphetane. The energy difference between the (*E*)- and the (*Z*)-transition state was 5 kJ/mol (C—C distance 200 pm); the (*E*)-oxaphosphetane was 5.4 kJ/mol more stable than the (*Z*)-isomer [7,9]. The calculations thus confirm the preference to the formation of the oxaphosphetane instead of the betain as the intermediate, but cannot explain the favored (*Z*)-configuration of the ligands of the oxaphosphetane four-membered ring.

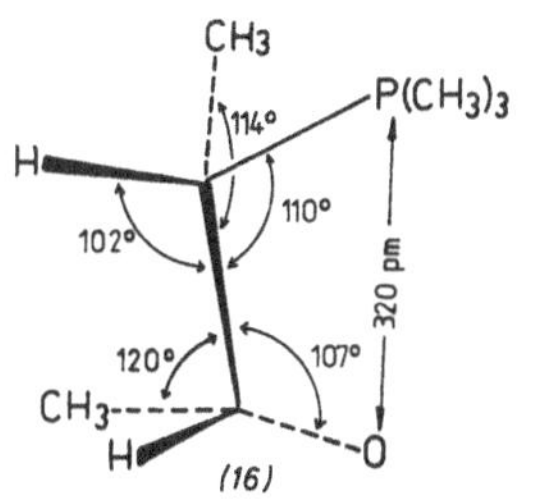

Scheme 4

According to their reactivity, the alkylidenephosphoranes (phosphorus ylides) used in Wittig reactions were divided into three groups [26]: ylides stabilizing the negative charge of their carbanionic centre by electron-attracting groups like —COR, —CO$_2$R, —CHO, —CN, etc., called stable ylides. They are stable against atmospheric oxygen and water. In contrast, unstable, reactive ylides possess no stabilizing α-substituents and carry electron-donating groups on their anionic centre, e.g. alkyl or O-alkyl. They are strong nucleophiles and must be handled under an inert gas. An intermediate position occupy the so-called moderate or metastable ylides, the methylene carbon of which can be substituted by halogen, vinyl, aryl, or alkynyl.

The above discussed mechanism of the Wittig reaction requires that the stereochemistry of the olefin formed strongly depends on the electronic character of the substituents attached to the ylidic carbon atom. This is in agreement with the experimental results. Reactive ylides preferentially give (*Z*)-olefins whereas stable ylides, as expected, show a contrary behavior yielding almost exclusively (*E*)-olefins. Metastable phosphoranes occupy an intermediate position and give about the same amount of (*E*)- and (*Z*)-olefin. Using alkylidene-tricyclohexylphosphoranes a high degree of (*E*)-stereoselectivity can also be obtained [27].

The addition of lithium salts during carbonyl olefination with reactive ylides causes a lowering of the (*Z*)-stereoselectivity. This effect is observed mainly in benzene or polar media such as ether (i.e. the more strongly the lithium cation is dissociated from its counterion [11]) (Table 1). On the other hand, the presence of lithium salts seems to increase the amount of the (*Z*)-isomer in the reaction of moderate [28], α-fluorosubstituted [29] or resonance-stabilized ylides [30] (Table 1).

When generating the ylide from the corresponding phosphonium salt, the choice of the method of formation is important for the stereochemistry of the reaction. With the original application of lithium alkyls as bases one equivalent of lithium halogenide is always formed; this lowers the stereoselectivity. Not before the development of methods for the preparation of "salt-free" ylide solutions, such as the sodium amide

Table 1. Influence of lithium salts on the (Z/E)-ratio of Wittig products (solvent benzene)

Added salt	$Ph_3P=CHC_2H_5 +$ $+ PhCH=O$ [11]	$Ph_3P=CHPh +$ $+ PhCH=O$ [28]
—	96:4 $(Z:E)$	18:82 $(Z:E)$
LiCl	90:10	21:79
LiBr	86:14	27:73
Li[BPh$_4$]	52:48	—

method [31, 32], the silazide technique [33] or the use of potassium in hexamethyl-phosphoric triamide [34, 35] (HMPA) new ways could be opened for carrying out stereoselective Wittig reactions.

As previously mentioned in the discussion of the mechanism of the Wittig reaction, the stationary ligands of the phosphorus effect a considerable influence on the geometry of the resulting alkenes. Electron-donating groups at the P-atom diminish the (Z)-stereoselectivity [27]. In the reaction of benzylidene-triphenylphos-phorane with benzaldehyde stilbene with an isomer ratio of $(Z/E) \sim 50/50$ is formed, whereas with benzylidene-tricyclohexylphosphorane (substitution of phenyl groups on the phosphorus by cyclohexyl groups) an (E)-stereoselectivity of about 95% is obtained [$(Z/E) = 5/95$] [27]. Similar effects can be observed in the successive substi-tution of the phenyl groups of benzylidene-triphenylphosphorane by methyl groups [36] as well as in the introduction of substituents in the phenyl ring of alkylidene-triaryl-phosphoranes [37].

The bulkiness of the groups attached to the carbon atoms, which are directly involved in the phosphetane formation, can affect the isomer ratio of the educts as well. Whereas the (Z)-stereoselectivity in the case of unstable phosphoranes bearing bulky groups next to the carbanionic centre (β-position to the P-atom) (e.g. branched alkyl substituents) is lowered [38, 39], the amount of the (Z)-isomer increases with the bulkiness of the alkyl groups of the aldehyde component [12, 40, 41]. In extreme cases, however, very bulky ligands on both of the reactants can suppress olefination completely [39].

α,β-Unsaturated and conjugated polyene aldehydes react with alkylidene-tri-phenylphosphoranes by the formation of a (Z)-double bond [4].

Protic solvents or the addition of proton-active compounds after oxaphosphetane formation shift the stereoselectivity of the reaction in the direction of the (E)-form. If the Wittig reaction is carried out in C_2H_5OD or if the oxaphosphetane solution, prepared at $-75\,°C$ in an aprotic solvent, is treated with deuterated ethanol, then deuterium is incorporated in high yield into the (E)-olefin formed, and the degree of deuterium labelling of the coexisting (Z)-olefin is lower. On the basis of these findings the mechanism discussed below has been established (Scheme 5).

At the stage of the betain *11*, a reaction with the alcohol involving attachment of a proton (deuterium) to the free electron pair occurs (*17*). The resulting unsolvated alkoxide anion of the complex now abstracts a proton from the carbon atom bearing the deuterium whereby a betain with conformation *18* is formed. To comply with the stereochemical conditions of the E_{1cb}-transition state, the free

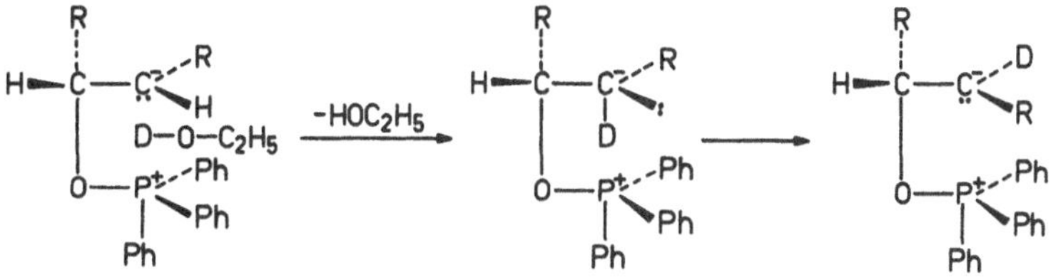

(17)   (18)   (19)

Scheme 5

electron pair must attain either the *syn*- or *anti*-position. For electronic and steric reasons rotation around the C—C axis to the *syn*-conformation is favored. Subsequent phosphine oxide elimination from *19* gives the (*E*)-olefin.

Aside from the depicted possibilities of shifting the carbonyl olefination towards the (*E*)-isomer a method for preparing (*E*)-olefins using oxido ylides has been developed [11,26,42–46]. In this method unstable lithiated phosphorus ylides *20*, giving the lithiated intermediate *22* on reaction with aldehyde *21*, are deprotonated to β-oxido ylides *23* with lithium alkyls and subsequently reprotonated. The resulting β-hydroxyphosphonium salt *24*, almost existing in the *threo*-form, is reacted with potassium *t*-butoxide to the (*E*)-olefin *25* [11,42,43] (Scheme 6). The β-oxido ylides *23* can be substituted by electrophiles like aldehydes [44,45], methyl iodide, FClO₃, iodobenzene dichloride, etc. [46] and subsequently converted into trisubstituted alkenes through protonation [44–46] (i.e. substitution + carbonyl olefination by means of β-oxido phosphonium ylides, the "scoopy"-variant of the Wittig reaction [46]).

This reaction, the pathway of which is still not fully clarified, can also be interpreted by a different mechanism.

At this point it should be noted, that olefination involving reaction of PO-stabilized carbanions like those from phosphonates, phosphinates, phosphine oxides, phosphonamides or thiophosphonates (Horner reaction [47], Horner-Wadsworth-Emmons reaction [48]) with carbonyl compounds almost exclusively yields (*E*)-olefins. This olefination is essentially restricted to PO-activated compounds carrying stabilizing groups on the carbanionic C-atom. However, these compounds are more reactive than the corresponding resonance-stabilized ylides. Therefore, this olefination method

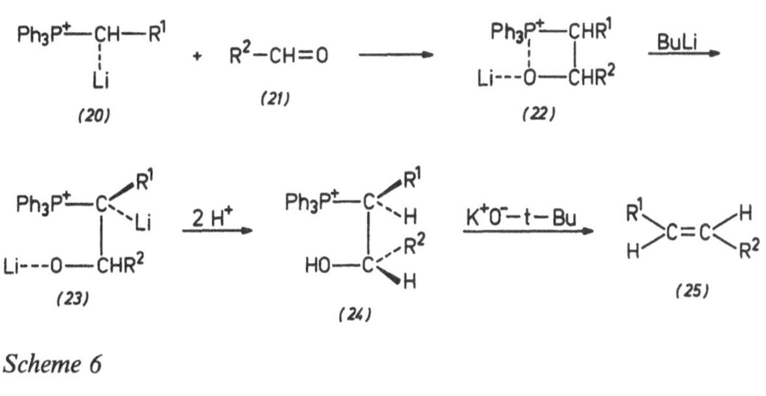

(20)   (21)   (22)

(23)   (24)   (25)

Scheme 6

represents a valuable supplement to the original Wittig reaction. However, since these reagents are carbanions or metallated compounds, respectively, and not ylides, they will not be dealt within this paper.

The Wittig reaction in general and the possibilities of preparing stereoselectively (Z)- or (E)-isomers gave great impulses to the synthesis of natural products. This is particularly true for fatty acids, prostaglandins, pheromones, juvenile hormones, carotenoids, and similar substances, the geometry of their double bonds being of paramount importance to their biological or physiological activity.

3 Synthesis of Fatty Acids and Fatty Acid Esters

Unsaturated fatty acids are essential structural components of lipids in the animal and plant kingdom. They normally possess between 12 and 28 carbon atoms, and one or more double bonds which are usually separated one from the other by one methylene group (methylene interruption). Animal lipids predominantly contain (Z)-unsaturated carboxylic acids. Whilst originally the chemistry of food and physiology of nutrition was dealing only with unsaturated fatty acids as essential components of food (e.g. vitamin F), in recent times interests in these compounds have become increasingly important in membrane research and in the chemistry of prostaglandins (the biochemical precursor of which they represent). Moreover, they are important synthesis intermediates in the synthesis of other natural products. The (Z)-stereoselective Wittig reaction may advantageously be applied to the synthesis of a great number of naturally occurring cis-unsaturated fatty acids, reviewed for the first time by Bergelson and Shemyakin [49, 50].

The reaction of the phosphonium salts 28, generated from the ω-halogencarboxylic esters 26, with sodium ethoxide in DMF or sodium hydride in DMSO yields the phosphoranes 29 which, on treatment with aldehydes 30, are converted into (Z)-alkenoic esters 31 [49, 51-53]. To synthesize Δ^5-unsaturated acids and higher homologous acids of type 31 (m ≥ 11) Bergelson et al. [49, 54] olefinated ethyl ω-formyl-alkanoates 34, obtained from ω-iodoalkanoic esters 32, with alkylidenephosphoranes 35 to the ethyl esters 31 (Scheme 7).

By this method (Z)-monounsaturated fatty acids and esters could be obtained with an (E)-isomer content of less than 10%; this stereoselectivity being however inferior to that of the commonly used "acetylenic approach" [55, 56]. However, the "salt-free" techniques used today in Wittig reactions allow (Z)-alkenoic acids to be synthesized with less than 2% of the (E)-isomers. Thus, Bestmann et al. prepared methyl and ethyl esters of (Z)-4,5,6,7,8,9,11- and 13-alkenoic acids of different chain lengths [35, 57-62], which served as intermediates in the synthesis of insect pheromones, both by reaction of ω-alkoxycarbonyl-substituted alkyl-triphenyl-phosphonium salts with simple alkanals and of ω-formylalkanoic esters with alkylidenephosphoranes. As the starting material for the synthesis of ω-substituted alkyl-phosphonium salts ω-chloro- and ω-bromocarboxylic esters were used. The corresponding ω-substituted aldehydes can usually be obtained by ozone cleavage of suitable olefin derivatives or by oxidation of alkohols [57, 58].

To prevent the base-catalyzed elimination of triphenylphosphine [63] 27 from the ester phosphonium salts, a frequently occurring competition reaction of ylide for-

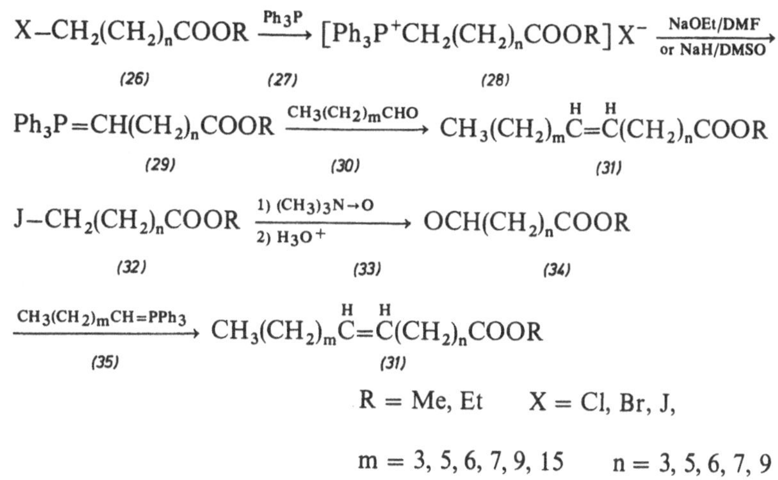

$$X-CH_2(CH_2)_nCOOR \xrightarrow{Ph_3P} [Ph_3P^+CH_2(CH_2)_nCOOR]X^- \xrightarrow[\text{or NaH/DMSO}]{\text{NaOEt/DMF}}$$

$$(26) \qquad\qquad (27) \qquad\qquad\qquad (28)$$

$$Ph_3P=CH(CH_2)_nCOOR \xrightarrow{CH_3(CH_2)_mCHO} CH_3(CH_2)_m\overset{H}{C}=\overset{H}{C}(CH_2)_nCOOR$$

$$(29) \qquad\qquad\qquad (30) \qquad\qquad\qquad (31)$$

$$J-CH_2(CH_2)_nCOOR \xrightarrow[\text{2) H}_3\text{O}^+]{\text{1) (CH}_3)_3\text{N}\to\text{O}} OCH(CH_2)_nCOOR$$

$$(32) \qquad\qquad\qquad (33) \qquad\qquad (34)$$

$$\xrightarrow{CH_3(CH_2)_mCH=PPh_3} CH_3(CH_2)_m\overset{H}{C}=\overset{H}{C}(CH_2)_nCOOR$$

$$(35) \qquad\qquad\qquad (31)$$

R = Me, Et   X = Cl, Br, J,

m = 3, 5, 6, 7, 9, 15   n = 3, 5, 6, 7, 9

Scheme 7

mation, it is recommended to use an excess of the salt and phosphonium iodides instead of the chlorides, because the latter are more susceptible to decomposition. In a synthesis of β,γ-unsaturated carboxylic acids described by Corey et al. phosphine elimination is prevented by using 2-carboxyethyl-triphenylphosphonium chloride and simultaneous addition of the salt and the carbonyl compound to sodium hydride [64]. Moreover, (Z)-Δ³-fatty acids are obtained by Michael addition of triphenylphosphine to acrylic esters in the presence of aldehydes [57]. (Z)-2-Dodecenoic, (Z)-2-tetradecenoic, (Z)-4-dodecenoic and 4,6-octadecadienoic acid were prepared by Jurasek et al. by means of the Wittig reaction and subsequent hydrolysis of the esters [65].

Saturated fatty acids with alkyl branching like e.g. (R,S)-tuberculostearic acid *39* are formed by olefination of *36* with ketone *37*, hydrogenation of the resulting methyl 10-methyl-9-octadecenoate *38* with Raney-nickel and subsequent hydrolysis [66]. α, β- and γ-branched unsaturated fatty acids were synthesized by Ucciani et al. by Wittig reaction of alkyl-branched aldehydes prepared by ion exchange-catalyzed condensation of alkanals [67].

Polyunsaturated fatty acids can also be prepared by Wittig reactions. Thus, Bergelson and Shemyakin synthesized α-eleostearic acid [(9Z,11E,13E)-9,11,13-octadeca-trienoic acid *42*] from (2E,4E)-2,4-nonadienal *40* and phosphorane *41* [49]. The intro-

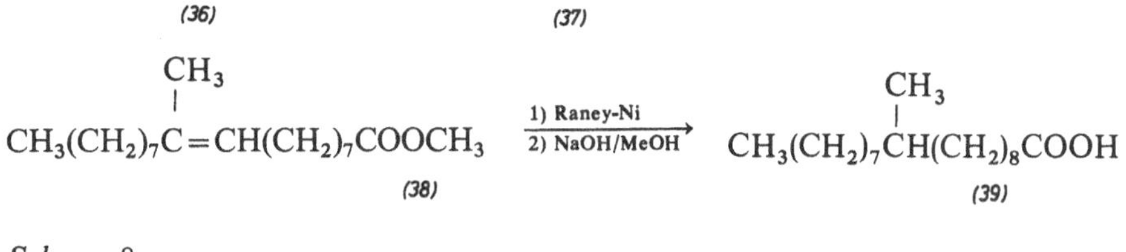

$$[Ph_3P^+(CH_2)_8COOCH_3]J^- \xrightarrow[\text{2) C}_8\text{H}_{17}\text{COCH}_3]{\text{1) NaOMe/DMF}}$$

$$(36) \qquad\qquad\qquad (37)$$

$$\underset{(38)}{CH_3(CH_2)_7\overset{\overset{\displaystyle CH_3}{|}}{C}=CH(CH_2)_7COOCH_3} \xrightarrow[\text{2) NaOH/MeOH}]{\text{1) Raney-Ni}} \underset{(39)}{CH_3(CH_2)_7\overset{\overset{\displaystyle CH_3}{|}}{CH}(CH_2)_8COOH}$$

Scheme 8

93

duction of an (E)-double bond by means of Wittig reaction is illustrated by the authors by the preparation of catalpa acid, [(9E,11E,13Z)-9,11,13-octadecatrienoic acid 46] [68], another member of naturally occurring conjugated-unsaturated fatty acids (Scheme 9).

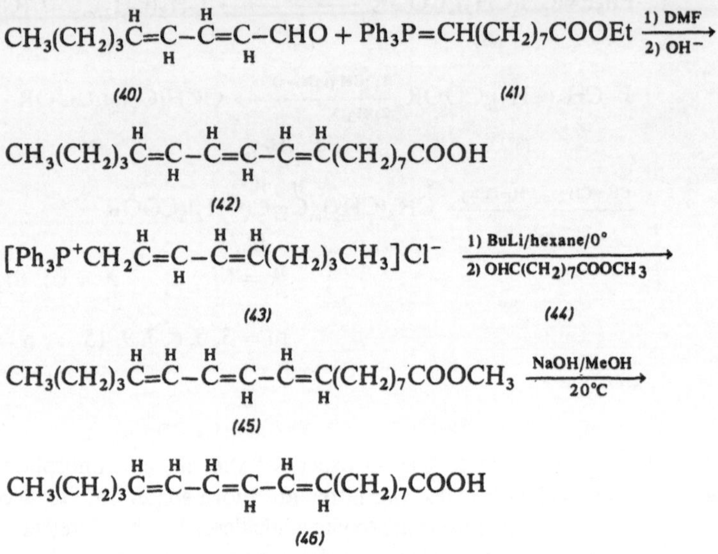

$$CH_3(CH_2)_3\overset{H}{C}=\overset{H}{C}-\overset{}{C}=\overset{}{C}-CHO + Ph_3P=CH(CH_2)_7COOEt \xrightarrow[\text{2) OH}^-]{\text{1) DMF}}$$

(40)   (41)

$$CH_3(CH_2)_3\overset{H}{C}=\overset{H}{C}-\overset{H}{C}=\overset{H}{C}-\overset{H}{C}=C(CH_2)_7COOH$$

(42)

$$[Ph_3P^+CH_2\overset{H}{C}=C-\overset{H}{C}=\overset{H}{C}(CH_2)_3CH_3]Cl^- \xrightarrow[\text{2) OHC(CH}_2)_7\text{COOCH}_3]{\text{1) BuLi/hexane/0°}}$$

(43)   (44)

$$CH_3(CH_2)_3\overset{H}{C}=\overset{H}{C}-\overset{H}{C}=\overset{H}{C}-\overset{}{C}=C(CH_2)_7COOCH_3 \xrightarrow[\text{20°C}]{\text{NaOH/MeOH}}$$

(45)

$$CH_3(CH_2)_3\overset{H}{C}=\overset{H}{C}-\overset{H}{C}=\overset{H}{C}-\overset{}{C}=C(CH_2)_7COOH$$

(46)

Scheme 9

Conjugated diunsaturated fatty acid methyl esters 50 with (E,Z)-configuration and 83–93 % stereochemical purity were synthesized by Bestmann et al. by (E)-selective olefination of ω-alkoxycarbonylalkanals such as 48 with formylmethylenephosphorane 47. The resulting (E)-α,β-unsaturated aldehydes 49 could be converted (Z)-stereoselectively with the ylides 35 into conjugated diunsaturated methyl esters 50 [69, 70] (Scheme 10). The same authors reacted 2-butynylphosphorane 51 with methoxycarbonylheptanal 52 to obtain methyl (E)-8-dodecen-10-ynoate 53 which could be partially hydrogenated to methyl (8E,10Z)-8,10-alkadienoate 55 by means of disiamylborane 54 (Scheme 10). This synthesis demonstrates that the Wittig reaction of 2-alkynylidenephosphoranes such as 51 with alkanals proceeds with an (E)-stereoselectivity of more than 90 % (c.f. p. 89).

Polyunsaturated carboxylic acids with methylene interruption (divinylmethane type), representing the major part of naturally occurring polyunsaturated fatty acids, can also be obtained via Wittig reaction. Methyl linoleate [methyl (9Z,12Z)-9,12-octadecadienoate 57] is obtained by olefination of ω-methoxycarbonylnonanal 44 with (Z)-3-nonenylidene-triphenylphosphorane 56 according to Scheme 11 [49]. After chromatographic purification with aluminium oxide the ester group is hydrolyzed to the corresponding carboxylic acid [3 % (E)-isomer]. By a combination of acetylene synthesis (58→59) and Wittig reaction (60→61), Bradshaw et al. synthesized crepeninic acid [71] [(Z)-9-octadecen-12-ynoic acid 62] occurring in essential oils of *Crepis* species (*Compositae*). Partial hydrogenation of the C-12-triple bond of 61 also yielded methyl linoleate 57 [71] (Scheme 11).

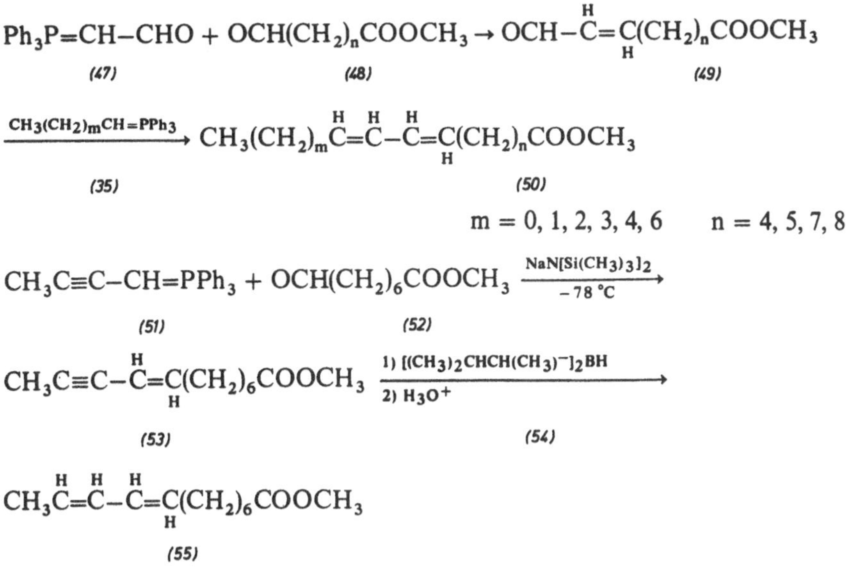

$$Ph_3P=CH-CHO + OCH(CH_2)_nCOOCH_3 \rightarrow OCH-\overset{H}{\underset{H}{C}}=C(CH_2)_nCOOCH_3$$

(47) (48) (49)

$$\xrightarrow{CH_3(CH2)_mCH=PPh_3} CH_3(CH_2)_m\overset{H}{C}=\overset{H}{C}-\overset{H}{\underset{H}{C}}=C(CH_2)_nCOOCH_3$$

(35) (50)

m = 0, 1, 2, 3, 4, 6   n = 4, 5, 7, 8

$$CH_3C\equiv C-CH=PPh_3 + OCH(CH_2)_6COOCH_3 \xrightarrow[-78\,°C]{NaN[Si(CH3)3]2}$$

(51) (52)

$$CH_3C\equiv C-\overset{H}{\underset{H}{C}}=C(CH_2)_6COOCH_3 \xrightarrow[2)\ H_3O^+]{1)\ [(CH_3)_2CHCH(CH_3)^-]_2BH}$$

(53) (54)

$$CH_3\overset{H}{C}=\overset{H}{C}-\overset{H}{\underset{H}{C}}=C(CH_2)_6COOCH_3$$

(55)

Scheme 10

$$CH_3(CH_2)_4\overset{H}{C}=\overset{H}{C}CH_2CH=PPh_3 + OCH(CH_2)_7COOCH_3 \xrightarrow{DMF}$$

(56) (44)

$$CH_3(CH_2)_4\overset{H}{C}=\overset{H}{C}-CH_2-\overset{H}{C}=\overset{H}{C}(CH_2)_7COOCH_3$$

(57)

$$CH_3(CH_2)_4C\equiv CH \xrightarrow[\text{8}]{Li/NH_3} CH_3(CH_2)_4C\equiv C-CH_2CH_2OH \xrightarrow[2)\ PPh_3]{1)\ PBr_3}$$

(58) (59) (27)

$$[CH_3(CH_2)_4C\equiv C(CH_2)_2P^+Ph_3]Br^- \xrightarrow[2)\ \ (44)]{1)\ BuLi/Et_2O}$$

(60)

$$CH_3(CH_2)_4C\equiv C(CH_2)\overset{H}{C}=\overset{H}{C}(CH_2)_7COOCH_3$$

(61)

$$\bigg| \begin{array}{l} H_2/Lindlar \\ \text{catalyst} \end{array}$$

$$CH_3(CH_2)_4\overset{H}{C}=\overset{H}{C}-CH_2-\overset{H}{C}=\overset{H}{C}(CH_2)_7COOCH_3$$

(57)

$$(61) \xrightarrow[EtOH]{KOH} CH_3(CH_2)_4C\equiv CCH_2\overset{H}{C}=\overset{H}{C}(CH_2)_7COOH$$

(62)

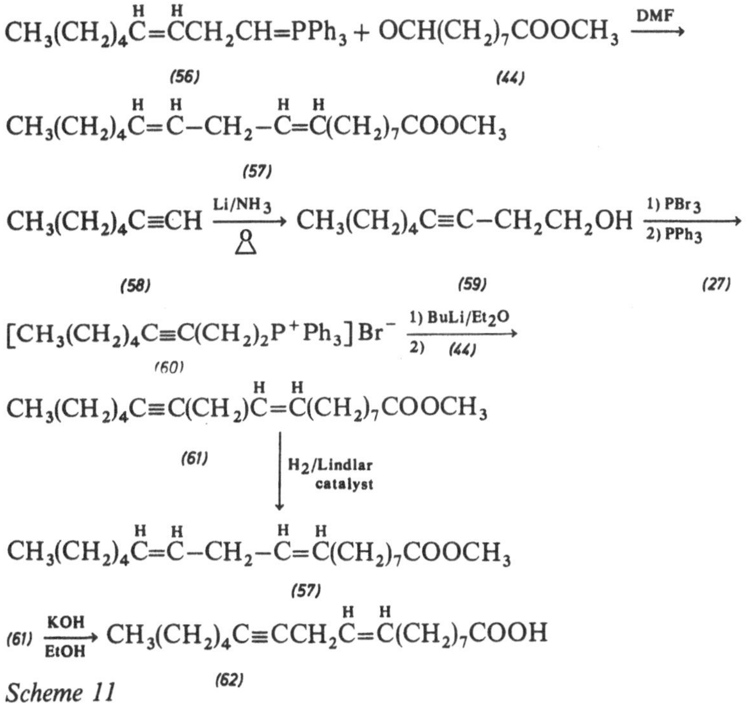

Scheme 11

An undesired side reaction, always involved in this reaction sequence, is the iso-merization of γ,δ-alkenylidenephosphoranes like e.g. *63* into the conjugated isomer *64* (Scheme 12). This complication can at least partially be prevented by slow addition of the base to a cooled solution of the phosphonium salt in the dark. The alternative olefination of (Z)-β,γ-unsaturated aldehydes with ω-alkoxycarbonylalkylidenephosphoranes normally gives less uniform products [72].

$$Ph_3P^+\diagdown\diagup\diagup\diagdown(CH_2)_n\,CH_3 \longrightarrow Ph_3P^+\diagdown\diagup\diagdown\diagup(CH_2)_{n+1}CH_3$$

(63)   (64)

Scheme 12

For the study of the physical and chemical properties of triply unsaturated fatty acids and for the comparison with γ-linolenoic acid [$6Z,9Z.12Z$)-6,9,12-octadecatrienoic acid] coworkers of the Unilever Research Centre in Vlaardingen (Netherlands) synthesized some methyl esters of (E,Z,Z)-trisunsaturated fatty acids [73]. For the preparation of the esters of $(2E,9Z,12Z)$-2,9,12-octadecatrienoic acid and of $(2E,11Z,14Z)$-2,11,14-eicosatrienoic acid (*68a* and *b*) they used the Wittig reaction to introduce the (E)-2-double bond into the starting diunsaturated aldehyde. Reduction of the acid chlorides of $(7Z,10Z)$-7,10-hexadecadienoic acid and of linolenoic acid (*65a* and *b*) with lithium tri-tert-butoxyaluminium hydride affords the corresponding aldehydes *66a* and *b* which can be olefinated with the stable ylide *67* to methyl (E,Z,Z)-alkatrienoates *68a, b* with a (Z)-2-isomer content of 4.6% [73] (Scheme 13).

$$CH_3(CH_2)_4\overset{H}{C}=\overset{H}{C}CH_2\overset{H}{C}=\overset{H}{C}(CH_2)_n COCl \xrightarrow{[(CH_3)_3CO]_3HAlLi}$$

(65)a,b

$$CH_3(CH_2)_4\overset{H}{C}=\overset{H}{C}-CH_2-\overset{H}{C}=\overset{H}{C}(CH_2)_n-CHO$$

(66)a,b

$$\xrightarrow{Ph_3P=CHCOOCH_3} CH_3(CH_2)_4\overset{H}{C}=\overset{H}{C}CH_2\overset{H}{C}=\overset{H}{C}(CH_2)_n-\overset{H}{C}=\overset{}{\underset{H}{C}}-CO_2CH_3$$

(67)   (68)a,b

$$n = \text{a) 5, b) 7}$$

Scheme 13

In the course of the structure elucidation of leucotrienes, which are open-chain metabolites of the arachidonic acid cascade, the synthesis of polyunsaturated fatty acid derivatives *via* the Wittig reaction has become, especially in recent years, highly important in preparative chemistry. Thus, e.g. for the synthesis of methyl 5,6-epoxy-7,9,11,14-eicosatetraenoate (*78*, leukotriene-A methyl ester), a substance representing a key compound in the synthesis as well as in the biogenesis of the other leukotrienes, Gleason et al. [74] (Smith Kline & French Labs.) carried out three olefinations with phosphoranes. The first two involve (E)-selective reactions of formylmethylenetriphenylphosphorane *47* with the aldehydes *69* and *71*, and give in 22% yield (based on *69*) the (E)-α,β-unsaturated epoxyaldehyde *72*. Synthon *77*, necessary for the third Wittig reaction, was generated by alkylation of propargyl alcohol *74*, partial hydrogenation to *76* and conversion into the phosphonium salt *77*. The conversion of the corresponding ylide (BuLi, $-78\,°C$) with 72 leads to a 2:1 mixture of the (E)- and (Z)-isomers with respect to the double bond in the 9-position [74] (Scheme 14).

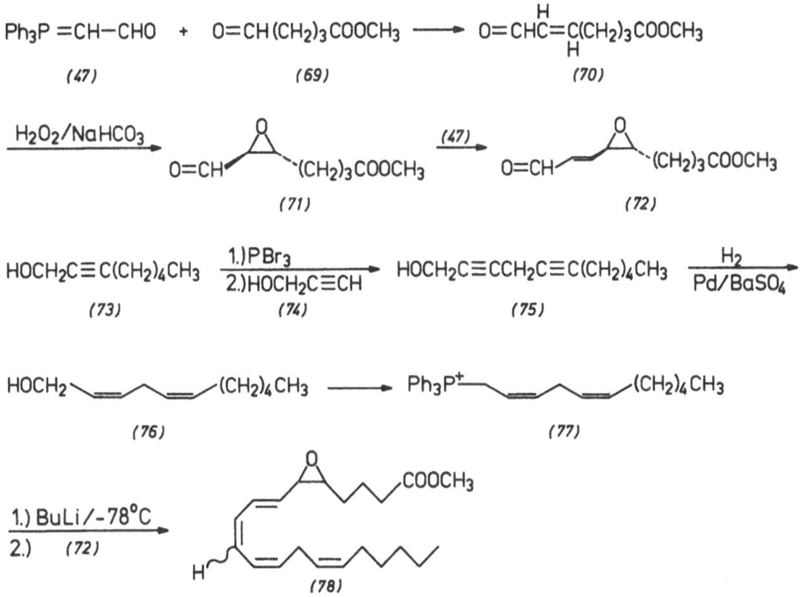

$Ph_3P=CH-CHO$ + $O=CH(CH_2)_3COOCH_3$ $\longrightarrow$ $O=\overset{H}{\underset{H}{CHC}}=C(CH_2)_3COOCH_3$

(47)   (69)   (70)

$\xrightarrow{H_2O_2/NaHCO_3}$ $O=CH$⎓$(CH_2)_3COOCH_3$ $\xrightarrow{(47)}$ $O=CH$⎓$(CH_2)_3COOCH_3$

(71)   (72)

$HOCH_2C\equiv C(CH_2)_4CH_3$ $\xrightarrow[2.)HOCH_2C\equiv CH]{1.)PBr_3}$ $HOCH_2C\equiv CCH_2C\equiv C(CH_2)_4CH_3$ $\xrightarrow{H_2}{Pd/BaSO_4}$

(73)   (74)   (75)

$HOCH_2$‿‿‿$(CH_2)_4CH_3$ $\longrightarrow$ Ph_3P^+‿‿‿$(CH_2)_4CH_3$

(76)   (77)

$\xrightarrow[2.) \ (72)]{1.) \ BuLi/-78\,°C}$

(78)

Scheme 14

Because the geometry of the 9-double bond was not clear at that time, Corey et al. [75] tried to prepare the (Z)-9-isomer as well as the (E)-9-isomer of leukotriene-A (78 and 86). In the synthesis of the former isomer the tribenzoyl derivative of D-$(-)$-ribose (79) was converted in 8 stepy into the optically active epoxyaldehyde 71 and the latter to 72. 72 was olefinated with ylide 82, generated by treatment of the corresponding phosphonium mesylate with lithium diisopropylamide in THF/HMPA [75] (Scheme 15). In the first olefination step $72+82\rightarrow78$, however, similar to the first method, a Δ^9-isomer mixture was formed. The loss of (Z)-selectivity of the Wittig reaction is due to the use of conjugated unsaturated, i.e. moderate ylides of type 82, and had to be expected because of the mechanism of the Wittig reaction (see Sect. 2).

Scheme 16 shows the preparation of the corresponding Δ^9-(E)-isomer, (E)-5(S), 6(S)-epoxy-$(7E,9E,11Z,14Z)$-7,9,11,14-eicosatetraenoic methyl ester 86 [75]: $(6E,7E)$-6,7-epoxyalkadienal 84, deduced from 71, was olefinated with the reactive ylide 85 (generated by reaction of the corresponding phosphonium iodide with BuLi in THF at $-78\,°C$) [75]. During this reaction the geometry of the double bonds of the aldehyde and the ylide remain unaffected and the newly generated double bond possesses (Z)-configuration, even in the presence of lithium iodide (Scheme 16). Reaction of 86 with glutathione or the corresponding N-trifluoroacetylglutathione dimethyl ester and subsequent hydrolysis gives 87, a SRS-active ("slow reacting substance") compound, identical with natural leukotriene-C-1 [75] (Scheme 16).

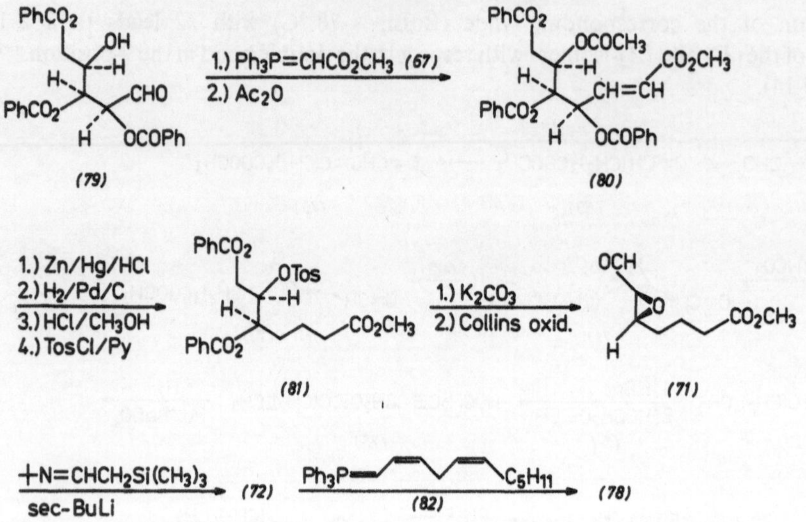

PhCO$_2$ ... OH, H–, H, CHO, PhCO$_2$, H, OCOPh **(79)**

1.) Ph$_3$P=CHCO$_2$CH$_3$ *(67)*
2.) Ac$_2$O

→ PhCO$_2$ ... OCOCH$_3$, CO$_2$CH$_3$, H–, CH=CH, PhCO$_2$, H, OCOPh **(80)**

1.) Zn/Hg/HCl
2.) H$_2$/Pd/C
3.) HCl/CH$_3$OH
4.) TosCl/Py

→ PhCO$_2$, OTos, H–, CO$_2$CH$_3$, PhCO$_2$ **(81)**

1.) K$_2$CO$_3$
2.) Collins oxid.

→ OCH, O, H, CO$_2$CH$_3$ **(71)**

$+$N=CHCH$_2$Si(CH$_3$)$_3$ *(72)*
sec-BuLi

Ph$_3$P= /\/\ C$_5$H$_{11}$ *(82)* → *(78)*

Scheme 15

OCH, O, H, CO$_2$CH$_3$ **(71)**

1.) LiCH=CH–CH=CHOEt *(83)*
2.) CH$_3$SO$_2$Cl/Et$_3$N
3.) pH 7.0 buffer

→

OCH, O, H, H, CO$_2$CH$_3$ **(84)**

Ph$_3$P= /\= C$_5$H$_{11}$ *(85)*

→

O, H, CO$_2$CH$_3$, C$_5$H$_{11}$, H **(86)**

glutathione derivative

→

H, OH, CO$_2$CH$_3$, C$_5$H$_{11}$, H, S–CH$_2$–CH–CONHCH$_2$CO$_2$H, NH–COCH$_2$CH$_2$CH–CO$_2$H, NH$_2$ **(87)**

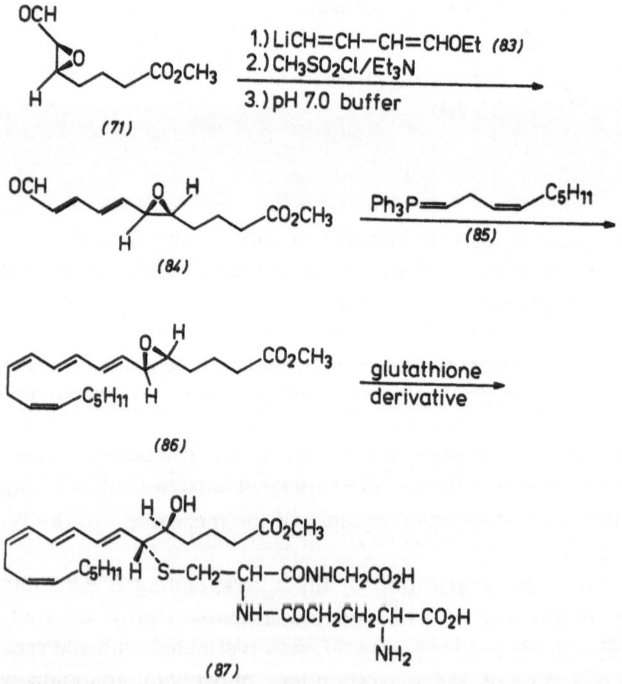

Scheme 16

Leukotriene-B could be obtained starting with the acetonide *88* of 2-deoxyribose [76]. Wittig olefination of *88* with methoxycarbonyl-methylenephosphorane *67* and subsequent hydrogenation affords hydroxy ester *89*. Tosylation of *89*, cleavage of the protective group and treatment with K$_2$CO$_3$ yields the epoxy ester *90*. Benzoylation of the hydroxy group at C-5, opening of the oxirane ring and oxidative glycol cleavage lead to the aldehyde ester *91*. As a second reaction partner for the Wittig reaction the phosphonium salt *98*, obtained from the D-(+)-mannose derivative *92*, was used. Wittig olefination of *92* with hexylidene-triphenylphosphorane *93* leads to (Z)-olefin *94*. Tosylation of *94* and subsequent cleavage of the protective group followed by reaction of the resulting tritosylate with phenyl chloroformate and diazabibyclo[4.3.0]nonane finally leads to the (Z)-epoxide *95*. Subsequent hydrolysis of the carbonate *95* and glycol cleavage affords an aldehyde which can be olefinated with allylidenephosphorane *96* to epoxytriene *97*. Treatment of *97* with HBr and triphenylphosphine yields, *via* the bromo alcohol, the phosphonium salt *98*. The following Wittig reaction of the 6-hydroxyalkadienylylide, obtained by treatment

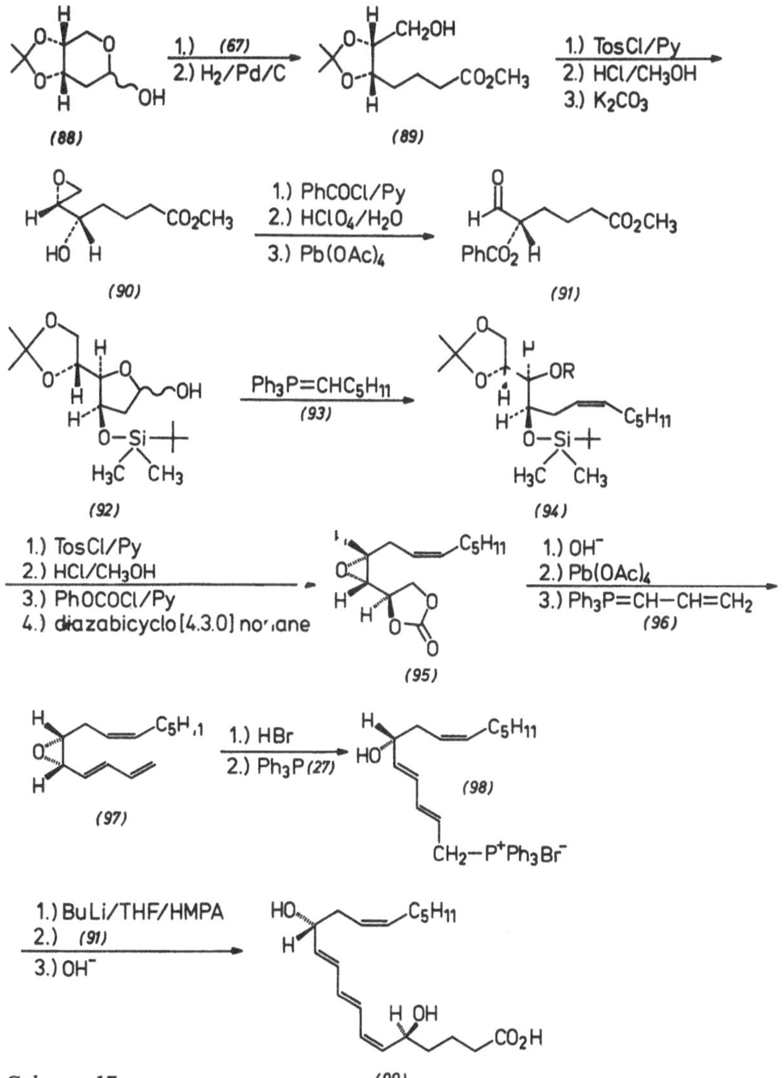

Scheme 17

99

Hans Jürgen Bestmann and Otto Vostrowsky

of the phosphonium salt *98* with two equivalents of BuLi in THF/HMPA, with *91* and subsequent hydrolysis of the ester groups gives 5(*S*),12(*R*)-dihydroxy-(6*Z*,8*E*, 10*E*,14*Z*)-6,8,10,14-eicosatetraenoic acid *99* (Scheme 17).

The retention time, determined by reserved phase HPLC, and spectroscopic data of *99* are identical with native leukotriene-B [76].

At about the same time, a synthesis of leukotriene-A, also termed SRS-A ("slow reacting substance of anaphylaxis"), which made use of the Wittig olefination was described [77]. The ylide of *100* is reacted with ethyl 5-formyl-2,4-pentadienoate *101* to give the (*E,E,Z,Z*)-tetraenoic ester *102*. Reduction and mesylation of *102*, subsequent conversion into the sulfonium salt, and treatment of the latter with a base yields a sulfonium ylide which is reacted with methyl 4-formylbutanoate *69* to the epoxytetraenoic ester *103*. After separation of the cis-epoxide by HPLC, *103* was treated with the *S*-trimethylsilyl derivative of glutathione dimethyl ester N-trifluoroacetamide. The diastereomeric products thus obtained were separated by means of HPLC and hydrolyzed to *104* [77] (Scheme 18).

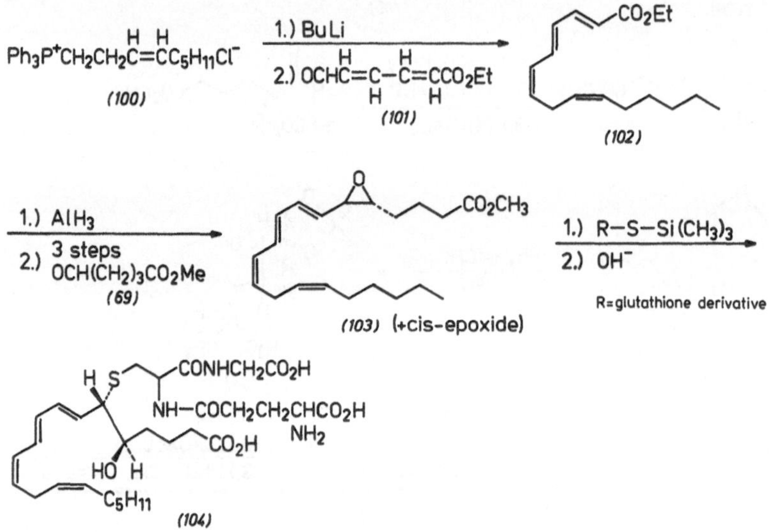

Scheme 18

4 Prostaglandins

Prostaglandins are cyclopentane derivatives of unsaturated C_{20}-fatty acids. They display hormonal as well as regulatory activities and were detected for the first time in spermatic fluid. They are obviously widely spread in animal tissue. Biosynthesis of prostaglandins involves cyclization of polyunsaturated carboxylic acids, e.g. arachidonic acid [(5*Z*,8*Z*,11*Z*,14*Z*)-5,8,11,14-eicosatetraenoic acid] is enzymatically converted directly into prostaglandin PGE_2. Prostaglandins stimulate and contract certain smooth muscles. At very low concentrations they lower the blood pressure and influence the heart frequency. Some of them can release labour and

100

be used for the interruption of pregnancy while others are antagonists of certain hormones, messenger substances in the hormone perception, or inhibit specific enzyme functions.

According to their functional groups on the cyclopentane ring prostaglandins are divided into three main groups, namely β-hydroxyketone prostaglandins PGE (*105* in Scheme 19), 1,3-cyclopentanediols PGF (*106*) and prostaglandins PGA *107* of the type of α,β-unsaturated ketones. The number of double bonds in the side chains are indicated by a subscript; the designations refer to the stereochemistry of the compounds (Scheme 19 gives three examples for the conventions of prostaglandin nomenclature).

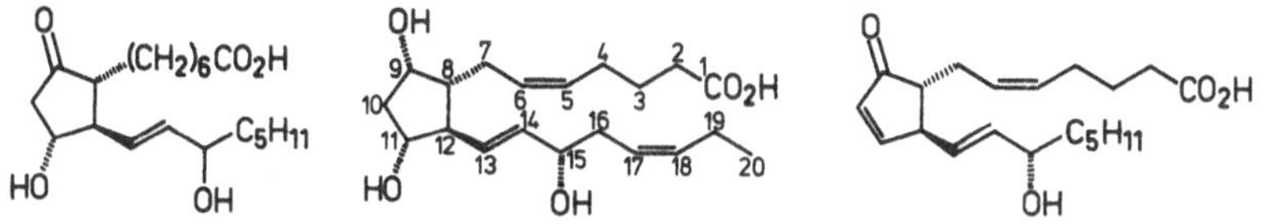

(105)

prostaglandin E_1
PGE_1

(106)

prostaglandin $F_{3\alpha}$
$PGF_{3\alpha}$

(107)

prostaglandin A_2
PGA_2

Scheme 19

The possibility of synthesizing pure (Z)-olefins by means of reactive "salt-free" ylides predestinates the (Z)-double bond at C-5 in e.g. *106* and *107* to be introduced into the corresponding aldehyde *via* the Wittig reaction. The (E)-configurated double bond at C-13 (*105, 106* and *107*) with its vicinal hydroxy group was frequently formed by the phosphonate method (cf. Chapter 2). In some cases, however, it could also be obtained by (E)-selective Wittig olefination using resonance-stabilized phosphoranes.

After the first syntheses of the prostanoic acid skeleton [78] and a few other prostaglandin derivatives [79–81], using Wittig olefinations, the first synthesis of a prostaglandin was published by Just and Simonovitch in 1967 [82]. These authors chose the epoxidation-solvolysis of the bicyclo[3.1.0]hexane derivative *113* and *114*, respectively, as the key reaction in their synthetic approach. The ester *108* was converted into the bicyclo[3.1.0]-hexane-6-carboxaldehyde *109* which was olefinated to *111*; the latter product was isolated in 60–80% yield from the isomer mixture. Hydrolysis, oxidation and alkylation of the resulting ketone gave *113* in poor yield; *113* exhibited the correct stereochemistry for the epoxidation-solvolysis step. Treatment of this product with H_2O_2 and formic acid yielded a mixture from which impure dl-PGE$_1$ methyl ester *105* could be isolated. The PGE$_1$ thus obtained displayed only 10–25% of the biological activity of natural prostaglandin. Reduction of *113* and the same reaction sequence afforded ester *115* of amorphous dl-PGF$_{1\alpha}$ [82] (Scheme 20).

Coworkers of the Smith Kline & French Laboratories later on could prove that under the conditions reported in the literature [82] no dl-PGE$_1$ *105* but only a mixture of isomeric PGF$_{1\alpha}$'s, is formed, none of these isomers corresponding to authentical PGF$_{1\alpha}$ [83]. Only through improvements of the chosen synthetic route developed by chemists of the McGill University [84] and of the Upjohn Company [85] not only dl-PGF$_{1\alpha}$ and dl-PGE$_1$ but also dl-PGE$_2$, dl-PGE$_3$, dl-PGF$_{2\alpha}$ and dl-PGF$_{3\alpha}$ [84,85]

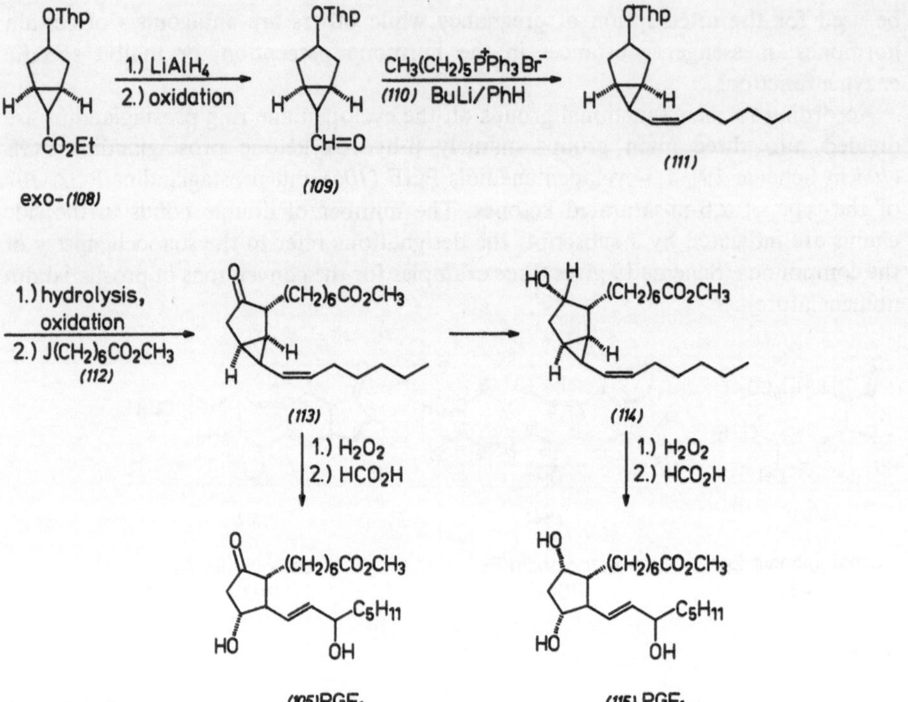

Hans Jürgen Bestmann and Otto Vostrowsky

Scheme 20

could be obtained. Axen et al. [86] prepared the bicyclo[3.1.0]hexane derivative *113* starting from norbornadiene. Hydroxylation of *113* led to the corresponding 1,2-diol which was hydrolyzed according to the method of Schneider [87]. Burdy et al. (Upjohn Co.) applied the same synthesis for the preparation of 15-methyl substituted prostaglandins. The methyl group was introduced *via* the phosphonium salt *116*. Reaction of *116* with *109* yielded *117* which was subsequently converted into 15-Methyl-PGF₁ and 15-methyl-PGE₁ [88] (Scheme 21).

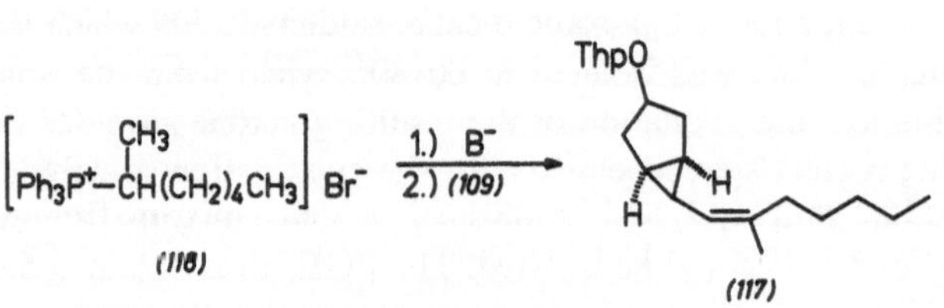

Scheme 21

In a synthesis of PGE₁ reported by Corey [89] for the preparation of the O,S-acetal-protected dienol thioether aldehyde *122*, the ylide generated from phosphonium salt *119* was used as an electrophile in the alkylation of *118*. The following Wittig reaction of the resulting phosphonium salt *120* with the thioenolether aldehyde *121* in the presence of phenyllithium as the base gave intermediate *122* in 35% yield.

102

Similar to the "cyclization route to PGE_1" [90], *122* could be converted into prostaglandin E_1 [89] (Scheme 22).

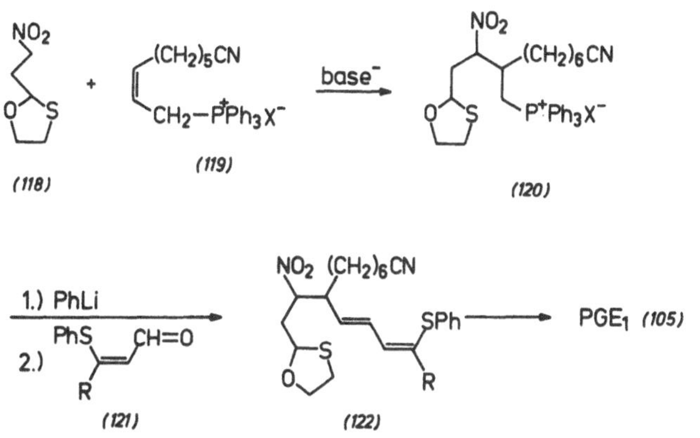

Scheme 22

In the "bicycloheptene route to prostaglandin" (also developed at the Harvard university [91,92]), which can be utilized for the preparation of all primary prostaglandins as well as non-natural prostaglandin analogs, the stereochemically uniform lactone aldehyde *127* is the key compound. It is obtained from the bicycloheptene derivate *123* as the starting compound [91,92] (Scheme 23). Olefination of *127* with phosphonate *128* leads to the introduction of an (*E*)-double bond into 13-position. Subsequent hydrolysis of the protective group and a further Wittig reaction with $[Ph_3P^+(CH_2)_4COOH]$ Br^- (*132*) in NaH/DMSO affords stereoselectively the (*Z*)-double bond in the second side chain of *133*. The bis-tetrahydropyranylether *133* can easily undergo hydrolysis, oxidation, hydrogenation, and hydrolysis and oxidation. These processes stereospecifically lead to $PGF_{2\alpha}$, PGE_2, $PGF_{1\alpha}$ and PGE_1 [91-93] (for summarizing literature see Refs. [94-96]).

The possibility of introducing a cisoid side chain into aldehydes and lactols like *131* in Scheme 23 also allows the access to a great number of prostaglandins and analogous compounds. Table 2 gives a survey of the reactions of certain aldehydes and lactols with the ylide generated from 5-triphenylphosphoniopentanoic acid *132*. Besides the ylide $Ph_3P=CH(CH_2)_3COO^-$ also structurally related phosphoranes were used for the introduction of the (*Z*)-Δ^5-double bond. Thus, dicarboxylic prostaglandin analogs [97], 2,3-dimethyl analogs [98], tetrazole analogs [99] and sulfonic acid prostaglandin analogs [100] could be obtained. For the synthesis of intermediates of thromboxanes, a new class of arachidonic acid metabolites which are structurally related to prostaglandins, also Wittig olefination was used for the introduction of the (*Z*)-5-unsaturated side chain according to Scheme 24 [101-103].

In many cases, the second side chain of the cyclopentane ring of prostaglandins possesses a double bond with (*E*)-geometry at C-13. The introduction of this side chain is usually achieved by olefination with phosphonate anions according to the Horner technique. However, because of the vicinal oxygen function at C-15 it is also recommended to introduce the side chain *via* (*E*)-stereoselective Wittig synthesis

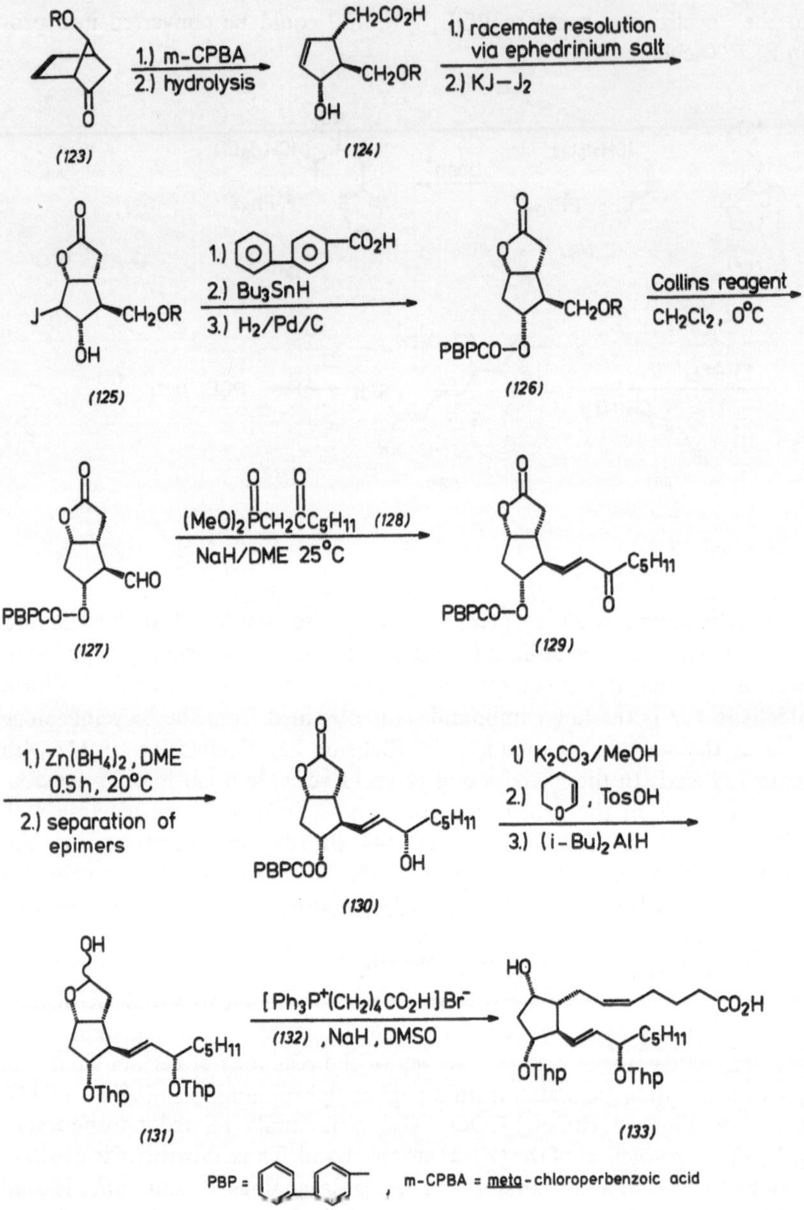

Scheme 23

of resonance-stabilized ylides. Thus, Finch and Fitt (Ciba Pharmac. Co.) used hexanoylmethylene-tributylphosphorane *140* for the construction of this molecular part. This is illustrated by the reaction scheme below where the *trans*-methoxime *139*, prepared from 7-(2-carboxy-5-oxocyclopent-1-enyl)heptanoic acid *138 via* several steps, was converted into the corresponding Thp-ether, the ring ester reduced, the carbinol group oxidized and subsequently olefinated with *140* [127)] (Scheme 25).

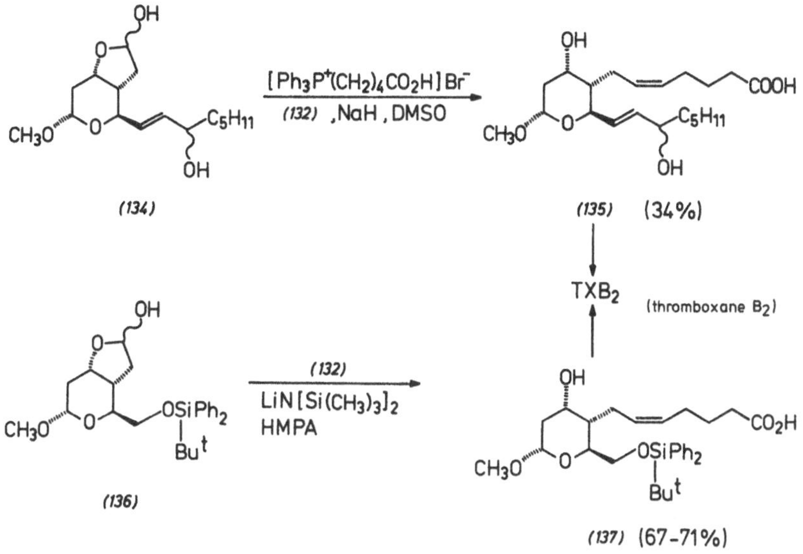

Scheme 24

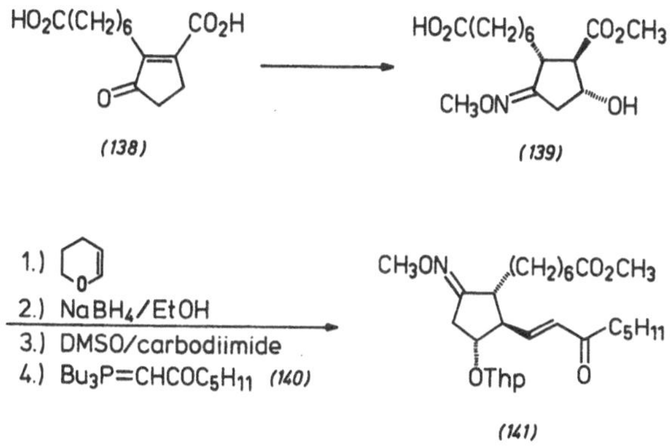

Scheme 25

Reduction of the carbonyl group at C-15 and liberation of the hydroxy group at C-11 give a mixture of dl-PGE$_2$- and dl-epi-PGE$_1$-methoxime methyl ester which could be separated chromatographically. However, the authors did not succeed in cleaving the methoxime group [127].

The synthesis of prostaglandin PGB$_1$, as described by Morin et al. (Eli Lily Research Labs.), starts with 7-(2-methoxyphenyl)heptanoic acid *142* which is converted into *143* through Birch reduction, esterification and acetalization. Ozone cleavage of *143* and cyclization of the resulting dialdehyde affords cyclopentene carbaldehyde *144* which was subjected to (E)-selective Wittig olefination with

105

Table 2. Introduction of the (Z)-5-unsaturated side chain of prostaglandin derivatives into aldehydes and lactols *via* Wittig reaction with $Ph_3P^+(CH_2)_4CO_2RX^-$

PG derivative	reactant	reaction conditions	product	yield (%)	Ref.
PGE$_2$	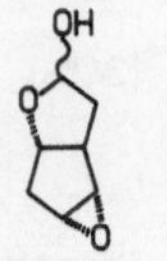	NaH/DMSO 2h, room temp.	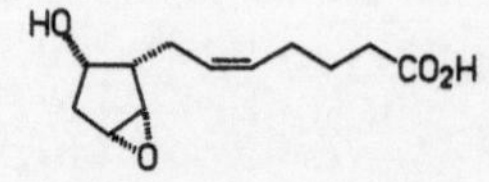	50–55	104)
PGA$_2$	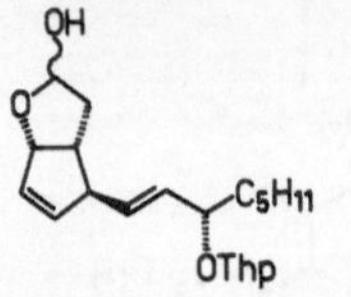	NaH/DMSO	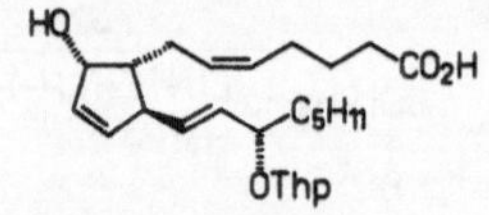		105)
PGC$_2$	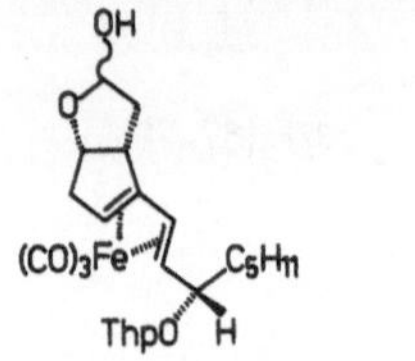	NaH/DMSO	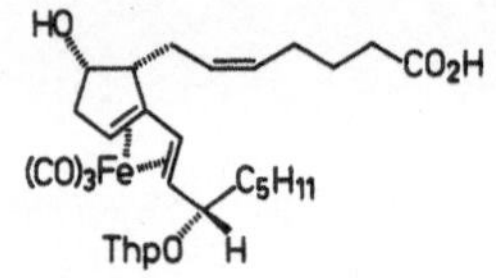	70	106)
11-deoxy-PGE$_2$	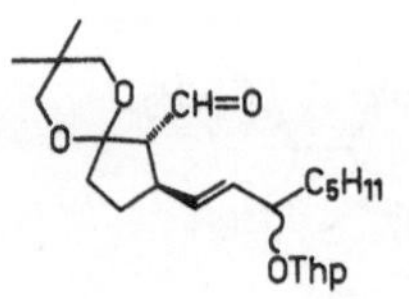	NaH/DMSO	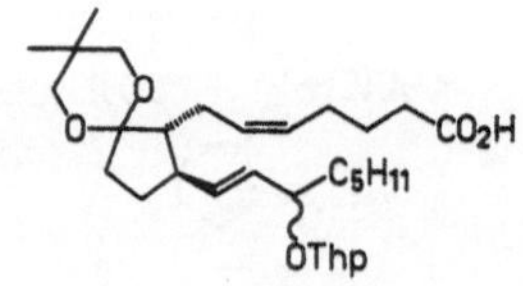		107)
12-α-fluoro-PGF$_{2α}$	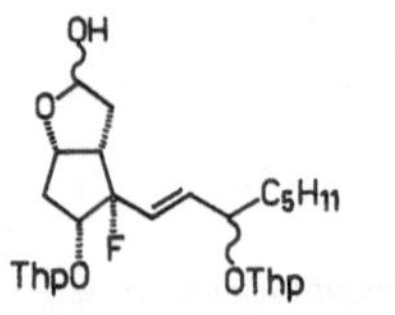	NaH/DMSO	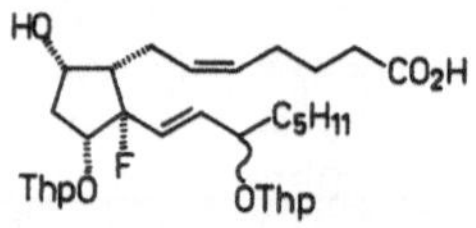	70	108)
13-dehydro-PGF$_{2α}$	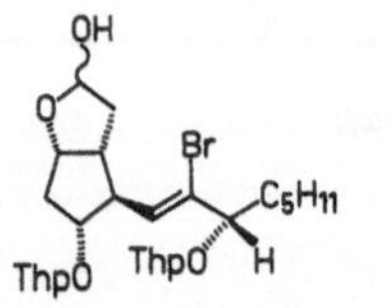		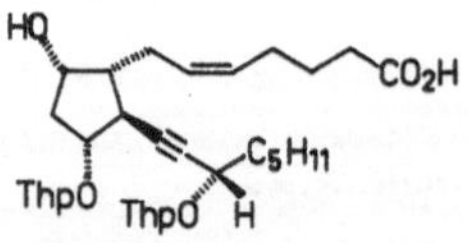		109)
	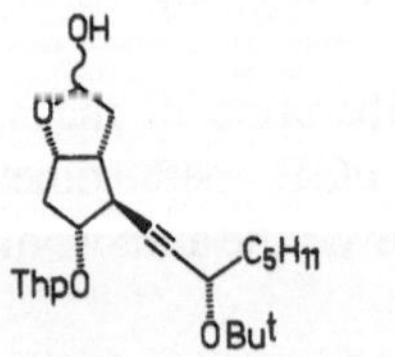		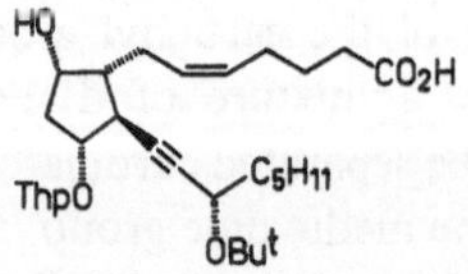	93	110)
16,16-ethano-PGF$_{2α}$	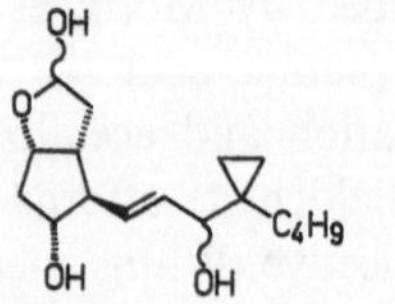	NaH/DMSO 2h, 25°C	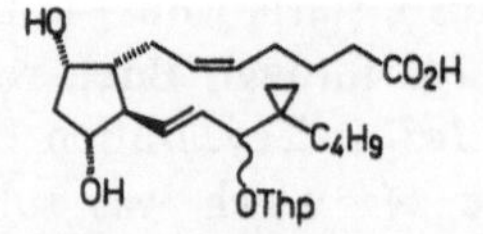	59	111)

Table 2. (continued)

PG derivative	reactant	reaction conditions	product	yield (%)	Ref.
11-oxa-PGF$_{2\alpha}$	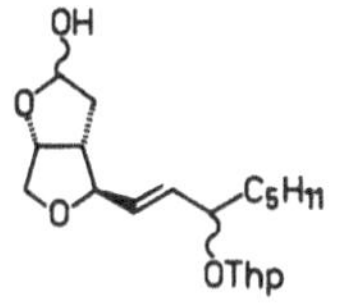	NaH/DMSO	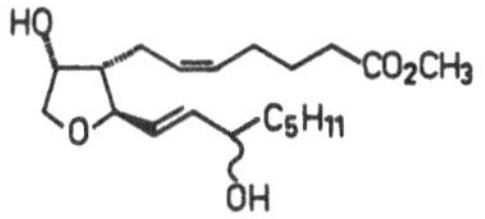		112)
pyrrolidinone analog PG	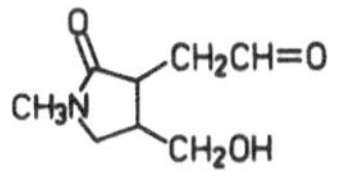		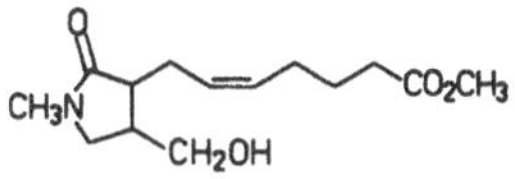		113)
benzo-type analog PG	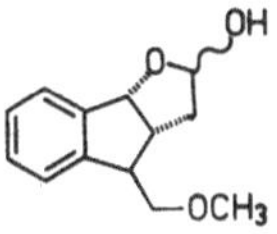	NaH/DMSO 25°C	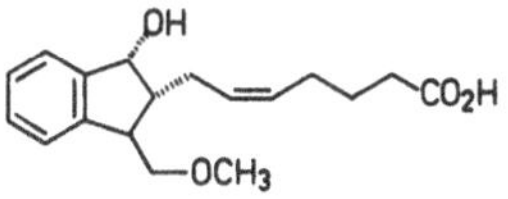		114)
9-deoxy-9α-hydroxy-methyl PGF$_{2\alpha}$	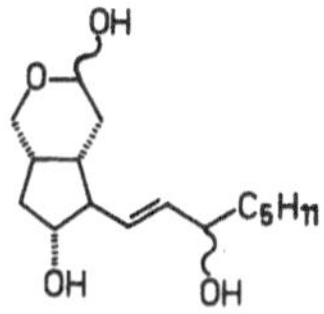		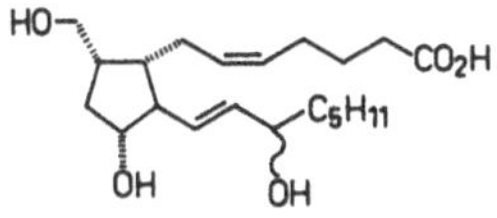	65	115)
6,9-aza-PGH$_2$-analog	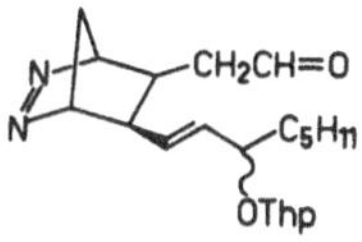	NaH/DMSO	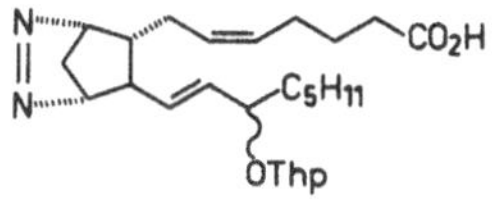	66	116)
12-substituted prostaglandins	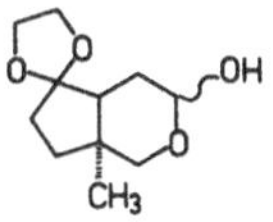	DMSO 3d, 25°C	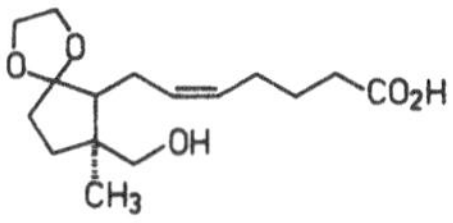	55	117)
16-methylene-prostaglandin F$_{2\alpha}$	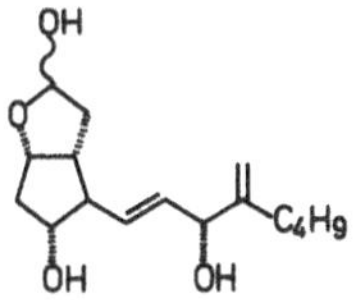	NaH/DMSO 2h, 25°C	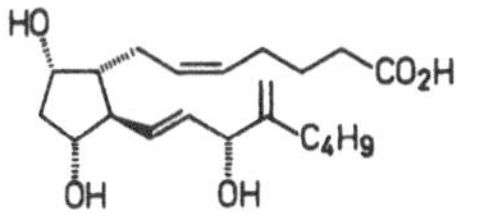		118)
12-methylprosta-glandin A$_2$	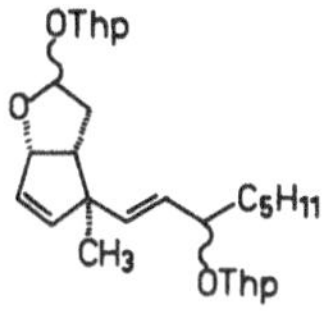	NaH/DMSO	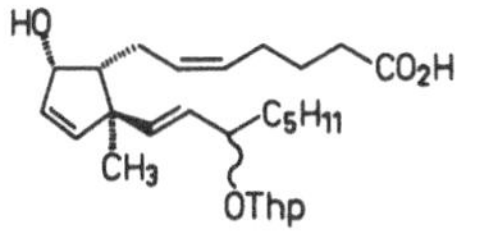	71	119)

107

Table 2. (continued)

PG derivative	reactant	reaction conditions	product	yield (%)	Ref.
PGF$_{2\alpha}$	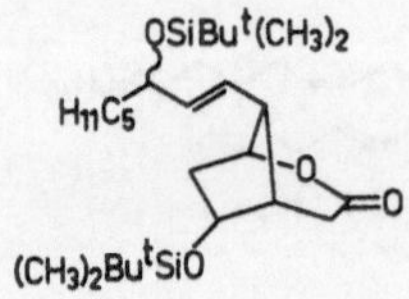	NaH/DMSO	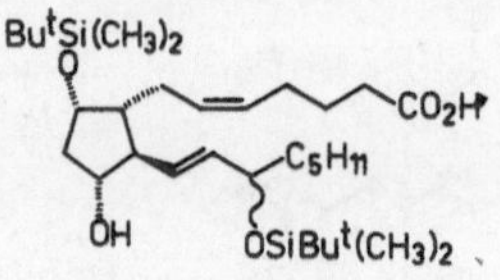	40	120)
11-norprostaglandins	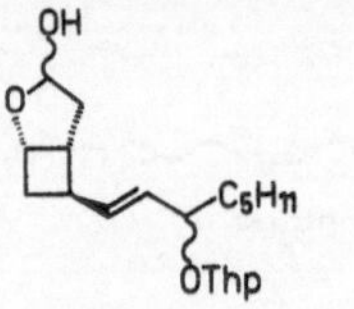	KH/DMSO	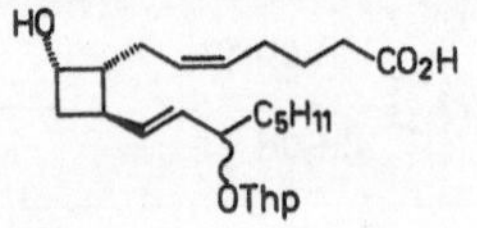	88	121)
9,11-etheno-PGH$_2$-analog	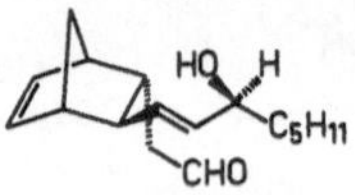	NaH/DMSO	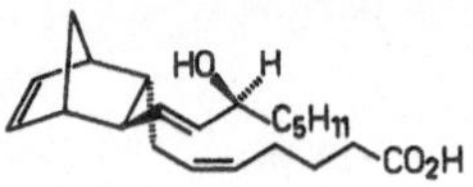	34	122)
six-membered ring analog PG	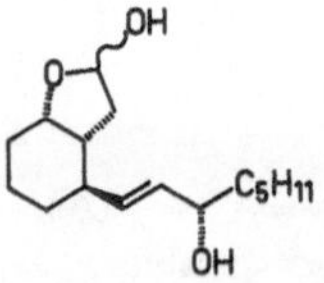		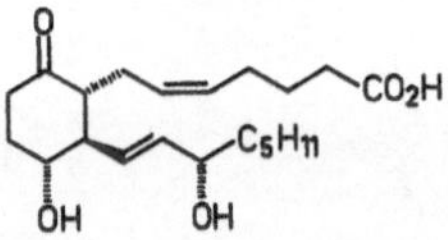		123)
dl-12,15-ethylene-13,14-dihydro-PGF$_{2\alpha}$ methyl ester	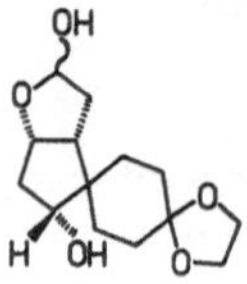	1.) NaH/DMSO 2.) CH$_2$N$_2$	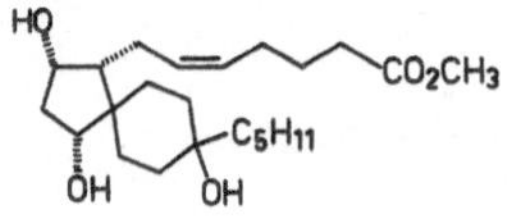		124)
carbocyclic thromboxane A$_2$	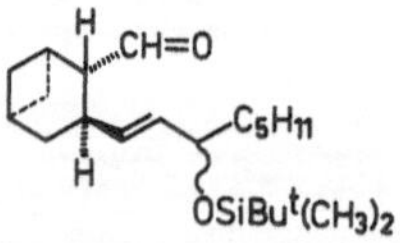	1.) NaH/DMSO 2.) CH$_2$N$_2$	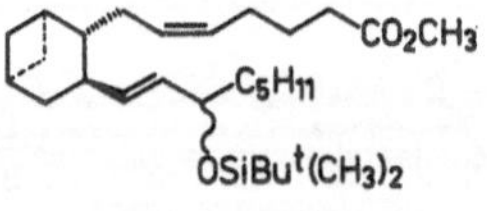		125)
stable thromboxane A$_2$-analog	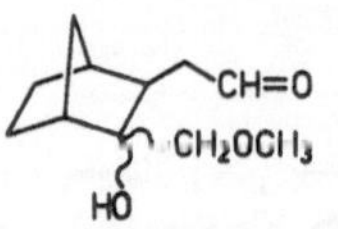	NaH/DMSO	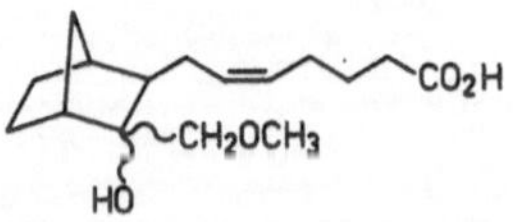	60	126)

hexanoylmethylene-triphenylphosphorane *145*. Hydrolysis and base-catalyzed isomerization of the resulting product *146* yields dl-PGB$_1$-ester *147* [128)] (Scheme 26).

Miyano and Dorn used the same ylide *145* for the introduction of the C-12 side chain. The aldehyde *150*, prepared by condensation of 3-oxoundecanedioic acid *148* with styrylglyoxal *149*, was reduced with Zn/acetic acid and olefinated (*E*)-selectively to dl-15-dehydro-PGE$_1$ *151* [129)] (Scheme 27).

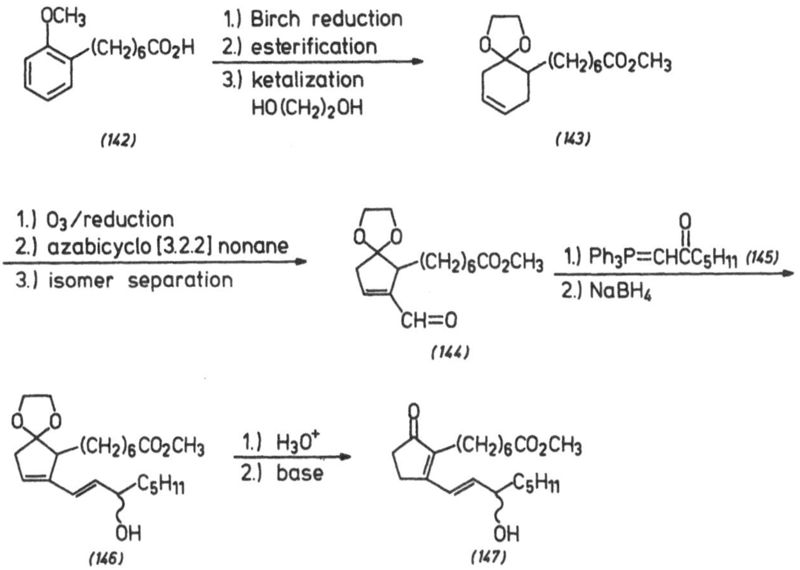

OCH$_3$

—(CH$_2$)$_6$CO$_2$H

1.) Birch reduction
2.) esterification
3.) ketalization
   HO(CH$_2$)$_2$OH

—(CH$_2$)$_6$CO$_2$CH$_3$

(142)

(143)

1.) O$_3$/reduction
2.) azabicyclo [3.2.2] nonane
3.) isomer separation

—(CH$_2$)$_6$CO$_2$CH$_3$

CH=O

(144)

1.) Ph$_3$P=CHCC$_5$H$_{11}$ (145)

2.) NaBH$_4$

—(CH$_2$)$_6$CO$_2$CH$_3$

—C$_5$H$_{11}$

OH

(146)

1.) H$_3$O$^+$
2.) base

—(CH$_2$)$_6$CO$_2$CH$_3$

—C$_5$H$_{11}$

OH

(147)

Scheme 26

HO$_2$CCH$_2$C(CH$_2$)$_7$CO$_2$H

(148)

+

CH=O

(149)

1.) condensation
2.) base
3.) OsO$_4$/NaJO$_4$

—(CH$_2$)$_6$CO$_2$H

CH=O

OH

(150)

1.) Zn/HOAc/H$_2$O
2.) (145)

—(CH$_2$)$_6$CO$_2$H

—C$_5$H$_{11}$

OH

(151)

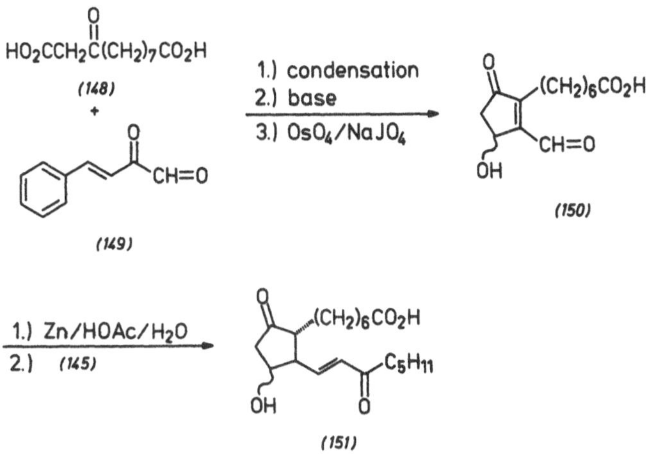

Scheme 27

According to Scheme 28, hexanoylmethylene-triphenylphosphorane *145* was used as the olefinating agent for the introduction of the C-12 side chain of a series of PG-analogs [130-133]. The fifth example in this scheme shows the (*E*)-selective condensation of *159* with formylmethylene-triphenylphosphorane *47*. The resulting (*E*)-α,β-unsaturated aldehyde *160* is a further starting compound for the synthesis of PG-derivatives with a hydroxy group at C-15 [134] (Scheme 28).

109

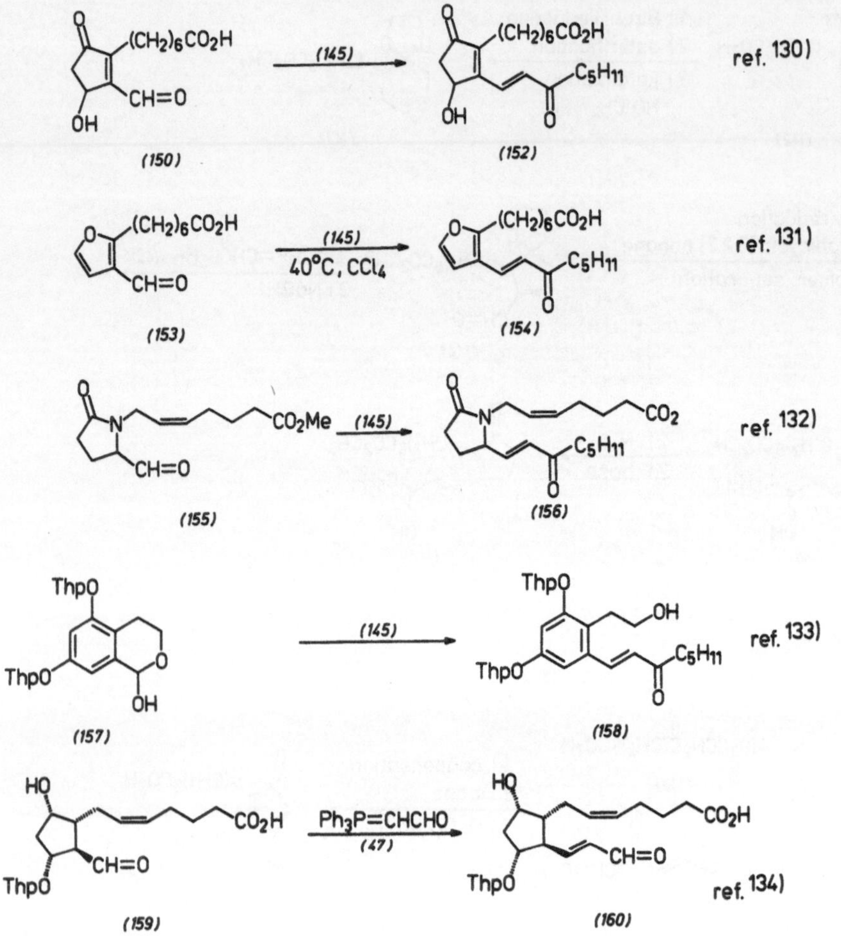

Scheme 28

A stereospecific total synthesis of prostaglandins E_3 and F_3, containing an additional double bond in this side chain, starts from the optically active phosphonium salt *161*. In this synthesis the (*E*)-13-double bond and the 15-hydroxy function are generated simultaneously by condensation of the chiral bicyclic aldehyde *163* with the β-oxido ylide *162* obtained by treatment of *161* with methyllithium. The corresponding phosphonium salt (*S*)(+)-*161*, already possessing the (*Z*)-configurated Δ^{17}-double bond of prostaglandins, was prepared from (*S*)(—)-tartaric acid [135] (Scheme 29).

At the Woodward Institute in Basle a synthesis of prostaglandins was developed which involves the introduction of two side chains into the starting aldehyde by stereoselective Wittig olefinations. The first olefinic bond is introduced (*E*)-selectively at C-13 using the stable phosphorane *145* and the second olefinic bond is introduced with (*Z*)-geometry at C-5 by the application of the reactive ylide $Ph_3P=CH(CH_2)_3$-COOR *168* [136] (Scheme 30 illustrates this synthesis using PGF_2 as an example).

110

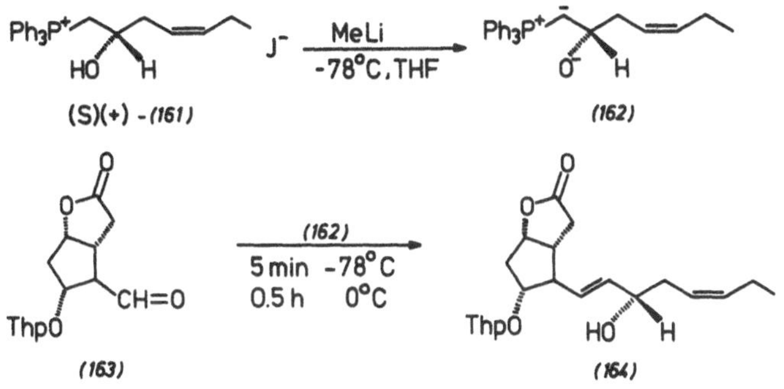
MeLi
-78°C, THF
(S)(+) -(161)
(162)
(162)
5 min -78°C
0.5 h 0°C
(163)
(164)

Scheme 29

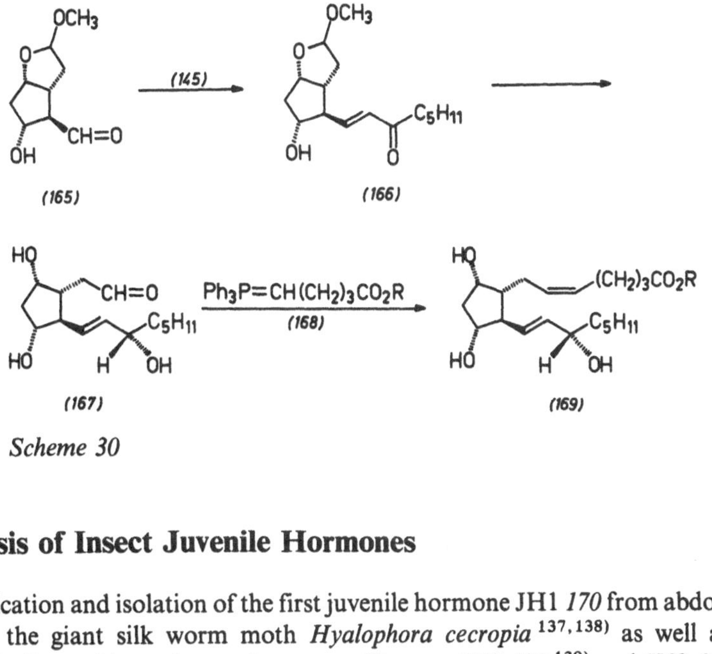
(145)
(165)
(166)
Ph3P=CH(CH2)3CO2R
(168)
(167)
(169)

Scheme 30

5 Synthesis of Insect Juvenile Hormones

The identification and isolation of the first juvenile hormone JH1 *170* from abdominal extracts of the giant silk worm moth *Hyalophora cecropia* [137,138] as well as the following findings of homologous hormones like e.g. JH2 *171* [139] and JH3 *172* [140] (Scheme 31) initiated numerous syntheses. Because of the possibilities of using these substances as agrochemicals (pesticides of the third generation) [141] they are not only of scientific interest.

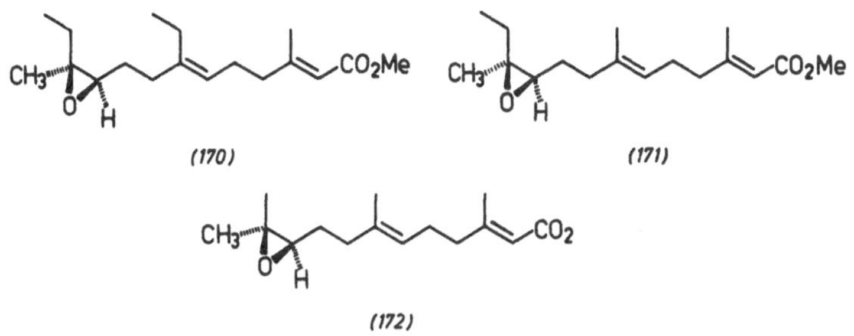
(170)
(171)
(172)

Scheme 31

The two double bonds of juvenile hormones as well as the double bond of the synthesis intermediates, giving the oxirane ring on epoxidation, usually are formed by Horner-Wadsworth-Emmons reactions. Although this is a reaction of phosphonate anions $(RO)_2POCHCO_2R$ and not of phosphorus ylides, we will examplarilly present one general juvenile hormone synthesis of the kind in Scheme 32. It was the first synthesis of racemic JH1, for which three phosphonate olefinations were used. In the examples following afterwards only juvenile hormone preparations will be reported where phosphorus ylides (Wittig reaction) were used.

Already in 1967 Dahm et al. reported the total synthesis of JH1 *170* [142]. The three olefination reactions each required separation of the geometric isomers; the

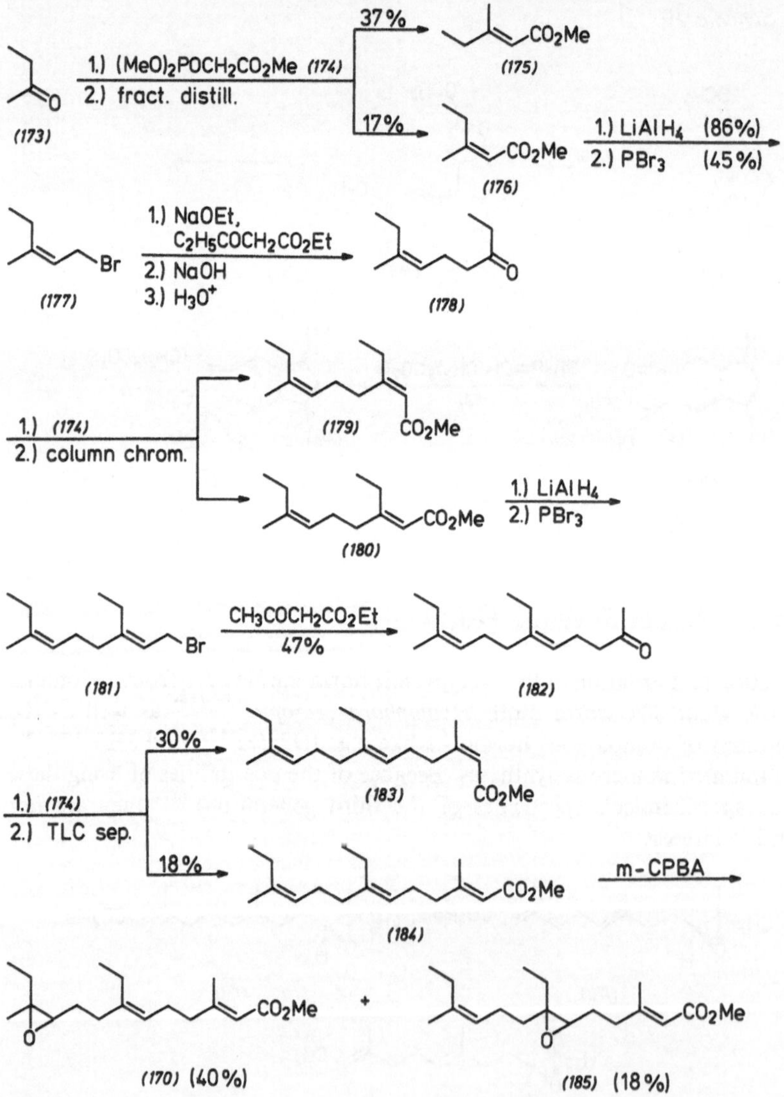

Scheme 32

$(2E,6E,10Z)$-2,6,10-isomer was identical with naturally occurring JH1. However, the total synthesis at the same time affords a large number of possible other isomers of JH1 (Scheme 32) [142].

A group of scientists of the Agricultural Research Service of the USDA in Beltsville reported two syntheses of JH1 *170*, the second of which (Scheme 33) included a Wittig olefination. 4-Methylhex-(E)-3-enyl bromide *187*, obtained from *186* via Julia synthesis, was reacted with propylidenephosphorane *188* to yield *189* [143]. Carbonyl olefination of *189* with aldehyde *190* gave the 2,6,10-alkatrienoic ester *191* which represents a precursor of JH1 [144] (Scheme 33).

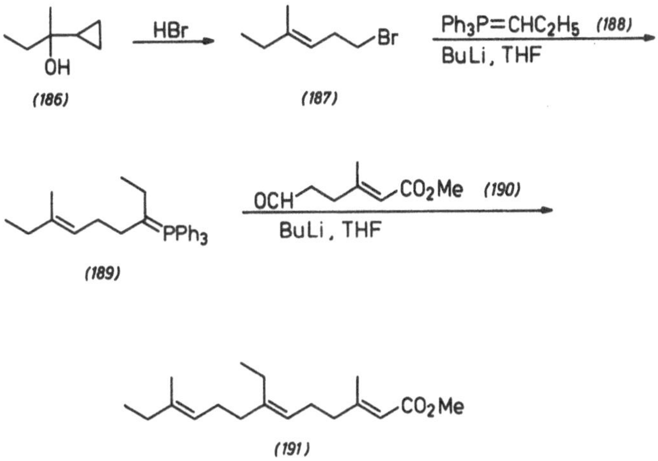

Scheme 33

An interesting synthesis of JH1 *170* starting from furan was published by Cavill et al. [145]. The use of a glycol precursor (*193*) prior to the oxirane formation guarantees the generation of a uniform 10,11-epoxide, in contrast to the usual m-chloroperbenzoic acid oxidation [145] (Scheme 34).

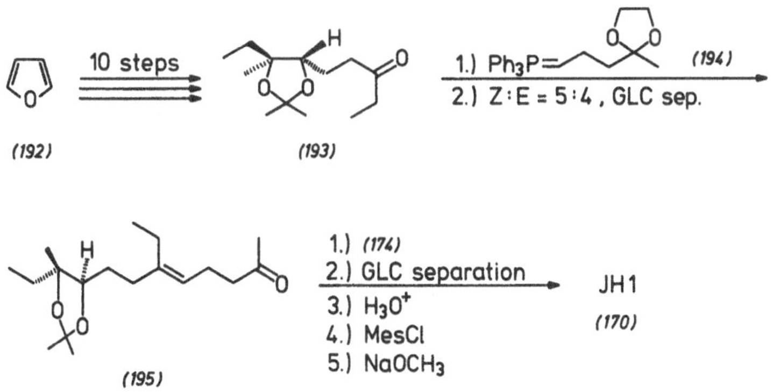

Scheme 34

113

Pommer et al. [146] reported a synthesis of a stereoisomer mixture of JH1 using a Horner-Wadsworth-Emmons olefination for introducing the double bond at C-2 of *184* [146]. Three Wittig olefinations were applied in the condensation of $C_4 + C_6 +$ $+ C_5 + C_2$ synthons [147]. The isomers obtained were separated by distillation (Scheme 35).

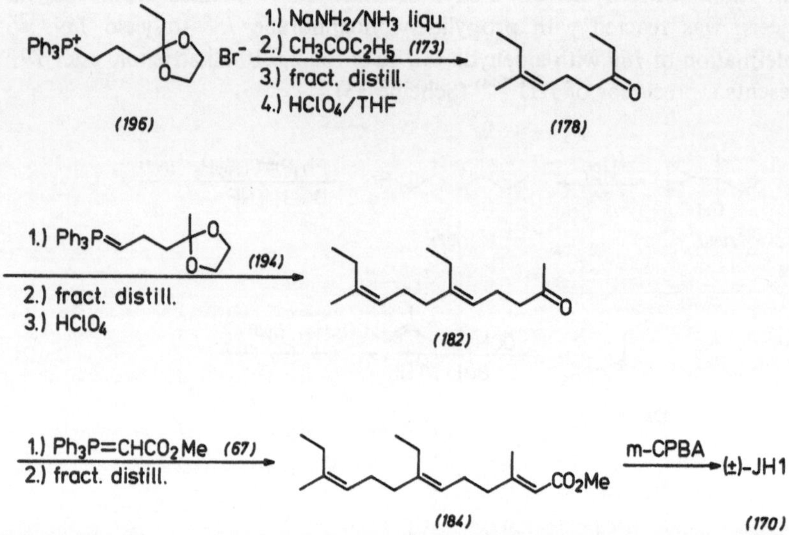
Ph3P+
Br-
1.) NaNH2/NH3 liqu.
2.) CH3COC2H5 (173)
3.) fract. distill.
4.) HClO4/THF
(196)
(178)
1.) Ph3P=
(194)
2.) fract. distill.
3.) HClO4
(182)
1.) Ph3P=CHCO2Me (67)
2.) fract. distill.
CO2Me
m-CPBA
(±)-JH1
(184)
(170)

Scheme 35

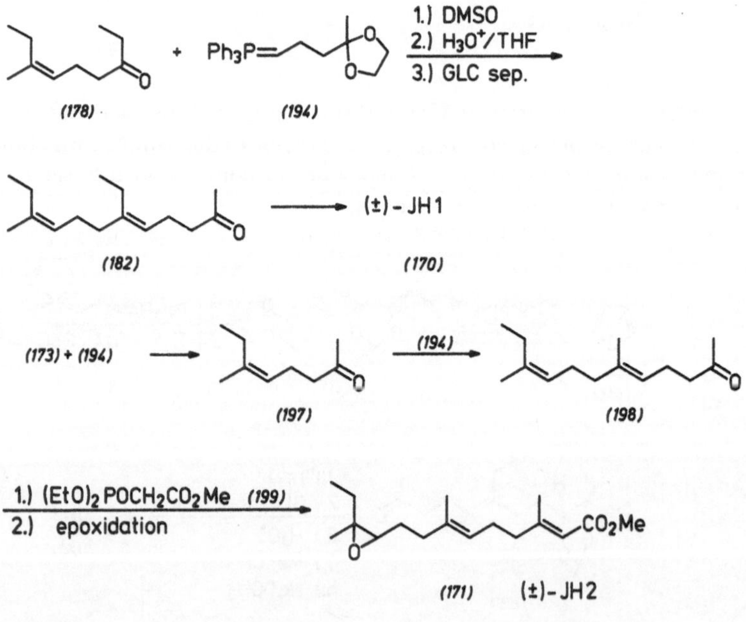
+ Ph3P=
1.) DMSO
2.) H3O+/THF
3.) GLC sep.
(178)
(194)
(±)-JH1
(182)
(170)
(173) + (194)
(194)
(197)
(198)
1.) (EtO)2POCH2CO2Me (199)
2.) epoxidation
CO2Me
(171)
(±)-JH2

Scheme 36

The (E)-double bond at C-6 of a racemic JH1 is formed by condensation of ethyl ketone *178* with phosphorane *194* in the New Brunswick synthesis. In this synthesis about 40% (Z)- and 60% (E)-isomer were formed which were separated by preparative gas chromatography. Compound *182* was converted into (±)-JH1 by use of the phosphonate method [148, 149] (Scheme 36). Applying two Wittig reactions and a phosphonate condensation the same authors could prepare racemic JH2 *171* [148, 149] (Scheme 36).

A combination of the Wittig reaction with an acetylene synthesis was applied by Cochrane and Hanson in the synthesis of JH2 [150]. Subsequently, they used Corey's stereoselective route [151] but obtained only 27% (Z)-isomer of the hormone, i.e. a poor stereoselectivity (Scheme 37).

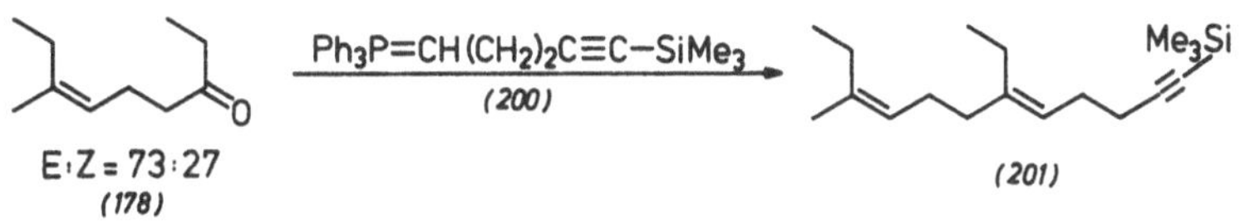

$$Ph_3P=CH(CH_2)_2C\equiv C-SiMe_3$$
$$(200)$$

E:Z = 73:27
(*178*)

(*201*)

Scheme 37

In view of the large number of stereoselective syntheses of juvenile hormones the Wittig reaction recedes into the background in comparison to the numerous other selective key reactions like e.g. acetylene synthesis, organo-copper coupling, stereoselective fragmentation processes, Julia syntheses, Claisen rearrangements, Biellmann coupling, and condensation with dihydropyranes. This is due to the relatively poor stereoselectivity of the carbonyl olefination in the formation of trisubstituted olefins, as may be inferred from the discussed mechanism of the Wittig reaction (Chapter 2) and as it could be demonstrated in many cases.

Corey and Yamamoto developed a highly stereoselective synthesis of (±)-JH1, JH2, JH3, and of a position-5-isomer of JH2 [152−155] using β-oxidophosphonium ylides as the real olefination agents (Scheme 38).

The aldehyde ester *212* prepared by two different paths is the key substance in a Zoecon synthesis of juvenile hormones. Wittig olefination of *212* with 1-methoxypropylidenephosphorane *213* yields *214* which is converted via three steps in (±)-JH1 *170* (Scheme 39) [156]. Application of the same Wittig reaction to the 4-methylderivative of *212* affords the corresponding (±)-JH2 [157].

The syntheses described above and the individual reaction steps, respectively, are the historical basis for the development of the synthesis of isoprenoid juvenile hormones. In addition to the latter, a great number of naturally occurring or synthetic analogous substances as well as non-isoprenoid compounds with juvenile hormone activity can also be obtained via Wittig reaction. However, since these syntheses do not offer basically new aspects with respect to the olefination step, they are not further discussed here.

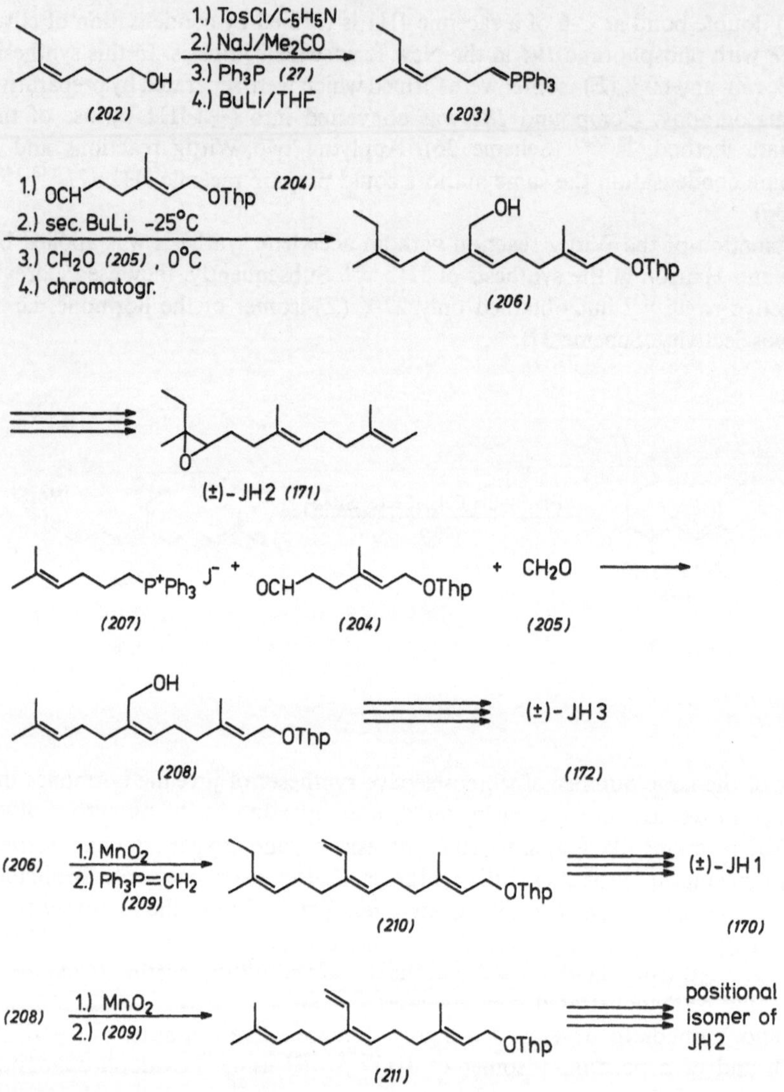
1.) TosCl/C5H5N
2.) NaJ/Me2CO
3.) Ph3P (27)
4.) BuLi/THF
OH
(202)
PPh3
(203)
1.) OCH OThp (204)
2.) sec. BuLi, -25°C
3.) CH2O (205), 0°C
4.) chromatogr.
OH
OThp
(206)
O
(±)-JH2 (171)
P+Ph3 J−
OCH
OThp
CH2O
(207)
(204)
(205)
OH
OThp
(208)
(±)-JH3
(172)
(206)
1.) MnO2
2.) Ph3P=CH2
(209)
OThp
(210)
(±)-JH1
(170)
(208)
1.) MnO2
2.) (209)
OThp
(211)
positional
isomer of
JH2

Scheme 38

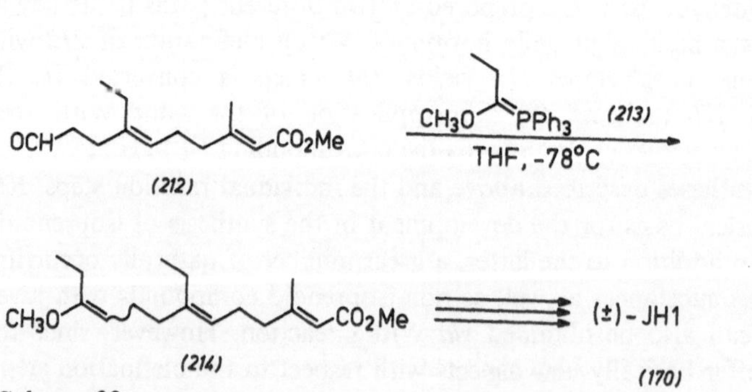
OCH
CO2Me
(212)
CH3O
PPh3
(213)
THF, -78°C
CH3O
CO2Me
(214)
(±)-JH1
(170)

Scheme 39

6 Aliphatic Fragrance and Aroma Substances

Among the fragrance and aroma substances a great variety of mono- and polyolefinic aliphatic alcohols, aldehydes, ketones, carboxylic acids, and their esters, as well as lactones are found [158]. Of these aroma substances, (E)-2-alkenals, (E)-2-alkenoic acids, (E)-2-alkenoic esters as well as (E)-2-alken-1-ols are predestinated for the synthesis *via* the Wittig reaction because of the (E)-stereoselectivity of the olefination using resonance-stabilized ylides.

(E)-2-Hexenal (leaf aldehyde) is a constituent responsible for the smell of green leafs, (E)-2-octenal a main component of the aroma of raw potatoes; (E)-2-nonenal is the organoleptic main constituent of the smell of cucumbers and is found in carot root oil, tomatoes, beef and raspberries [158]. (E)-2-Decenal and (E)-2-dodecenal are components of some essential oils, (E)-2-tridecenal is responsible for the bug-like smell of coriander seed oil [158].

(E)-2-alkenals are readily prepared in 35–50% yield by carbonyl olefination of alkanals *215* using formylmethylene-triphenylphosphorane *47* (stereoselectivity of the (E)-isomer ~95%) (Scheme 40). (E)-2-nonenal, (E)-2-undecenal and (E)-2-tridecenal have been synthesized in our laboratories [69] to be used for systematic synthesis of insect pheromones.

$$CH_3-(CH_2)_n-CHO + Ph_3P=CH-CHO \rightarrow CH_3-(CH_2)_n-\overset{H}{\underset{H}{C}}=C-CHO$$

$$(215) \qquad\qquad (47) \qquad\qquad\qquad (216)$$

$$n = 0, 1, 2, \ldots .$$

Scheme 40

(E)-2-Alkenoic acids are formed by heating of oils and fats of animal or plant origin. They can also be prepared by hydrolysis of the corresponding esters [158]. The Wittig reaction of alkanals *215* with alkoxycarbonylmethylene-triphenylphosphorane *217* affords the (E)-2-alkenoic esters *218* in about 95% yield of the (E)-isomer. Alkaline hydrolysis of the esters *218* gives the (E)-2-alkenoic acids *220* (Scheme 41).

$$CH_3-(CH_2)_n-CHO + Ph_3P=CH-COOR \rightarrow CH_3-(CH_2)_n-\overset{H}{\underset{H}{C}}=C-COOR$$

$$(215) \qquad\qquad (217) \qquad\qquad\qquad (218) \quad\downarrow OH^-$$

$$\xrightarrow{LiAlH_4} CH-(CH_2)_n\overset{H}{\underset{H}{C}}=CCH_2OH \qquad\qquad CH_3(CH_2)_n\overset{H}{\underset{H}{C}}=C-COOH$$

$$(219) \qquad\qquad\qquad (220)$$

Scheme 41

(E)-2-Alkenols *219* like (E)-2-hexenol, (E)-2-octenol and (E)-2-nonenol were found in the smell of apples, in the aroma of potatoes, in aroma concentrates of mushrooms, and in Cucurbitacae [158]. They are best obtained by reduction of the corresponding (E)-2-carboxylic acid esters *218* (Scheme 41).

The vinyl ketone *222* which displays an unpleasant smell is found in the volatiles of cooked mushrooms; it is an aroma component of clover. The racemic alcohol *223* called "mushroom alcohol" because of its typical smell was isolated from soya beans, corn oil and oxidized butter. (—)-1-Octen-3-ol *223* is the most important aroma component of the mushroom Agaricus bisporus [158]. For the preparation of the ketone *222* as well as for the racemic alcohol *223* a synthesis using almost exclusively phosphorus ylides was developed [159]. Thus, methylene-triphenylphosphorane *209* is acylated with hexanoyl chloride *221*, to yield hexanoylmethylene-triphenylphosphorane *145* through transylidation [160]. *145* is olefinated with formaldehyde *205* to ketone *222*. Reduction of the carbonyl group yields racemic 1-octen-3-ol *223*. (Scheme 42) [159].

$$H_2C=PPh_3 + ClCO(CH_2)_4CH_3 \rightarrow CH_3(CH_2)_4\overset{\overset{\displaystyle O}{\|}}{C}-CH=PPh_3$$

$$(209) \qquad\qquad (221) \qquad\qquad\qquad (145)$$

$$\xrightarrow{CH_2O} CH_3(CH_2)_4\overset{\overset{\displaystyle O}{\|}}{C}-CH=CH_2 \xrightarrow{\text{reduction}} CH_3(CH_2)_4\overset{\overset{\displaystyle OH}{|}}{C}H-CH=CH_2$$

$$(205) \qquad\qquad (222) \qquad\qquad\qquad (223)$$

Scheme 42

Double bonds with (Z)-geometry having a higher positional number than C-2 are characteristic of aliphatic aroma compounds. They can easily be generated by (Z)-stereoselective Wittig reactions. (Z)-3-Hexenol (leaf alkohol) is, apart from other alkenols, responsible for the smell of green leaves, and (Z)-3-hexenal is responsible for that of apples and soya bean oil. At higher concentrations (>2 ppm) (Z)-4-heptenal is responsible for the rancid smell of butter and oils and contributes to the off flavour of frozen, tainted cod (*Gadus morlina*) [158]. (Z)-3-Nonenal, (Z)-3-nonenol, (Z)-6-nonenal and (Z)-6-nonenol were found in Cucurbitae. These fragrance substances can easily be prepared by the Wittig reaction. Thus, reaction of the ylides of the phosphonium salts *224*, generated according to the silazide technique [33], with the aldehydes *215* gives the carboxylic esters *225*, the tetrahydropyranyl ethers *226* and alkadienes *227* with almost 98% stereochemical purity (Scheme 43). From these compounds (Z)-alkenols and (Z)-alkenals can be obtained, using well known reactions (cf. Scheme 47 in Chapter 7).

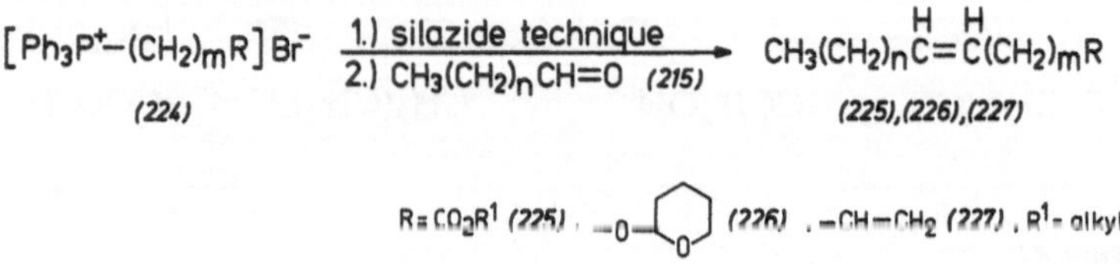

$$[Ph_3P^+-(CH_2)_mR]\,Br^- \xrightarrow[\text{2.) } CH_3(CH_2)_nCH=O \ (215)]{\text{1.) silazide technique}} CH_3(CH_2)_n\overset{H}{C}=\overset{H}{C}(CH_2)_mR$$

$$(224) \qquad\qquad\qquad\qquad\qquad\qquad (225),(226),(227)$$

$$R = CO_2R^1 \ (225) \ , \ -O-\underset{O}{\bigcirc} \ (226) \ , \ -CH=CH_2 \ (227) \ , \ R^1 = \text{alkyl}$$

Scheme 43

Remarkable are the relations between the chemical structure of aliphatic compounds and their odour threshold for the human olfactory system and hence for their importance in aroma chemistry. Thus, e.g. the odour threshold for alkanals decreases with increasing number of carbon atoms (C_5 to C_{10}), the introduction of a double bond generally lowering the threshold. In general, it can be assumed that alkenals with (Z)-geometry smell more strongly and aromatically then the corresponding (E)-isomers [158].

Polyunsaturated aliphatic alcohols, aldehydes, ketones, and esters occur as fragrance components in fats, oils, fruits and plants [158]. As an example the synthesis of ethyl (2E,4Z)-2,4-decadienoate (*232*, "pear ester"), which is responsible for the aroma of bartlett pears [161] is given. 2,4-Diunsaturated ester *232* may be obtained by a number of highly stereoselective syntheses, a lot of them making use of the Wittig reaction. Ohloff and Pawlak condensed 4,5-epoxy-(E)-2-pentenal *228* with the ylide generated from *229* (butyllithium/ether) to the alkadiene epoxide *230* which was oxidized with periodic acid to the 2,4-decadienal *231*. *231* is subsequently converted with MnO_2/NaCN in ethanol to the pear ester *232* [162] [75% (Z)-amount of the C-4 double bond] (Scheme 44).

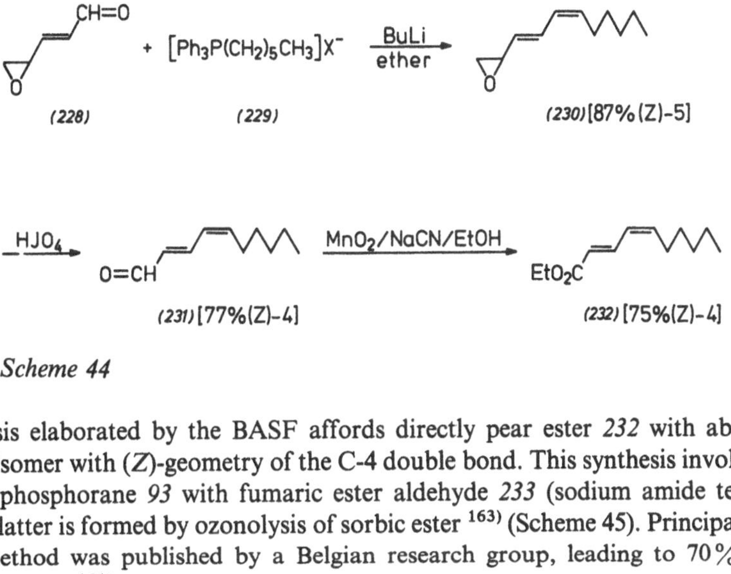

Scheme 44

A synthesis elaborated by the BASF affords directly pear ester *232* with about 85% of the isomer with (Z)-geometry of the C-4 double bond. This synthesis involves reaction of phosphorane *93* with fumaric ester aldehyde *233* (sodium amide technique). The latter is formed by ozonolysis of sorbic ester [163] (Scheme 45). Principally, the same method was published by a Belgian research group, leading to 70% of (2E,4Z)-2,4-isomer [164].

$$\underset{(93)}{Ph_3P{=}CH(CH_2)_4CH_3} + \underset{(233)}{OCH{-}\overset{H}{C}{=}\underset{H}{C}{-}COOEt} \xrightarrow[\text{NaNH}_2\text{-method}]{\text{"salt free"}}$$

$$\underset{(232)}{CH_3(CH_2)_4{-}\overset{H}{C}{=}\overset{H}{C}{-}\overset{H}{C}{=}\underset{H}{C}{-}COOEt}$$

Scheme 45

A synthesis of the pear ester worked out in our laboratory uses the (Z)- and the (E)-stereoselective Wittig olefination for the generation of the two double bonds in the molecule *232* [165]. Reaction of 2-ethoxyvinyl-triphenylphosphonium bromide *234* with one equivalent sodium ethanolate yields 2,2-ethoxyethylene-triphenylphosphorane *235* [166,167]. (Z)-Stereoselective olefination of *235* with hexanal *236* followed by mild hydrolysis affords (Z)-2-octenal *237*. The latter can be converted (E)-stereoselectively with ethoxycarbonylmethylenephosphorane *238* into *232* [90% (E,Z), 5% (E,E), 4% (Z,Z) and 1% (Z,E) isomer] (Scheme 46) [165].

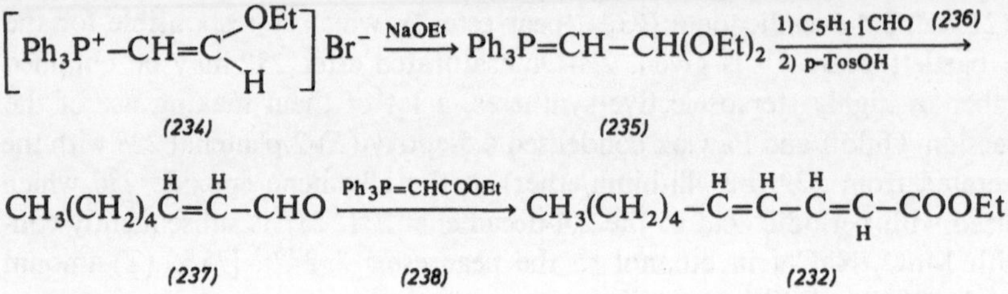

$$\left[Ph_3P^+-CH=C\overset{OEt}{\underset{H}{\diagdown}} \right] Br^- \xrightarrow{NaOEt} Ph_3P=CH-CH(OEt)_2 \xrightarrow[\text{2) p-TosOH}]{\text{1) }C_5H_{11}CHO\ \textit{(236)}}$$

$$\textit{(234)} \qquad\qquad\qquad\qquad \textit{(235)}$$

$$CH_3(CH_2)_4\overset{H}{C}=\overset{H}{C}-CHO \xrightarrow{Ph_3P=CHCOOEt} CH_3(CH_2)_4-\overset{H}{C}=\overset{H}{C}-\overset{H}{C}=C-COOEt$$
$$\underset{H}{}$$

$$\textit{(237)} \qquad\qquad \textit{(238)} \qquad\qquad\qquad \textit{(232)}$$

Scheme 46

7 Insect Pheromones

Among insect pheromones a great number of mono- and polyolefinic compounds are found; a lot of them can be obtained by the Wittig reaction. Especially the syntheses of sex attractants of female butterflies and moths, which are mostly mono- and bisunsaturated alcohols, acetates or aldehydes [168], offer a broad field for the application of the Wittig reaction and have stimulated the development of many new stereoselective variants. Thus, the methods of "salt-free" Wittig reactions (Chapter 2) like the sodium amide method [11,31,32], the silazide technique [33], potassium in HMPA [34,35] or the use of dipolar aprotic solvents like dimethyl formamide [169], dimethyl sulfoxide [51,170] or hexamethylphosphoric triamide [51,170] were often used.

7.1 Butterflies — Lepidoptera

For the preparation of (Z)-unsaturated lepidopterous pheromones *via* Wittig reaction four different methods are mainly applied.
a) Reaction of ω-substituted aldehydes *239* with alkylidene-triphenylphosphoranes *240* to (Z)-alkenoic esters, (Z)-alkenyl esters and (Z)-alkenyl ethers *241*, respectively;
b) reaction of ω-substituted phosphoranes *242* with simple alkanals *243* to *241*;
c) condensation of terminally unsaturated alkenals *244* with alkylidenephosphoranes *240* to 1,(Z)-m-alkadienes *245* (m = 5, 6, 7, ...);
d) (Z)-stereoselective carbonyl olefination of alkanals *243* with terminally unsaturated alkenylidenephosphoranes *246* to the same alkadienes *245* (Scheme 47).

Scheme 47 also shows consecutive reactions of the corresponding Wittig olefinations and the conversions of functional groups involved, respectively. Thus, the (Z)-1,n-alkadienes *245* can be hydroborated and oxidized regioselectively to the (Z)-alken-1-ols *248*. Compounds *248* are also obtained by reduction of the (Z)-alkenoic esters *247* with LiAlH₄ or by alkaline hydrolysis of the alkenyl carboxylates *246*, acetates *249* and the (Z)-alkenyl tetrahydropyranyl ethers *252*. Esterification of

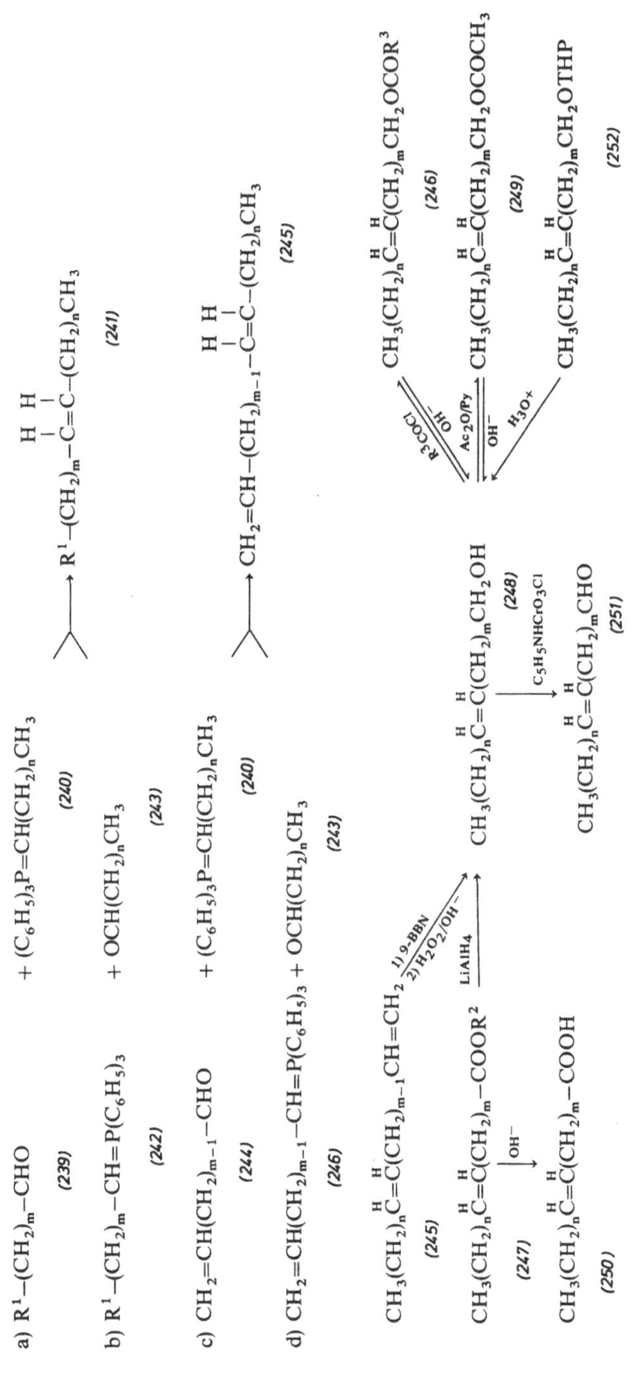

a) $R^1-(CH_2)_m-CHO$ $+ (C_6H_5)_3P=CH(CH_2)_nCH_3$
   (239)   (240)

$$\longrightarrow R^1-(CH_2)_m-\overset{H}{\underset{|}{C}}=\overset{H}{\underset{|}{C}}-(CH_2)_nCH_3$$
                                                      (241)

b) $R^1-(CH_2)_m-CH=P(C_6H_5)_3$ $+ OCH(CH_2)_nCH_3$
   (242)   (243)

c) $CH_2=CH(CH_2)_{m-1}-CHO$ $+ (C_6H_5)_3P=CH(CH_2)_nCH_3$
   (244)   (240)

$$\longrightarrow CH_2=CH-(CH_2)_{m-1}-\overset{H}{\underset{|}{C}}=\overset{H}{\underset{|}{C}}-(CH_2)_nCH_3$$
                                                          (245)

d) $CH_2=CH(CH_2)_{m-1}-CH=P(C_6H_5)_3 + OCH(CH_2)_nCH_3$
   (246)   (243)

$CH_3(CH_2)_n\overset{H}{C}=\overset{H}{C}(CH_2)_{m-1}CH=CH_2 \xrightarrow[2)\,H_2O_2/OH^-]{1)\,9\text{-BBN}}$
(245)

$CH_3(CH_2)_n\overset{H}{C}=\overset{H}{C}(CH_2)_m-COOR^2 \xrightarrow{LiAlH_4} CH_3(CH_2)_n\overset{H}{C}=\overset{H}{C}(CH_2)_mCH_2OH$
(247)   (248)

$\Big\downarrow OH^-$

$CH_3(CH_2)_n\overset{H}{C}=\overset{H}{C}(CH_2)_m-COOH \xrightarrow{C_5H_5NHCrO_3Cl} CH_3(CH_2)_n\overset{H}{C}=\overset{H}{C}(CH_2)_mCHO$
(250)   (251)

$$\begin{array}{c} \nearrow \xrightarrow{R^2COCl} CH_3(CH_2)_n\overset{H}{C}=\overset{H}{C}(CH_2)_mCH_2OCOR^3 \\ Ac_2O/Py \quad\quad\quad (246) \\ \\ \overset{HO^-}{\Big\Vert} \quad\quad\quad\quad\quad\quad CH_3(CH_2)_n\overset{H}{C}=\overset{H}{C}(CH_2)_mCH_2OCOCH_3 \\ OH^- \quad\quad\quad\quad\quad (249) \\ \\ \searrow \xrightarrow{H_3O^+} CH_3(CH_2)_n\overset{H}{C}=\overset{H}{C}(CH_2)_mCH_2OTHP \\ \quad\quad\quad\quad\quad\quad (252) \end{array}$$

$R^1 = COOR^2$, CH_2OCOR^3, CH_2OTHP, ...
$R^2 = CH_3, C_2H_5, ...$ $R^3 = H, CH_3, C_2H_5, ...$
$n = m = 0, 1, 2, 3, ...$
THP = tetrahydropyranyl-

Scheme 47

121

alkenols *248* with the usual methods gives (*Z*)-alkenyl acetates *249* or other (*Z*)-alkenyl carboxylates *246*, the oxidation with pyridinium chlorochromate (PCC) yields the (*Z*)-alkenals *251*. By alkaline hydrolysis of the esters *247* the (*Z*)-unsaturated fatty acids *250* are obtained (Scheme 47).

By means of the Wittig reaction unsaturated moth sex attractants can be synthesized according to a generalized reaction sequence [58]. Thus, numerous (*Z*)-5-, (*Z*)-6-, (*Z*)-7-, (*Z*)-8-, (*Z*)-9-, (*Z*)-11- and (*Z*)-13-unsaturated pheromones and pheromone analogs like alkenols, alkenyl acetates, alkenyl esters, alkenals, and alkyl branched analogs were synthesized with a stereochemical purity of up to 98 % [33,35,58,59,60,61,62,171]. These compounds were required for investigations concerning the structure-activity relationships in the series of lepidoptera pheromones [60,172,173,174].

(*Z*)-7,8-Epoxy-2-methyloctadecane *258* (disparlure) is the sex pheromone of the femaly gypsy moth *Lymantria dispar* [175], a serious forest pest in Europe and the New World. With its oxirane ring this pheromone structure can directly be deduced from the (*Z*)-olefinic precursor *257*, again representing a Wittig product. The preparation of *257*, which has proved to be a biosynthetic intermediate in the insect [176], has often been reported [175,176b,177]. Thus, isooctyl-triphenylphosphonium bromide *255* is obtained by reaction of the Grignard reagent *253* of the isoamyl bromide with oxetane *254*, hydrobromic acid and triphenylphosphine. The ylide generated by treatment of *255* with sodium silazide was olefinated with undecanal *256* to (*Z*)-2-methyl-7-octadecene *257*. (*Z*)-stereoselective epoxidation of *257* with m-chloroperbenzoic acid (m-CPBA) (~95%) gives the racemic oxirane *258* in good yield and with high stereochemical purity [177] (Scheme 48). So far *258* is the only known chiral lepidoptera sex attractant. The direction of the rotatory power of the natural product is still unknown. Independent syntheses of the two enantiomeric forms and biological tests resulted in a higher attractivity of the (+)(7*R*,8*S*)-isomer for *Lymantria* males.

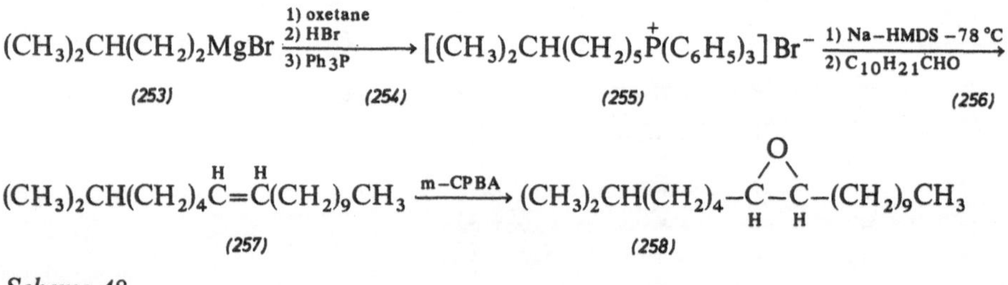

$$(CH_3)_2CH(CH_2)_2MgBr \xrightarrow[\substack{2)\ HBr \\ 3)\ Ph_3P}]{1)\ oxetane} [(CH_3)_2CH(CH_2)_5\overset{+}{P}(C_6H_5)_3]\ Br^- \xrightarrow[\substack{2)\ C_{10}H_{21}CHO}]{1)\ Na-HMDS\ -78\ ^\circ C}$$

$$(253) \qquad\qquad (254) \qquad\qquad\qquad (255) \qquad\qquad\qquad\qquad (256)$$

$$(CH_3)_2CH(CH_2)_4\overset{H}{C}=\overset{H}{C}(CH_2)_9CH_3 \xrightarrow{m-CPBA} (CH_3)_2CH(CH_2)_4-\overset{O}{\overset{/\backslash}{\underset{H}{C}}}-\overset{}{\underset{H}{C}}-(CH_2)_9CH_3$$

$$(257) \qquad\qquad\qquad\qquad\qquad (258)$$

Scheme 48

Amongst the sex attractants of months also a great number of diunsaturated compounds are found, without showing any regularity regarding the geometry and position of the double bonds of the pheromones within the lepidoptera families, subfamilies and species. For the synthesis of these attractants again the (*Z*)-stereoselective carbonyl olefination can be applied. (*E*)-Double bonds were formed by (*E*)-stereoselective Wittig reaction using resonance-stabilized ylides or other olefin-generating reactions. Thus a general synthetic route to bisolefinic pheromones has become available [57,58,70,178,179].

Some typical representatives of conjugated unsaturated alkadiene attractants include (10E,12Z)-10,12-hexadecadien-1-ol (bombykol), the first sex pheromone isolated by A. Butenandt et al. [180] from the silkworm moth *Bombyx mori*, and (10E,12Z)-10,12-hexadecadienal (bombykal), a second pheromone component of the same species [181]. Further biologically active members of this class of compounds are (9Z,11E)-9,11-tetradecadienyl acetate, (7E,9Z)-7,9-dodecadienyl acetate as well as (8E,10E)-8,10-dodecadien-1-ol, which are the sex attractants and pheromone components, respectively, of the noctuids *Spodoptera littoralis* [182] and *S. litura* [183], of the European grape berry moth *Lobesia botrana* [184] and the worldwide occurring apple pest *Laspeyresia pomonella* [185].

Bombykol was for the first time prepared by Butenandt using the Wittig reaction of 2-hexynylidene-triphenylphosphorane with 9-formylnonanoic ester. The resulting (Z)- resp. (E)-10-hexadecen-12-ynoic acid esters were resolved by urea, the (E)-isomer hydrogenized with Lindlar catalyst and reduced with lithium aluminium hydride. Further purification *via* the urea inclusion compound and low-temperature crystallization gave (10E,12Z)-10,12-hexadecadien-1-ol [186,187].

In a further synthesis of bombykol and bombykal Bestmann and coworkers [69,70] used the stereochemistry of the Wittig reaction. 9-Formylnonanoic ester *259* was reacted (E)-selectively with formylmethylene-triphenylphosphorane *47* and the resulting α,β-unsaturated aldehyde *260* subsequently converted (Z)-stereoselectively into (10E,12Z)-10,12-hexadecadienoic ester *262* according to the silazide technique [33]. Reduction of *262* with LiAlH₄ gave bombykol *263* which could be oxidized with pyridinium chlorochromate (PCC) to bombykal *264*. Acetylation of *263* yielded (10E,12Z)-10,12-hexadecadienyl acetate *265* [69,70] (Scheme 49).

Following the same reaction sequence, (6E,8Z)-, (7E,9Z)-, (8E,10Z)-, (9E,11Z)-, (10E,12Z)- and (11E,13Z)-alkadienyl acetates and alkadienols of various chain

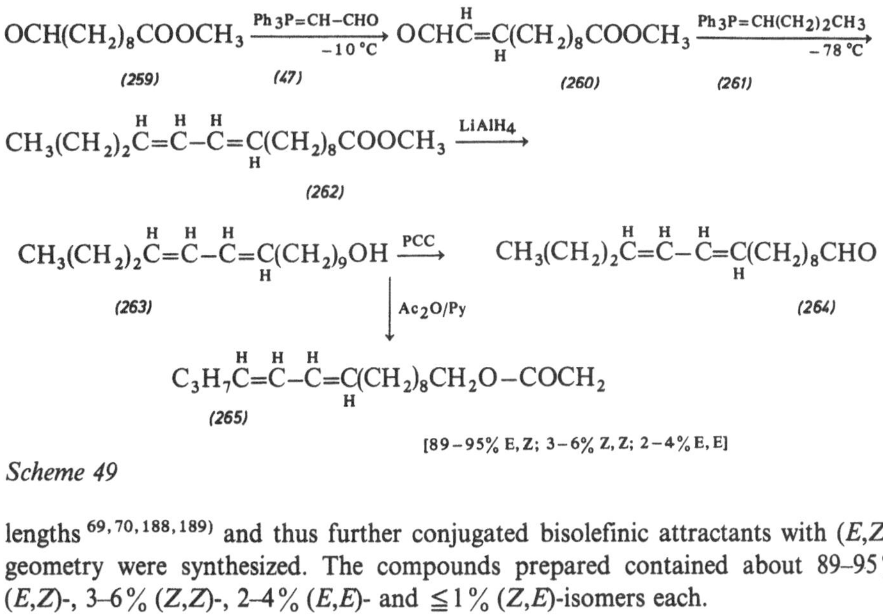

$$
OCH(CH_2)_8COOCH_3 \xrightarrow[-10\,°C]{Ph_3P=CH-CHO} OCHC\overset{H}{\underset{H}{=}}C(CH_2)_8COOCH_3 \xrightarrow[-78\,°C]{Ph_3P=CH(CH_2)_2CH_3}
$$

(259)　　　　　　(47)　　　　　　　　　　　(260)　　　　　(261)

$$
CH_3(CH_2)_2\overset{H}{C}=\overset{H}{C}-\overset{H}{C}=C(CH_2)_8COOCH_3 \xrightarrow{LiAlH_4}
$$

(262)

$$
CH_3(CH_2)_2\overset{H}{C}=\overset{H}{C}-\overset{H}{C}=C(CH_2)_9OH \xrightarrow{PCC} CH_3(CH_2)_2\overset{H}{C}=\overset{H}{C}-\overset{H}{C}=C(CH_2)_8CHO
$$

(263)　　　　　　　　　　│ Ac₂O/Py　　　　　　　　　　(264)

$$
C_3H_7\overset{H}{C}=\overset{H}{C}-\overset{H}{C}=C(CH_2)_8CH_2O-COCH_2
$$

(265)

[89–95% E, Z; 3–6% Z, Z; 2–4% E, E]

Scheme 49

lengths [69,70,188,189] and thus further conjugated bisolefinic attractants with (E,Z)-geometry were synthesized. The compounds prepared contained about 89–95% (E,Z)-, 3–6% (Z,Z)-, 2–4% (E,E)- and ≤1% (Z,E)-isomers each.

(7E,9Z)-7,9-Dodecadienyl acetate *271*, the sex attractant of the grape berry moth *Lobesia botrana* was prepared analogously. Originally, Roelofs et al. [185a)] olefinated the aldehyde *266* with diethyl-2-(cyclohexylimino)ethylphosphonate *267*, and the resulting (E)-2-unsaturated aldehyde *268* with the phosphorane *188*. Hydrolysis, reduction and acetylation of the obtained *270* gave the pheromone *271* with a content of about 88% (E,Z)-isomer (Scheme 50). Similar syntheses of *271* using formylmethylene-triphenylphosphorane *47* or diethylethoxycarbonylmethylphosphonate and carried out under "salt-free" conditions (see p. 89) have been reported [188,190,191].

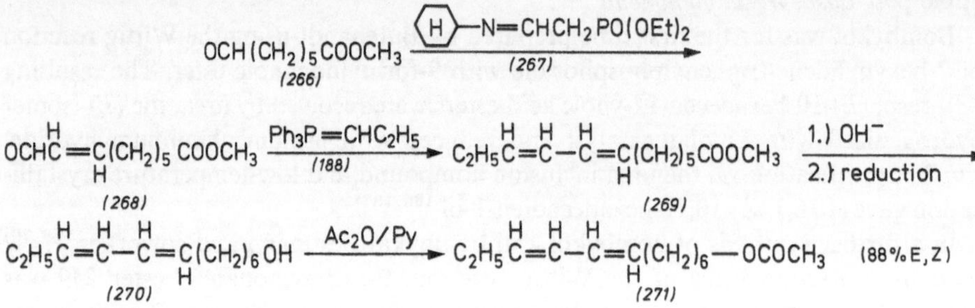

$$OCH(CH_2)_5COOCH_3 \xrightarrow{\langle H \rangle - N = CHCH_2PO(OEt)_2 \atop (267)}$$
$$(266)$$

$$OCHC\overset{H}{=}C(CH_2)_5COOCH_3 \xrightarrow[\text{(188)}]{Ph_3P=CHC_2H_5} C_2H_5\overset{H}{C}=\overset{H}{C}-\overset{H}{C}=C(CH_2)_5COOCH_3 \xrightarrow[\text{2.) reduction}]{1.) OH^-}$$
$$(268) \qquad\qquad\qquad\qquad (269)$$

$$C_2H_5\overset{H}{C}=\overset{H}{C}-\overset{H}{C}=C(CH_2)_6OH \xrightarrow{Ac_2O/Py} C_2H_5\overset{H}{C}=\overset{H}{C}-\overset{H}{C}=C(CH_2)_6-OCOCH_3 \quad (88\%\,E,Z)$$
$$(270) \qquad\qquad\qquad\qquad (271)$$

Scheme 50

For the synthesis of (9Z,11E)-9,11-tetradecadienyl acetate the molecule, because its "reversed" geometry compared with the examples described above, its appropriately synthesized by a reversed sequence of olefination steps. The reaction of (E)-2-pentenal with ylides of 9-substituted alkyl-triphenylphosphonium salts gives (9Z,11E)-9,11-tetradecadiene derivatives with 80–95% isomeric purity [188,190,192,193].

The females of the codling moth *Laspeyresia pomonella* use (8E,10E)-8,10-dodecadien-1-ol *276* as the attractive component of their pheromone system. Although, strictly speaking, this compound is not suitable for the Wittig synthesis because of its (E,E)-geometry it was for the first time obtained by Roelofs et al. [185a)] by Wittig condensation of the ylide of (E)-2-butenyl-triphenylphosphonium bromide *272* and 7-formylheptanoic ester *52* (Scheme 51). The isomeric esters *273* were hydrolyzed, the resulting acids *274* purified by crystallization and subsequently reduced to the alkadienols *275*. The isomeric mixture of *275* was exposed to sun light/iodine and thus the (E,E)-isomer enriched. The corresponding acetate *277* was separated into its isomers by means of silver nitrate/silica gel and again hydrolyzed to *276* [185a)] (Scheme 51). A similar synthesis uses the same phosphorane and the corresponding ester aldehyde, olefination being carried out in the presence of lithium bromide [194] and giving a different product distribution of the isomers with approximately 45% (E,E)-isomer [194].

(E)-9,11-Dodecadienyl acetate *279* is the sex attractant of the red bollworm moth *Diparopsis castanea*, a cotton pest of South Africa. Wittig reaction of allylidenephosphorane *96* with 8-formyloctyl acetate *278* yielded *279* with an isomer distribution of E:Z ranging from 4:1 to 3:7, depending on the polarity of the solvent used. By sulfolen formation (*281*) and subsequent stereospecific thermolysis, the (E)-isomer *279* could be isolated with 99% stereochemical purity [195] (Scheme 52).

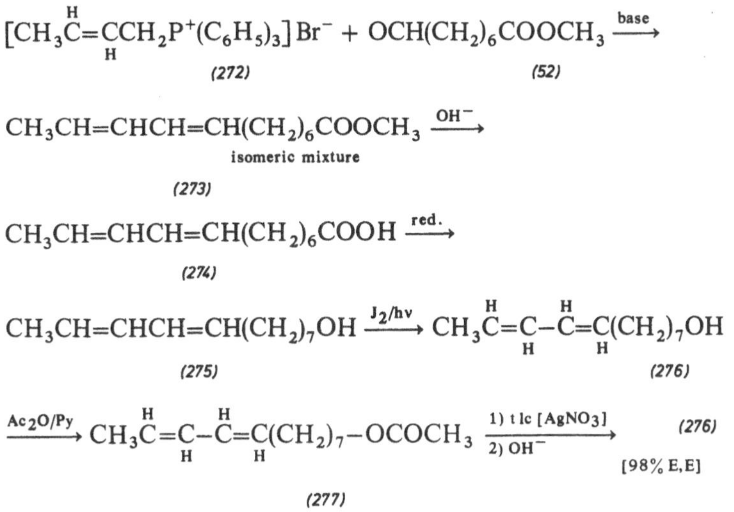

$$[CH_3\overset{H}{\underset{H}{C}}=CCH_2P^+(C_6H_5)_3]\,Br^- + OCH(CH_2)_6COOCH_3 \xrightarrow{\text{base}}$$

(272)   (52)

$$CH_3CH=CHCH=CH(CH_2)_6COOCH_3 \xrightarrow{OH^-}$$
isomeric mixture

(273)

$$CH_3CH=CHCH=CH(CH_2)_6COOH \xrightarrow{\text{red.}}$$

(274)

$$CH_3CH=CHCH=CH(CH_2)_7OH \xrightarrow{J_2/h\nu} CH_3\overset{H}{\underset{H}{C}}=C-\overset{H}{\underset{H}{C}}=C(CH_2)_7OH$$

(275)   (276)

$$\xrightarrow{Ac_2O/Py} CH_3\overset{H}{\underset{H}{C}}=C-\overset{H}{\underset{H}{C}}=C(CH_2)_7-OCOCH_3 \xrightarrow[\text{2) OH}^-]{\text{1) t lc [AgNO3]}}$$
(276)

[98% E,E]

(277)

Scheme 51

(3Z,5E)-3,5-Tetradecadienyl acetate *284* was recently presumed to be the pheromone of *Prionoxystus robiniae* on the basis of electronantennogram data; it could be synthesized *via* Wittig reaction of (E)-2-undecenal *282* and the ylide *283*. (3Z,5E)-3,5-tetradecadienyl acetate *284* was separated from its distillation residue [98% (3E,5E)-3,5-isomer] with 95% purity by means of spinning ribbon-column distillation [196] (Scheme 52).

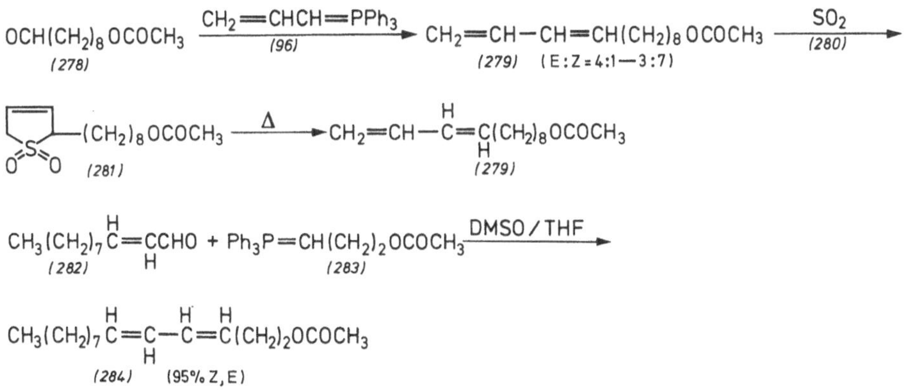

$$OCH(CH_2)_8OCOCH_3 \xrightarrow[(96)]{CH_2=CHCH=PPh_3} CH_2=CH-CH=CH(CH_2)_8OCOCH_3 \xrightarrow[(280)]{SO_2}$$
(278)   (279)   (E:Z=4:1—3:7)

$$\overset{\text{(structure)}}{\underset{O^{\!\!\nearrow S}\!\!\searrow O}{}}-(CH_2)_8OCOCH_3 \xrightarrow{\Delta} CH_2=CH-\overset{H}{\underset{H}{C}}=C(CH_2)_8OCOCH_3$$
(281)   (279)

$$CH_3(CH_2)_7\overset{H}{\underset{H}{C}}=CCHO + Ph_3P=CH(CH_2)_2OCOCH_3 \xrightarrow{DMSO/THF}$$
(282)   (283)

$$CH_3(CH_2)_7\overset{H}{\underset{H}{C}}=C-\overset{H}{\underset{}{C}}\overset{H}{=}C(CH_2)_2OCOCH_3$$
(284)   (95% Z,E)

Scheme 52

5,7-Dodecadiene derivatives were identified as sex attractants of *Lasiocampidae* species [168]. While Canadian authors synthesized all four isomeric 5,7-dodecadiene alcohols, acetates and aldehydes by nonstereoselective Wittig reactions and subsequent isomer separation [197], Bestmann et al. [198] described in a recently published paper the stereoselective preparation of the corresponding (Z,Z)-, (Z,E)- and (E,Z)-compounds. Therefore, the (Z)- resp. (E)-unsaturated aldehydes 285a–c were olefinated with the ylides of the corresponding salts 286a, b formed as reported in ref. [33]. Subsequent hydrolysis and hydroboration/oxidation of respectively 287a, b and 287c gave (5Z,7Z)-, (5Z,7E)- and (5E,7Z)-dodecadien-1-ols 288a–c. Acetylation or oxidation with pyridinium chlorochromate (PCC) yielded the corresponding alkadienyl acetates 289a–c and alkadienals 290a–c, respectively (Scheme 53). (5Z,7E)-5,7-Dodecadienal turned out to be a weak attractant for males of the pine moth *Dendrolimus pini* [198] (Scheme 53).

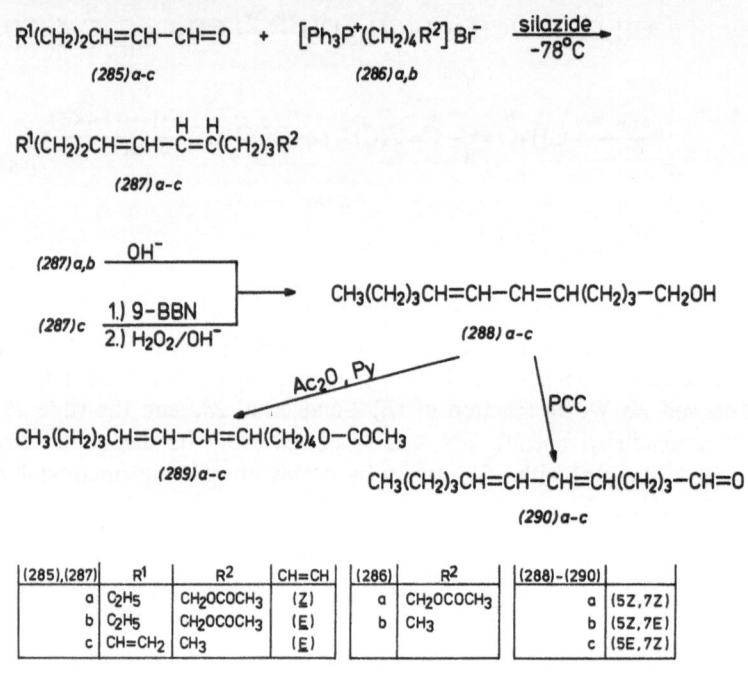

(285),(287)	R¹	R²	CH=CH	(286)	R²	(288)–(290)	
a	C₂H₅	CH₂OCOCH₃	(Z)	a	CH₂OCOCH₃	a	(5Z,7Z)
b	C₂H₅	CH₂OCOCH₃	(E)	b	CH₃	b	(5Z,7E)
c	CH=CH₂	CH₃	(E)			c	(5E,7Z)

Scheme 53

(9Z,12E)-9,12-Tetradecadienyl acetate 295, a sex attractant component of the pest insects *Cadra cautella, Plodia interpunctuella, Spodoptera eridiana, Anagasta kühniella, Ephestia elutella, Spodoptera exigua, Cadra figulella* and *Spodoptera litura* [168] contains two olefinic double bonds separated by a methylene group. It was synthesized by carbonyl olefination. Julia synthesis [199] of 1-cyclopropylethanol 291 leads to (E)-3-pentenyl bromide 292. Treatment of the phosphonium salt 293 with potassium/HMPA [200] or sodium/DMSO [201] affords the ylide 294 which is olefinated with 8-formyloctyl acetate 278 to the acetate 295 with an (E,E)-isomer content of 5–9 % [178,200,201] (Scheme 54).

The gelechid moth *Pectinophora gossypiella* uses (7Z,11Z)-7,11-hexadecadienyl acetate 300 and (7Z,11E)-7,11-hexadecadienyl acetate 313 as its sex pheromones. Another moth of the same family, the angouis grain moth *Sitotroga cerealella* uses only the (Z,E)-compound 313. Sonnet synthesized the (Z,Z)-7,11-alkadienyl acetate

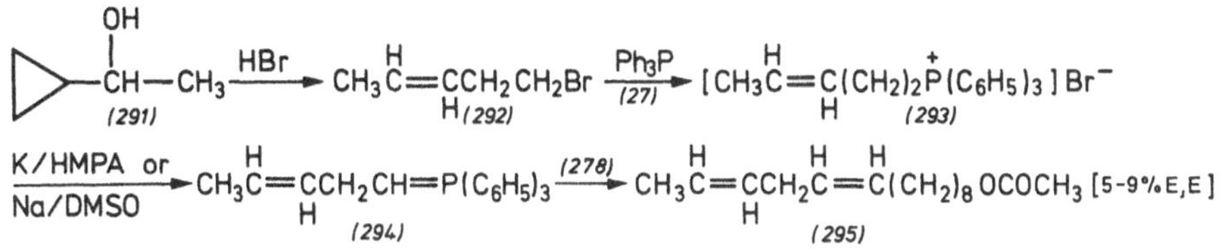

Scheme 54

300 by alkylation of the acetylide *296* with 1-bromo-3-chloropropane and subsequent formation of the phosphonium salt. The corresponding ylide *297* was olefinated with valeric aldehyde *298* and subsequently acetylated to (Z)-11-hexadecen-7-ynyl acetate *299*. Partial (Z)-selective hydrogenation with the Lindlar catalyst gives (7Z,11Z)-7,11-hexadecadienyl acetate *300* [202] (Scheme 55).

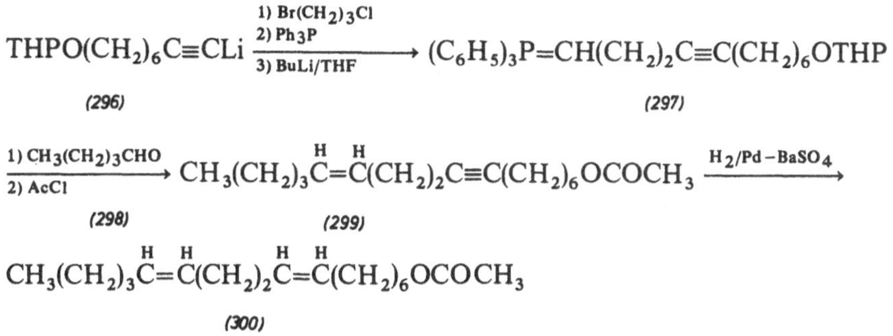

$$\text{THPO(CH}_2)_6\text{C}\equiv\text{CLi} \xrightarrow[\substack{\text{1) Br(CH}_2)_3\text{Cl} \\ \text{2) Ph}_3\text{P} \\ \text{3) BuLi/THF}}]{} (C_6H_5)_3P\text{=CH(CH}_2)_2\text{C}\equiv\text{C(CH}_2)_6\text{OTHP}$$

$$(296) \qquad\qquad\qquad\qquad\qquad (297)$$

$$\xrightarrow[\substack{\text{1) CH}_3\text{(CH}_2)_3\text{CHO} \\ \text{2) AcCl}}]{} CH_3(CH_2)_3\overset{H}{C}\text{=}\overset{H}{C}(CH_2)_2C\equiv C(CH_2)_6OCOCH_3 \xrightarrow{H_2/Pd-BaSO_4}$$

$$(298) \qquad\qquad (299)$$

$$CH_3(CH_2)_3\overset{H}{C}\text{=}\overset{H}{C}(CH_2)_2\overset{H}{C}\text{=}\overset{H}{C}(CH_2)_6OCOCH_3$$

$$(300)$$

Scheme 55

By the reaction of pentanal *298* with 3-ethoxycarbonylpropylidene-triphenylphosphorane *301* the corresponding ester *302* is obtained (Z)-stereoselectively. This ester is reduced with LiAlH₄ and converted *via* the bromocompound *303* into the corresponding phosphonium salt. The ylide *304* generated from this salt according to the sodium silazide method [33] reacts (Z)-stereoselectively with ethyl 6-formylhexanoate *305* to *306*. Reduction and subsequent acetylation yields the pheromone *300* [203]. To prepare the (7Z,11E)-7,11-hexadecadienyl acetate *313* (E)-4-nonenyl bromide *309* synthesized by a Crombie cleavage (*308→309*) is converted into the corresponding phosphonium salt. (Z)-stereoselective olefination of the ylide *310*, generated by treatment of the phosphonium salt with sodium silazide, with *305* gives the ester *311*. The latter is reduced with LiAlH₄ and converted into the second pheromone component *313* in the same way as described above for the isomer *300* [203] (Scheme 56).

An elegant route, directly leading to a 1:1 mixture of the two acetates *300* and *313* is reported by Anderson and Henrick [204], starting their synthesis with (1Z,5Z)-1,5-cyclooctadiene *314*. Cleavage of one double bond in *314* results in the (Z)-configurated synthon *316* which, in a stereocontrolled Wittig reaction in the presence of lithium bromide and ethanol, is olefinated with the ylide pentylidene-triphenylphosphorane

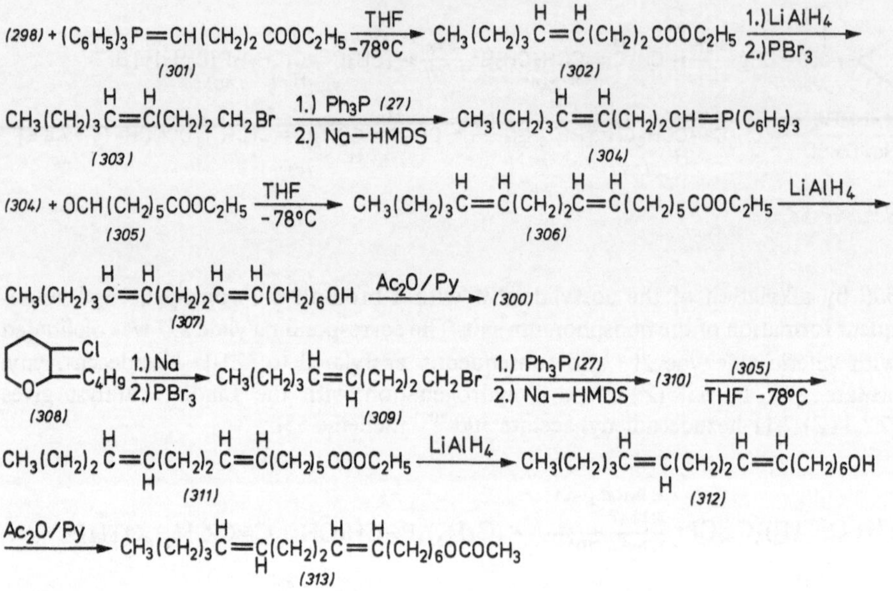

(298) + $(C_6H_5)_3P$=$CH(CH_2)_2 COOC_2H_5$ $\xrightarrow[-78°C]{THF}$ $CH_3(CH_2)_3\overset{H}{C}$=$\overset{H}{C}(CH_2)_2 COOC_2H_5$ $\xrightarrow[2.)PBr_3]{1.)LiAlH_4}$

(301)   *(302)*

$CH_3(CH_2)_3\overset{H}{C}$=$\overset{H}{C}(CH_2)_2 CH_2 Br$ $\xrightarrow[2.)\ Na-HMDS]{1.)\ Ph_3P\ (27)}$ $CH_3(CH_2)_3\overset{H}{C}$=$\overset{H}{C}(CH_2)_2 CH$=$P(C_6H_5)_3$

(303)   *(304)*

(304) + $OCH(CH_2)_5 COOC_2H_5$ $\xrightarrow[-78°C]{THF}$ $CH_3(CH_2)_3\overset{H}{C}$=$\overset{H}{C}(CH_2)_2\overset{H}{C}$=$\overset{H}{C}(CH_2)_5 COOC_2H_5$ $\xrightarrow{LiAlH_4}$

(305)   *(306)*

$CH_3(CH_2)_3\overset{H}{C}$=$\overset{H}{C}(CH_2)_2\overset{H}{C}$=$\overset{H}{C}(CH_2)_6 OH$ $\xrightarrow{Ac_2O/Py}$ *(300)*

(307)

$\overset{Cl}{\underset{O}{\bigcirc}}$—$C_4H_9$ $\xrightarrow[2.)PBr_3]{1.)Na}$ $CH_3(CH_2)_3\overset{H}{C}$=$\overset{}{\underset{H}{C}}(CH_2)_2 CH_2 Br$ $\xrightarrow[2.)\ Na-HMDS]{1.)\ Ph_3P\ (27)}$ *(310)* $\xrightarrow[THF\ -78°C]{(305)}$

(308)   *(309)*

$CH_3(CH_2)_2\overset{H}{C}$=$\overset{}{\underset{H}{C}}(CH_2)_2\overset{H}{C}$=$\overset{H}{C}(CH_2)_5 COOC_2H_5$ $\xrightarrow{LiAlH_4}$ $CH_3(CH_2)_3\overset{H}{C}$=$\overset{}{\underset{H}{C}}(CH_2)_2\overset{H}{C}$=$\overset{H}{C}(CH_2)_6OH$

(311)   *(312)*

$\xrightarrow{Ac_2O/Py}$ $CH_3(CH_2)_3\overset{H}{C}$=$\overset{}{\underset{H}{C}}(CH_2)_2\overset{H}{C}$=$\overset{H}{C}(CH_2)_6OCOCH_3$

(313)

Scheme 56

317 to a 1:1 mixture of the (*Z,Z*)- and the (*E,Z*)-alkadiene ester *318*. Coupling of the latter with the C_3-synthon, followed by hydrolysis and acetylation afford the two pheromone acetates *300* and *313* [204] (Scheme 57).

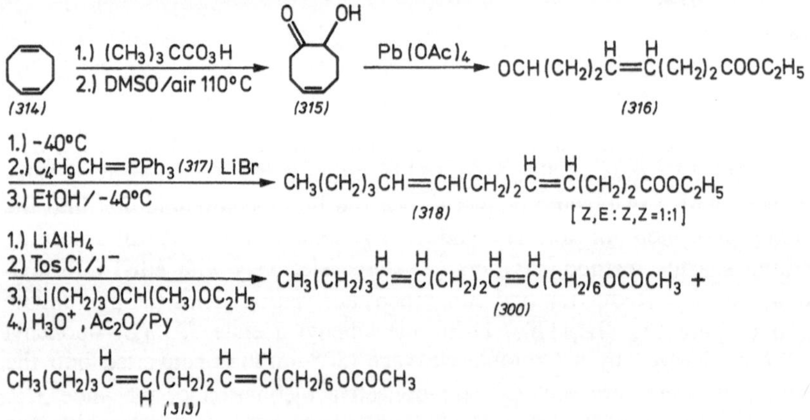

$\bigcirc$ $\xrightarrow[2.)\ DMSO/air\ 110°C]{1.)\ (CH_3)_3 CCO_3H}$ $\overset{O}{\underset{}{\bigcirc}}\overset{OH}{}$ $\xrightarrow{Pb(OAc)_4}$ $OCH(CH_2)_2\overset{H}{C}$=$\overset{H}{C}(CH_2)_2 COOC_2H_5$

(314)   *(315)*   *(316)*

$\begin{matrix}1.)\ -40°C\\2.)\ C_4H_9 CH=PPh_3\ (317)\ LiBr\\3.)\ EtOH/-40°C\end{matrix}$ $\longrightarrow$ $CH_3(CH_2)_3 CH$=$CH(CH_2)_2\overset{H}{C}$=$\overset{H}{C}(CH_2)_2 COOC_2H_5$

                                      (318)   [Z,E : Z,Z =1:1]

$\begin{matrix}1.)\ LiAlH_4\\2.)\ TosCl/J^-\\3.)\ Li(CH_2)_3OCH(CH_3)OC_2H_5\\4.)\ H_3O^+,\ Ac_2O/Py\end{matrix}$ $\longrightarrow$ $CH_3(CH_2)_3\overset{H}{C}$=$\overset{H}{C}(CH_2)_2\overset{H}{C}$=$\overset{H}{C}(CH_2)_6OCOCH_3$ +

                                      (300)

$CH_3(CH_2)_3\overset{H}{C}$=$\overset{}{\underset{H}{C}}(CH_2)_2\overset{H}{C}$=$\overset{H}{C}(CH_2)_6 OCOCH_3$

                         (313)

Scheme 57

For the synthesis of *313* Hammond and Descoins also used a Wittig reaction [205]. Thus, starting from 1-hepten-3-ol *319* the phosphonium salt *320* is formed via several steps. The ylide of this salt is olefinated with aldehyde *305* to *311*. Reduction of *311* and subsequent acetylation afford the pheromone *313* (Scheme 58) [205].

Very recently, polyenic hydrocarbons were found as a new class of sex pheromones in the family of *Geometridae* (*Lepidoptera*) [206,207]. From this class (3Z,6Z,9Z)-

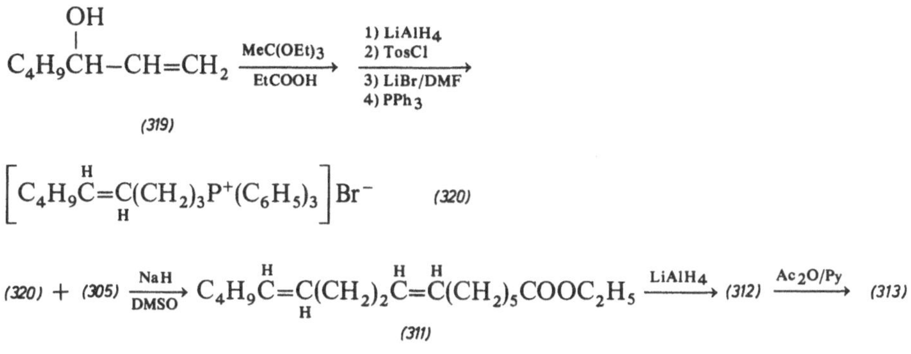

$$\underset{(319)}{\overset{\overset{\displaystyle OH}{\displaystyle |}}{C_4H_9CH-CH=CH_2}} \xrightarrow[\text{EtCOOH}]{\text{MeC(OEt)}_3} \xrightarrow[\substack{\text{3) LiBr/DMF}\\ \text{4) PPh}_3}]{\substack{\text{1) LiAlH}_4\\ \text{2) TosCl}}}$$

$$\left[C_4H_9\overset{H}{\underset{H}{C}}{=}C(CH_2)_3P^+(C_6H_5)_3 \right] Br^- \qquad (320)$$

$$(320) + (305) \xrightarrow[\text{DMSO}]{\text{NaH}} \underset{(311)}{C_4H_9\overset{H}{\underset{H}{C}}{=}C(CH_2)_2\overset{H}{C}{=}\overset{H}{C}(CH_2)_5COOC_2H_5} \xrightarrow{\text{LiAlH}_4} (312) \xrightarrow{\text{Ac}_2\text{O/Py}} (313)$$

Scheme 58

1,3,6,9-nonadecatetraene *325* was isolated from the winter moth *Operophthera brumata* and synthesized by a combination of acetylene synthesis and Wittig olefination [206]. Thus the starting 1-bromononane *321* is subjected to a twofold alkylation with lithium acetylide and to a Grignard reaction to yield *322*. The latter is partially hydrogenated and converted into the alkadienylphosphonium salt *323*. The corresponding ylide is olefinated (Z)-selectively with acrolein *324* to the tetraene pheromone *325* [206] (Scheme 59).

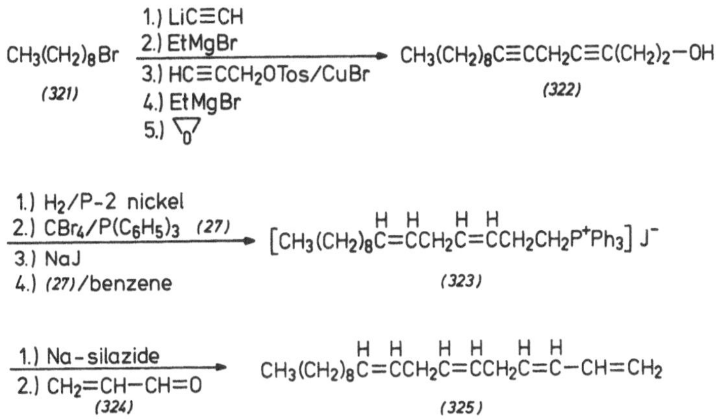

$$\underset{(321)}{CH_3(CH_2)_8Br} \xrightarrow[\substack{\text{3.) HC}\equiv\text{CCH}_2\text{OTos/CuBr}}]{\substack{\text{1.) LiC}\equiv\text{CH}\\ \text{2.) EtMgBr}}}\begin{array}{c}\text{4.) EtMgBr}\\ \text{5.) }\end{array} \underset{(322)}{CH_3(CH_2)_8C\equiv CCH_2C\equiv C(CH_2)_2{-}OH}$$

$$\xrightarrow[\substack{\text{3.) NaJ}\\ \text{4.) (27)/benzene}}]{\substack{\text{1.) H}_2\text{/P-2 nickel}\\ \text{2.) CBr}_4\text{/P(C}_6\text{H}_5)_3 \;(27)}} \underset{(323)}{\left[CH_3(CH_2)_8\overset{H}{\underset{}{C}}{=}\overset{H}{C}CH_2\overset{H}{C}{=}\overset{H}{C}CH_2CH_2P^+Ph_3 \right] J^-}$$

$$\xrightarrow[\substack{\text{2.) CH}_2{=}\text{CH}{-}\text{CH}{=}\text{O}\\ (324)}]{\text{1.) Na-silazide}} \underset{(325)}{CH_3(CH_2)_8\overset{H}{C}{=}\overset{H}{C}CH_2\overset{H}{C}{=}\overset{H}{C}CH_2\overset{H}{C}{=}\overset{H}{C}{-}CH{=}CH_2}$$

Scheme 59

From females of the forest pest *Bupalus piniarius*, another *Geometridae* species, a 6,9-nonadecadiene with unknown geometry of the double bonds, was isolated as their sex attractant. For the synthesis of three of the four possible isomers Wittig reactions were used according to Scheme 60 [207]. To obtain (6Z,9Z)-6,9-nonadecadiene *327* methyl linoleate *57* is reduced to the corresponding aldehyde which is simultaneously olefinated with *209* to 1,10-13-nonadecatriene *326*. Regioselective hydroboration of *326* with subsequent protonolysis gives the (Z,Z)-isomer *327* (Scheme 60). For the preparation of the corresponding (Z,E)- (*332*) and (E,Z)-isomer

(335), respectively, similar reaction sequences were chosen. Thus, from the Grignard compound of 1-undecyne 328 and oxirane 3-tridecyn-1-ol 329 is obtained which is subsequently reduced (E)-stereospecifically to (E)-3-tridecen-1-ol 330. Conversion of the latter into the corresponding tosylate and treatment with triphenylphosphine yield phosphonium salt 331 the corresponding ylide of which is olefinated with hexanal to give (6Z,9E)-6,9-nonadecadiene 332. The reversed isomer 335 is prepared by the same sequence starting from 1-heptyne 333 and using principally the same individual steps (Scheme 60) [207].

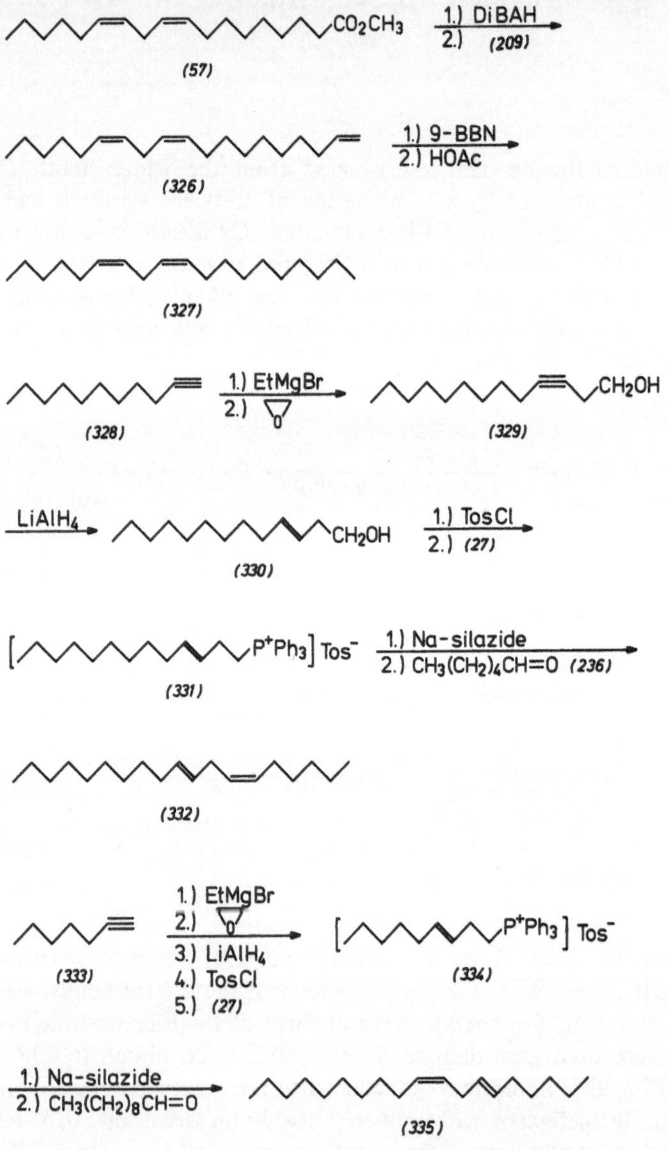

CO_2CH_3 (57) 1.) DiBAH 2.) (209)

(326) 1.) 9-BBN 2.) HOAc

(327)

(328) 1.) EtMgBr 2.) → CH_2OH (329)

LiAlH₄ → CH_2OH (330) 1.) TosCl 2.) (27)

[P^+Ph_3] Tos⁻ (331) 1.) Na-silazide 2.) $CH_3(CH_2)_4CH=O$ (236)

(332)

(333) 1.) EtMgBr 2.) 3.) LiAlH₄ 4.) TosCl 5.) (27) → [P^+Ph_3] Tos⁻ (334)

1.) Na-silazide 2.) $CH_3(CH_2)_8CH=O$ → (335)

Scheme 60

7.2 Beetles — Coleoptera

The pheromones of the beetles, the greatest order in the class of insects, are, compared to those of the butterflies, a structurally inhomogeneous class of substances. Especially in the family of bark beetles (*Scolytidae*), representing dangerous forest pests, the sex attractants and aggregants have been the object of numerous investigations. Ipsenol [*344*, (—)-2-methyl-6-methylene-7-octen-4-ol] and ipsdienol [*350*, (—)-2-methyl-6-methylene-2,7-octadien-4-ol] are two chiral pheromone components of the bark beetle *Ips paraconfusus* and of some other *Scolytidae* species which have been synthesized by Wittig reactions. Only recently, did Mori succeed in determining the absolute configuration of the naturally occurring *344* to be (S)(—)-*344* by a stereo-selective synthesis, starting from (S)-leucine *336*. Via tosylate *338* and the optically active epoxide *339* the lactone *340* is obtained which is converted into (S)-*341*. The exomethylene group of *341* is protected by selenoid formation, the lactone function reduced and the resulting semiacetal *343* olefinated with the ylide *209* with cleavage of the protective group to (S)-*344* ([α]$_D^{24}$ = 16.5, ethanol) [208, 209] (Scheme 61).

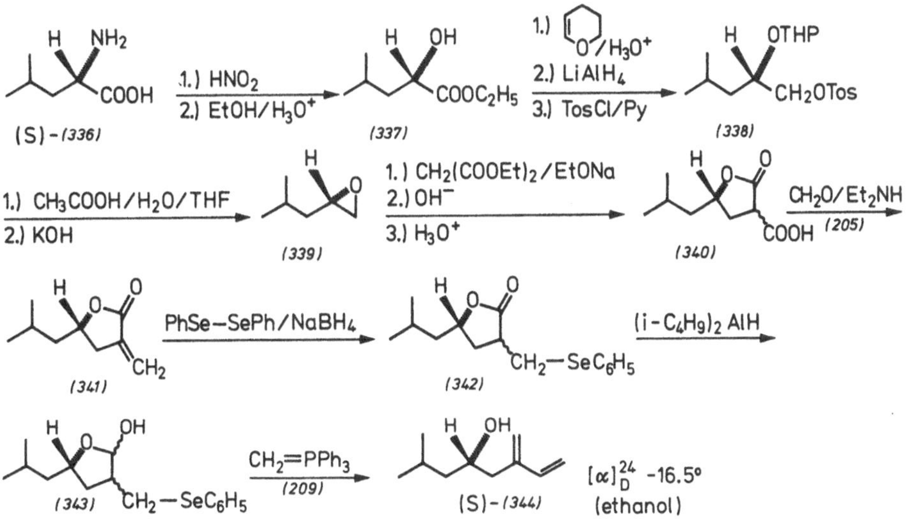

Scheme 61

Mori also succeeded in synthesizing the optically active ipsdienol *350* via carbonyl olefination of the acetonide *345* of (R)(+)-glycerine aldehyde with isopropylidene-phosphorane to form *346*. Methoxymercuration of *346* followed by the splitting of the protective group, tosylation and treatment with alkali afford the epoxide *349* which is converted into the optically active (R)-*350* via the same reaction sequence as in the synthesis of *344* according to Scheme 61 [210] (Scheme 62).

α-Multistriatin *356* is one of the three aggregation pheromone components of the bark beetle *Scolytus multistriatus*, the originator of the "dutch elm disease". The synthesis of *356*, identified as 2,4-dimethyl-5-ethyl-6,8-dioxabicyclo[3.2.1]octane with (1S,2R,4S,5R)-configuration, starts from glycerine aldehyde acetonide *345* which is converted via the alcohol *351* into the ketone *352*. The latter is olefinated

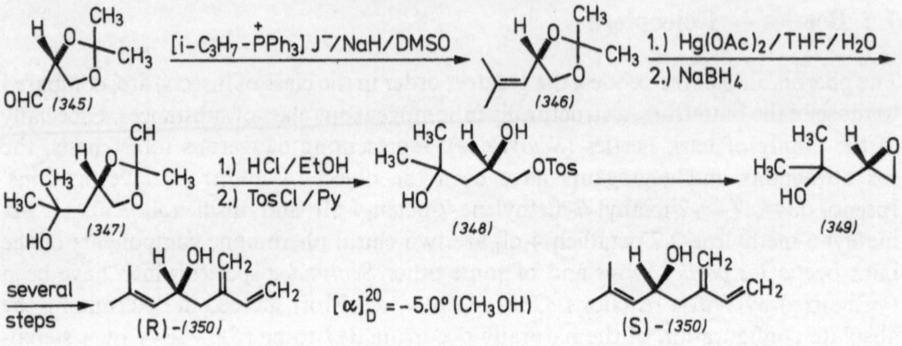

Scheme 62

to the methylene compound *353*. *Via* several steps the crude product *355* is formed which gives a mixture of four possible Multistriatin isomers on heating. Pure (—)-Multistriatin *356* can be isolated by preparative gas chromatography [211)] (Scheme 63) and thus the configuration of the natural product determined.

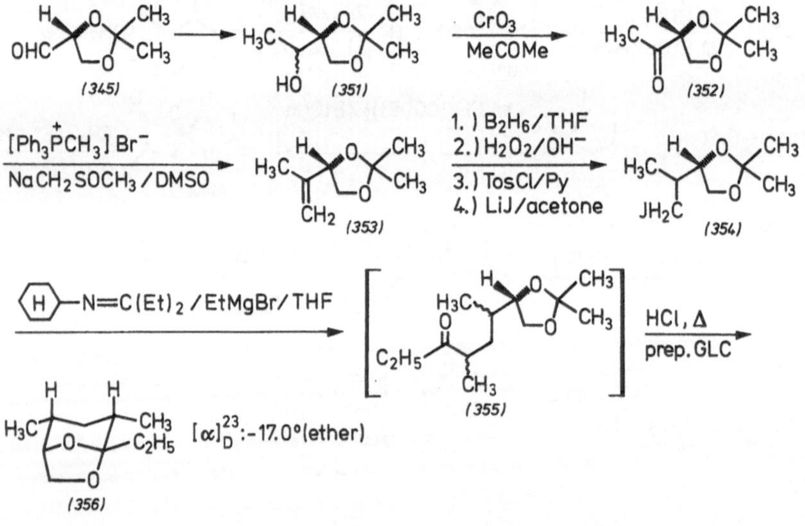

Scheme 63

Sulcatol *362*, the aggregating pheromone of the ambrosia beetle *Gnathotrichus sulcatus* consists of 65% (S)(+)- and 35% (R)(—)-6-methyl-5-hepten-2-ol. Mori [212)] synthesized the two optical antipodes starting from (R)- and (S)-glutamic acid *357*, respectively. Tosylation, halogenation and two hydrogenation steps yield the optically active semiacetal *361* the carbonyl olefination of which gives (R)-*362* (and (S)-*362*, respectively, when starting from (R)-*357*) [212)] (Scheme 64).

Grandisol *373*, (+)-cis-1-methyl-1-(2-hydroxymethyl)-2-isopropenylcyclobutane, is the main component of the pheromone complex of the curculionid beetle *Anthono-*

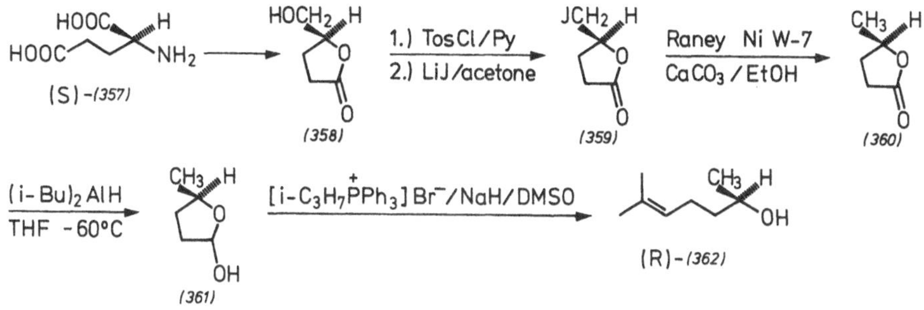

Scheme 64

mus grandis, responsible for almost one third of all insect cotton damage of the United States. Hobbs and Magnus [213,214] synthesized optically active grandisol starting from β-pinan-2-ol *363*. The key reaction is thereby the preparation of the olefin *367* via *364* and *365* from the semiacetal *366* by means of a Wittig reaction. Hydroboration/oxidation of *367* followed by acetylation and dehydrogenation of the resulting *368* affords a mixture of the acetates *369* and *370*. Subsequent oxidation, hydrogenation, photolysis and decarbonylation lead *via* acetate *372* to (+)(1R,2S)-grandisol *373* [213,214] (Scheme 65).

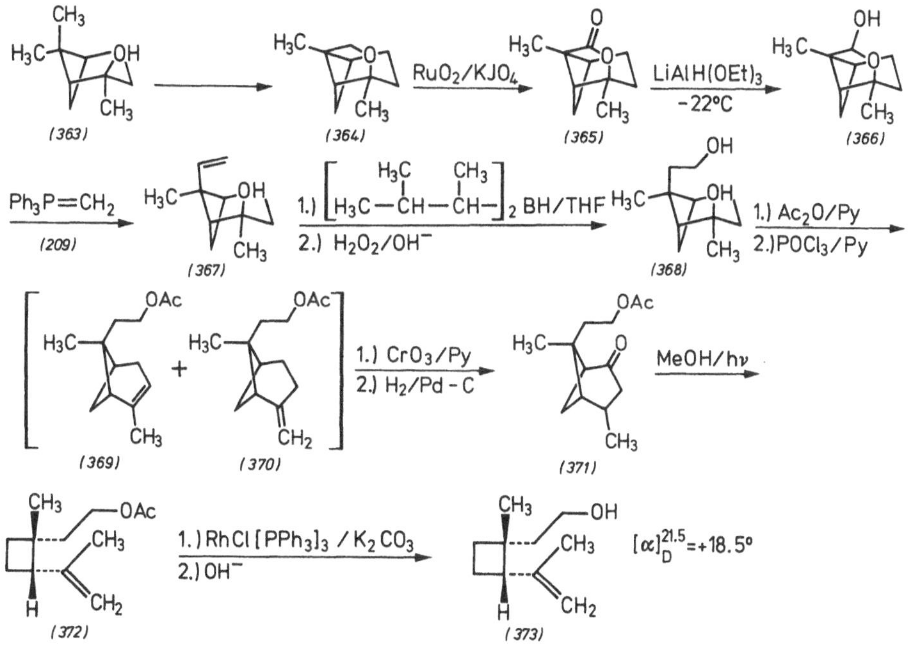

Scheme 65

Wittig methylenations similar to those described in Scheme 65 were also used by Zurflüh et al. [215], Kosugi et al. [216] and Stork and Cohen [217] in the synthesis of racemic grandisol (Scheme 66). In the first two cases (refs. [215,216]) the cyclobutane

133

system was prepared by photochemical cycloaddition. In the latter case stereo-selective cyclization of an epoxynitril was applied [217].

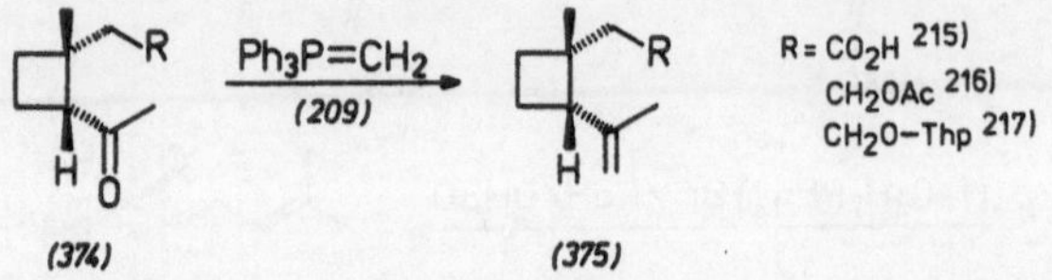

R = CO_2H [215]
CH_2OAc [216]
CH_2O-Thp [217]

(374)　　(375)

Scheme 66

The sex pheromone of the scarabaeid beetle *Popillia japonica* is a chiral lactone (*379*) with a (Z)-unsaturated side chain [218]. Optically active (R)(−)-glutamic acid *357*, which is converted by retention of configuration *via* three steps into the Wittig synthon *377*, serves as the starting material. Reaction of the latter with the ylide *378* yields the lactone *379* with the desired (R)(Z)-configuration [218] (Scheme 67).

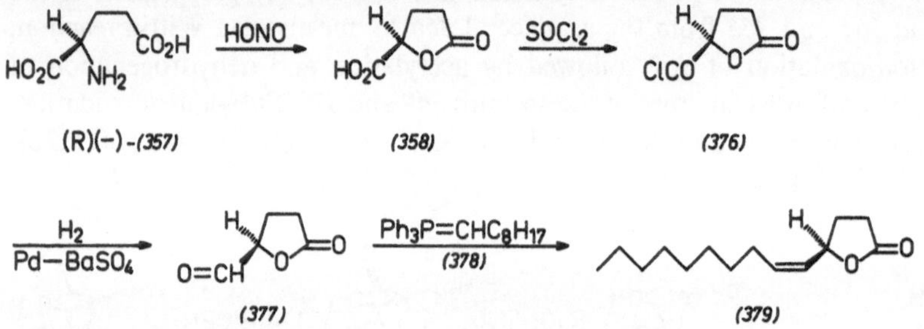

(R)(−)-(*357*)　　(*358*)　　(*376*)

(*377*)　　(*379*)

Scheme 67

7.3 Flies — Diptera

(Z)-9-Tricosene *382* (muscalure) together with a series of homologous alkenes was isolated as the sex-specific attractive basic component from faecal and cuticular lipids of the house fly *Musca domestica* [219]. These alkenes, all possessing (Z)-configuration, were, in addition to other methods, synthesized by selective Wittig olefinations [219,220,221].

Carlson et al. obtained (Z)-9-tricosene *382* by reaction of nonanal *380* with the ylide generated from the phosphonium salt *381* and butyllithium in DMSO [219,220] (Scheme 68). Bestmann et al. [221] synthesized the (Z)-olefin *382* by olefination of tetradecanal *384*, with the ylide generated by treatment of the corresponding phosphonium salt *383* with potassium in HMPA [34] (Scheme 68).

$CH_3(CH_2)_7CHO$
(*380*)

$+ [(C_6H_5)_3P^+(CH_2)_{13}CH_3] Br^-$ $\xrightarrow[10-20\,°C]{BuLi/DMSO}$ $CH_3(CH_2)_7\overset{H}{C}=\overset{H}{C}(CH_2)_{12}CH_3$
(*381*)　　　　　　　　　　　　　　　　　　　(*382*)　　15% E, 73% yield

$[CH_3(CH_2)_8P^+(C_6H_5)_3] Br^- + OCH(CH_2)_{12}CH_3$ $\xrightarrow{K/HMPA}$ (*382*)　4% E, 32% yield
(*383*)　　　　　　　　　(*384*)

Scheme 68

7.4 Hymenoptera

One of the reported syntheses of (*E*)-9-oxodec-2-enoic acid *392*, the queen substance of the honey bee *Apis mellifera*, uses two ylide reactions [222]. Starting from pimelic acid *385* the resonance-stabilized ylide *386* is prepared by alkylation of methylene-triphenylphosphorane *209* and the former hydrolyzed to 7-oxooctanoic acid *387*. Reduction of the corresponding thiol ester *389* and olefination of the resulting aldehyde *390* with phosphorane *67* gives the (*E*)-2-unsaturated ester *391*. The latter was hydrolyzed to pheromone *392* [222] (Scheme 69).

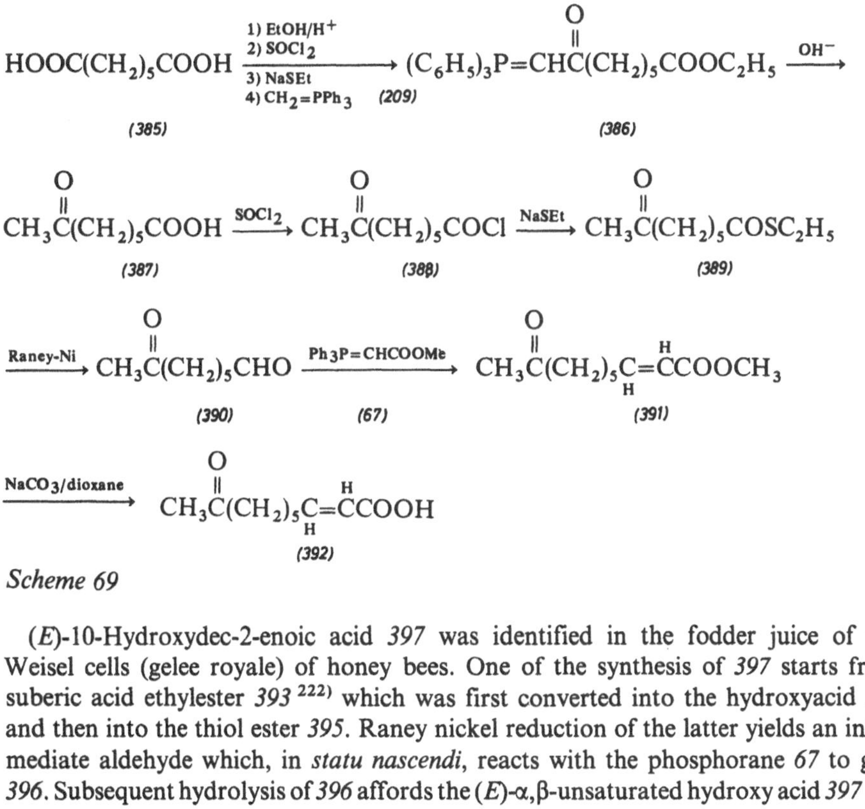

$$HOOC(CH_2)_5COOH \xrightarrow[\substack{3)\ NaSEt \\ 4)\ CH_2=PPh_3}]{\substack{1)\ EtOH/H^+ \\ 2)\ SOCl_2}} (C_6H_5)_3P=CHC(CH_2)_5COOC_2H_5 \xrightarrow{OH^-}$$

(385)   (209)   (386)

$$CH_3\overset{O}{\overset{\|}{C}}(CH_2)_5COOH \xrightarrow{SOCl_2} CH_3\overset{O}{\overset{\|}{C}}(CH_2)_5COCl \xrightarrow{NaSEt} CH_3\overset{O}{\overset{\|}{C}}(CH_2)_5COSC_2H_5$$

(387)   (388)   (389)

$$\xrightarrow{Raney-Ni} CH_3\overset{O}{\overset{\|}{C}}(CH_2)_5CHO \xrightarrow{Ph_3P=CHCOOMe} CH_3\overset{O}{\overset{\|}{C}}(CH_2)_5\overset{H}{\underset{H}{C}}=CCOOCH_3$$

(390)   (67)   (391)

$$\xrightarrow{NaCO_3/dioxane} CH_3\overset{O}{\overset{\|}{C}}(CH_2)_5\overset{H}{\underset{H}{C}}=CCOOH$$

(392)

Scheme 69

(*E*)-10-Hydroxydec-2-enoic acid *397* was identified in the fodder juice of the Weisel cells (gelee royale) of honey bees. One of the synthesis of *397* starts from suberic acid ethylester *393* [222] which was first converted into the hydroxyacid *394* and then into the thiol ester *395*. Raney nickel reduction of the latter yields an intermediate aldehyde which, in *statu nascendi*, reacts with the phosphorane *67* to give *396*. Subsequent hydrolysis of *396* affords the (*E*)-α,β-unsaturated hydroxy acid *397* [222] (Scheme 70).

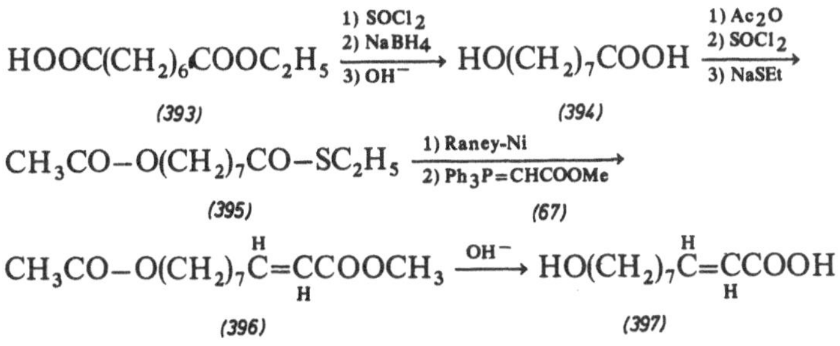

$$HOOC(CH_2)_6COOC_2H_5 \xrightarrow[\substack{3)\ OH^-}]{\substack{1)\ SOCl_2 \\ 2)\ NaBH_4}} HO(CH_2)_7COOH \xrightarrow[\substack{3)\ NaSEt}]{\substack{1)\ Ac_2O \\ 2)\ SOCl_2}}$$

(393)   (394)

$$CH_3CO-O(CH_2)_7CO-SC_2H_5 \xrightarrow[\substack{2)\ Ph_3P=CHCOOMe}]{\substack{1)\ Raney-Ni}}$$

(395)   (67)

$$CH_3CO-O(CH_2)_7\overset{H}{\underset{H}{C}}=CCOOCH_3 \xrightarrow{OH^-} HO(CH_2)_7\overset{H}{\underset{H}{C}}=CCOOH$$

(396)   (397)

Scheme 70

Different esters of 3,7-dimethylpentadecan-2-ol are the main components of the pheromone of some diprionids [168]. With their methyl branching they possess structures principally accessible by Wittig methylenation. Thus, Magnusson [223] obtained the ketone *399* by Baeyer-Villiger oxidation and alkylation, starting from pure *trans*-2,3-dimethylcyclohexanone *398*. *399* is olefinated with *209* to *400*. Reduction (H$_2$, Pd/C) and acetylation give the acetate *401* of the *erythro*-isomer of 3,7-dimethylpentadecan-2-ol [223], one of the pheromone components of diprionid species (Scheme 71).

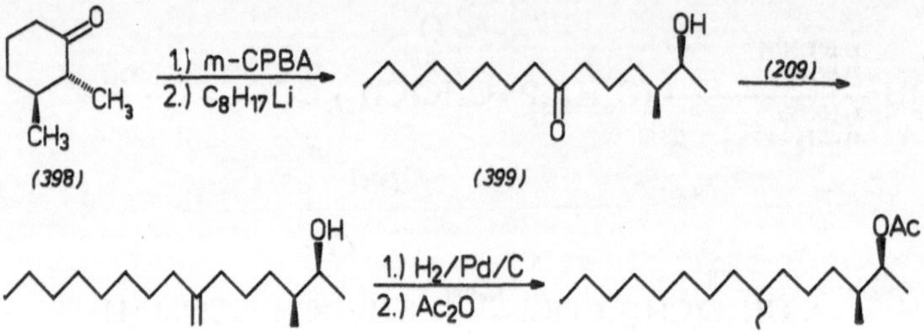

Scheme 71

7.5 Orthoptera

The females of the German cockroach *Blatella germanica* secrete a pheromone which is received by the males through antennal contact. One of the two pheromone components has been identified as 3,11-dimethylnonacosan-2-one *405* [224,225] and synthesized by Schwarz and coworkers [226] using a phosphorus ylide as an assisting agent. The anion of the stable phosphorane *403* is alkylated with *402* and the resulting acyl ylide *404* hydrolyzed to *405* [226]. In another synthesis of *405*, published in 1976, coupling of the *S*-aryl thioester *406* with lithium dialkylcuprate *407* is the key reaction [227]. The Wittig reaction of the resulting ketone *408* with *209* followed by catalytical hydrogenation, hydrolysis, halogenation and coupling with 3-(1-ethoxy)-ethoxy-2-methylbutyllithium afford the acetal protected alcohol *413* the hydrolysis of which and subsequent oxidation give the cockroach pheromone *405* [227] (Scheme 72).

For the preparation of the second component *421* of the cockroach pheromone Burgsthaler et al. [228] also used a Wittig reaction. Lithium acetylide is alkylated with the two halides *414* and *416* and the resulting alkynyl bromide *417* converted into the phosphonium salt. Olefination of the corresponding ylide with 9-bromo-2-nonanone *418* gives a (Z)/(E)-mixture of olefin *419* which is converted into the pheromone 3,11-dimethyl-29-hydroxynonacosan-2-one *421* by acetoacetate synthesis, hydrogenation, hydrolysis, and decarboxylation [228] (Scheme 73).

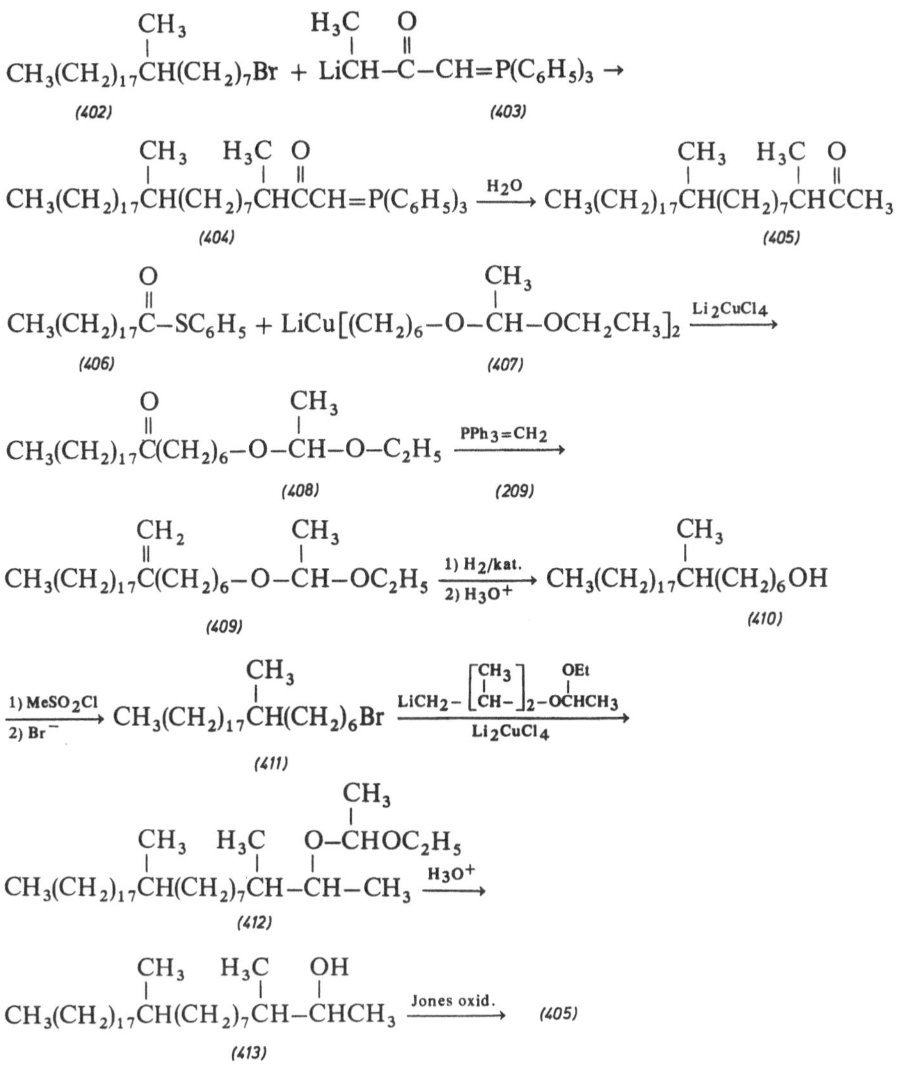

$$\underset{(402)}{CH_3(CH_2)_{17}\overset{\overset{\displaystyle CH_3}{|}}{C}H(CH_2)_7Br} + \underset{(403)}{Li\overset{\overset{\displaystyle H_3C}{|}}{C}H-\overset{\overset{\displaystyle O}{\|}}{C}-CH=P(C_6H_5)_3} \rightarrow$$

$$\underset{(404)}{CH_3(CH_2)_{17}\overset{\overset{\displaystyle CH_3}{|}}{C}H(CH_2)_7\overset{\overset{\displaystyle H_3C}{|}}{C}H\overset{\overset{\displaystyle O}{\|}}{C}CH=P(C_6H_5)_3} \xrightarrow{H_2O} \underset{(405)}{CH_3(CH_2)_{17}\overset{\overset{\displaystyle CH_3}{|}}{C}H(CH_2)_7\overset{\overset{\displaystyle H_3C}{|}}{C}H\overset{\overset{\displaystyle O}{\|}}{C}CH_3}$$

$$\underset{(406)}{CH_3(CH_2)_{17}\overset{\overset{\displaystyle O}{\|}}{C}-SC_6H_5} + \underset{(407)}{LiCu[(CH_2)_6-O-\overset{\overset{\displaystyle CH_3}{|}}{C}H-OCH_2CH_3]_2} \xrightarrow{Li_2CuCl_4}$$

$$\underset{(408)}{CH_3(CH_2)_{17}\overset{\overset{\displaystyle O}{\|}}{C}(CH_2)_6-O-\overset{\overset{\displaystyle CH_3}{|}}{C}H-O-C_2H_5} \xrightarrow{PPh_3=CH_2}$$

$$\underset{(409)}{CH_3(CH_2)_{17}\overset{\overset{\displaystyle CH_2}{\|}}{C}(CH_2)_6-O-\overset{\overset{\displaystyle CH_3}{|}}{C}H-OC_2H_5} \xrightarrow[\text{2) } H_3O^+]{\text{1) } H_2/kat.} \underset{(410)}{CH_3(CH_2)_{17}\overset{\overset{\displaystyle CH_3}{|}}{C}H(CH_2)_6OH}$$

$$\xrightarrow[\text{2) Br}^-]{\text{1) MeSO}_2Cl} \underset{(411)}{CH_3(CH_2)_{17}\overset{\overset{\displaystyle CH_3}{|}}{C}H(CH_2)_6Br} \xrightarrow[\text{Li}_2\text{CuCl}_4]{LiCH_2-\left[\overset{\overset{\displaystyle CH_3}{|}}{C}H-\right]_2-\overset{\overset{\displaystyle OEt}{|}}{C}HCH_3}$$

$$\underset{(412)}{CH_3(CH_2)_{17}\overset{\overset{\displaystyle CH_3}{|}}{C}H(CH_2)_7\overset{\overset{\displaystyle H_3C}{|}}{C}H-\overset{\overset{\displaystyle O-\overset{\overset{\displaystyle CH_3}{|}}{C}HOC_2H_5}{|}}{C}H-CH_3} \xrightarrow{H_3O^+}$$

$$\underset{(413)}{CH_3(CH_2)_{17}\overset{\overset{\displaystyle CH_3}{|}}{C}H(CH_2)_7\overset{\overset{\displaystyle H_3C}{|}}{C}H-\overset{\overset{\displaystyle OH}{|}}{C}HCH_3} \xrightarrow{\text{Jones oxid.}} (405)$$

Scheme 72

7.6 Kairomones by Wittig Synthesis

Hydrocarbons with methyl branching at C-11 to C-13 were isolated from *Heliothis virescens (Noctuidae, Lepidoptera)* [229]. These compounds release a host-seeking behavior in the braconid *Cardiochiles nigriceps*, a *Heliothis* parasit. These kairomones were prepared in 1975 by Vinson et al. [229] by Wittig methylenation of the corresponding methyl ketones and subsequent catalytic hydrogenation [229].

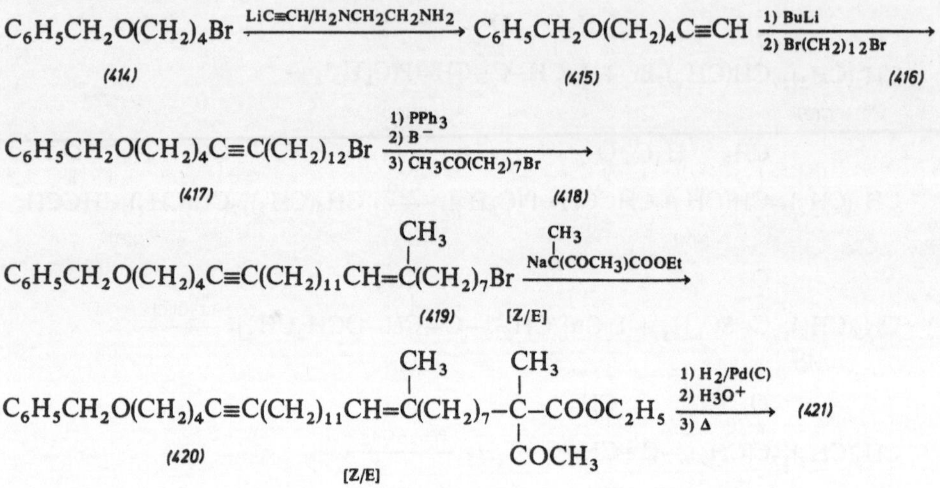

$$C_6H_5CH_2O(CH_2)_4Br \xrightarrow{\text{LiC}\equiv\text{CH/H}_2\text{NCH}_2\text{CH}_2\text{NH}_2} C_6H_5CH_2O(CH_2)_4C\equiv CH \xrightarrow[\text{2) Br(CH}_2)_{12}\text{Br}]{\text{1) BuLi}}$$

(414)   *(415)*   *(416)*

$$C_6H_5CH_2O(CH_2)_4C\equiv C(CH_2)_{12}Br \xrightarrow[\text{3) CH}_3\text{CO(CH}_2)_7\text{Br}]{\substack{\text{1) PPh}_3 \\ \text{2) B}^-}}$$

(417)   *(418)*

$$C_6H_5CH_2O(CH_2)_4C\equiv C(CH_2)_{11}CH=\overset{\overset{\displaystyle CH_3}{|}}{C}(CH_2)_7Br \xrightarrow{\text{Na}\overset{\overset{\displaystyle CH_3}{|}}{C}(COCH_3)COOEt}$$

(419)   [Z/E]

$$C_6H_5CH_2O(CH_2)_4C\equiv C(CH_2)_{11}CH=\overset{\overset{\displaystyle CH_3}{|}}{C}(CH_2)_7-\overset{\overset{\displaystyle CH_3}{|}}{\underset{\underset{\displaystyle COCH_3}{|}}{C}}-COOC_2H_5 \xrightarrow[\text{3) }\Delta]{\substack{\text{1) H}_2\text{/Pd(C)} \\ \text{2) H}_3\text{O}^+}} \textit{(421)}$$

(420)   [Z/E]

Scheme 73

8 Terpenoids

Terpenes and terpenoids may, regarding the structure of their carbon skeleton, formally be formed by a regular joining of isoprenic units, as it is shown by the isoprene rule of Ruzicka [230]. In the case of mono-, sesqui- and diterpenes alternating head-to-tail linking of these C_5-units occurs whereas tri- and tetraterpenes, mostly, in the middle of the molecule, contain tail-to-tail units, thus consisting of two mirror image molecular halves. In spite of this systematic structure terpenes and terpenoids exhibit, however, only little regularities concerning the position and configuration of their double bonds which can appear both in the carbon chain and carbocyclic rings. Therefore, there are no general syntheses known based on the Wittig reaction; the latter is however often used for the construction of single structural units.

Monoterpene hydrocarbons, alcohols, aldehydes and ketones are commercially available with sufficient purity. Thus, total syntheses are not required but only conversions of commercial product and synthetic derivatives.

However, for some sesquiterpenes which occur more rarely in nature in higher concentrations and which can be isolated in pure form only with great difficulties and are hardly commercially available, total syntheses making use of the Wittig carbonyl olefination are applied. Thus, e.g. olefination of the ylide, generated from the acetylenic phosphonium salt *423*, with the ketone *422* allows the construction of an isoprenoid chain by direct incorporation of a C_5-synthon (Scheme 74). The addition of water to the terminal triple bond of *424* yields the ketone *425* [231], an isoprenologue of *422* (as a mixture of *cis*- and *trans*-isomers) which may be converted further. Similarily, the ylide *426* was chosen as a C_{10}-synthon and used for the synthesis of chlorobiumquinone [232]. The Wittig reaction of *426* with aldehyde *427* gives the acetal *428*, after the hydrolysis of which the process can be repeated [232] (Scheme 74).

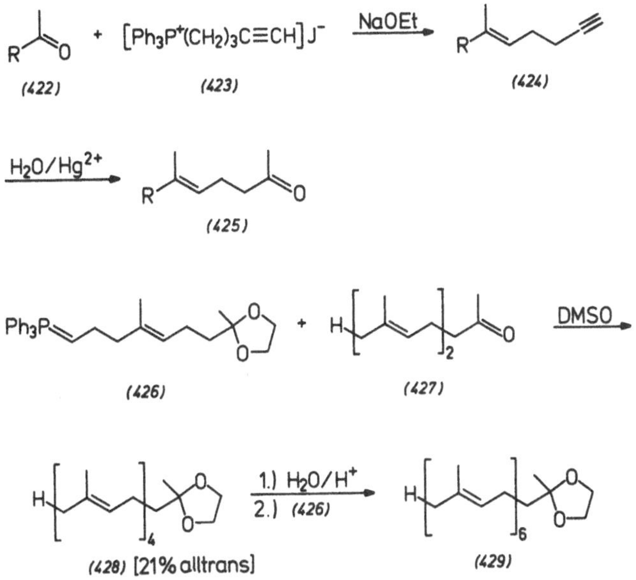

$$R \overset{\text{(422)}}{\underset{O}{\bigvee}} + [Ph_3P^+(CH_2)_3C{\equiv}CH]J^- \xrightarrow{\text{NaOEt}} \underset{\text{(424)}}{R} $$

(422)   (423)   (424)

$$\xrightarrow{H_2O/Hg^{2+}} \underset{\text{(425)}}{R}$$

(425)

$$Ph_3P{=} \underset{\text{(426)}}{\bigvee} + H \underset{\text{(427)}}{\bigvee} \xrightarrow{\text{DMSO}}$$

(426)   (427)

$$H \underset{\text{(428) [21\% alltrans]}}{\bigvee_4} \xrightarrow[\text{2.) (426)}]{\text{1.) } H_2O/H^+} H \underset{\text{(429)}}{\bigvee_6}$$

(428) [21% alltrans]   (429)

Scheme 74

Corey and Yamamoto [233] used the β-oxido synthesis [234] of trisubstituted olefins for the preparation of the acyclic sesquiterpene farnesol *433*. In this preparation the isoheptenylphosphonium salt *430* is converted into the hydroxyfarnesol derivative *432* by reaction with the tetrahydropyranyl ether — protected hydroxy aldehyde *431* and formaldehyde *205*. *432* is converted into farnesol *433* via several steps. Other reactions of *432* likewise proceeding via several steps lead to *434* which is a positional isomer of a C_{17}-juvenile hormone [233] (Scheme 75).

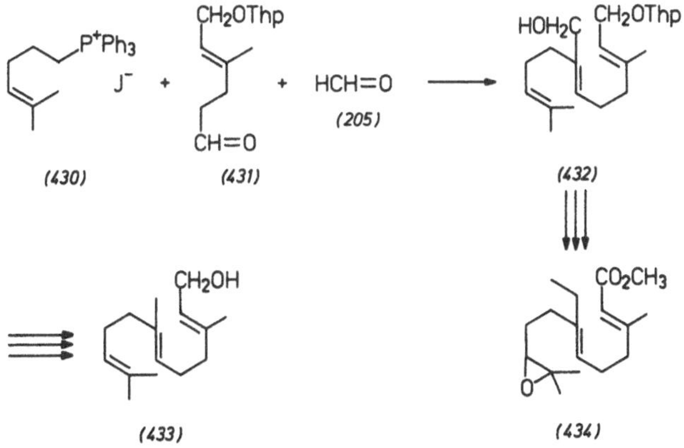

(430)   (431)   (205)   (432)

(433)   (434)

Scheme 75

The same synthetic route was applied to the preparation of the tricyclic sesquiterpene santalol [235]. The reaction of the tricyclic aldehyde *435* with ethylidenetriphenylphosphorane and paraformaldehyde *205* gives the α,β-unsaturated alcohol *436*, elimination of phosphorus being probably directed by the thermodynamical stability of *436* [235] (Scheme 76).

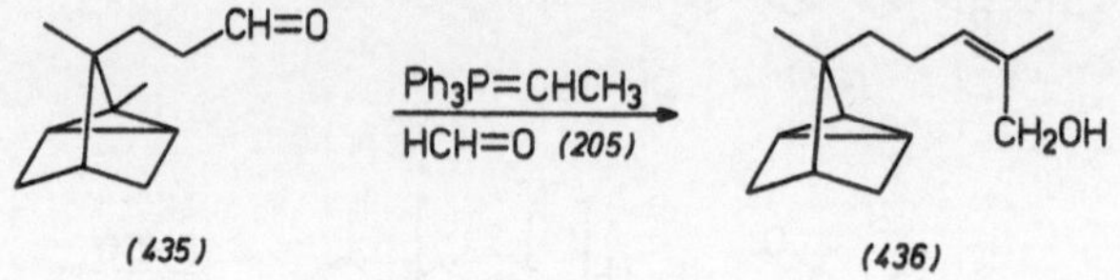

(435)   (436)

Scheme 76

As a further Wittig synthon isopropylidene-triphenylphosphorane was used for the preparation of trans 2,6-farnesol and trans-nerolidol [236] which are structurally related to one another like linalool and geraniol.

Trans β-sinensal *441* is obtained from the aldehyde *438* prepared by selective ozonolysis of trans-farnesene *437*. Wittig olefination of *438* with 1-formylethylidenephosphorane *440* gives trans β-sinensal *441*. Reaction of *438* with 1-ethoxycarbonylethylidene-triphenylphosphorane *439* yields the ethyl ester *442*, from which *441* can also be obtained [237] (Scheme 77). Furthermore, a working group of the BASF synthesized α-sinensal *443* and β-sinensal *441* using Wittig reactions [238].

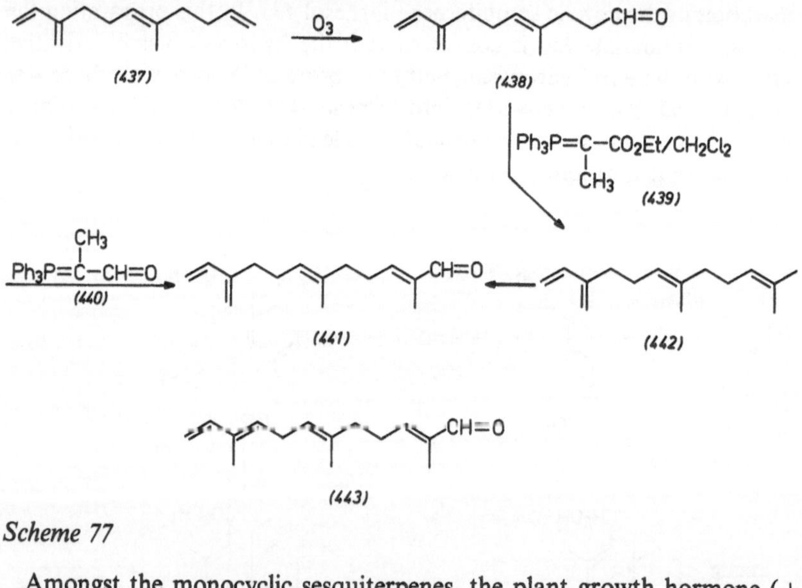

(437)   (438)

$$Ph_3P=C-CO_2Et/CH_2Cl_2$$
$$\overset{|}{C}H_3 \quad (439)$$

$$Ph_3P=\overset{CH_3}{\overset{|}{C}}-CH=O$$
(440)

(441)   (442)

(443)

Scheme 77

Amongst the monocyclic sesquiterpenes, the plant growth hormone (+)-abscisic acid *446* and α-humulene *451* are found which both can be synthesized by Wittig reactions. To synthesize (+)-abscisinic acid *446* D. L. Roberts et al. [239] converted α-ionone *444* into the allylic alcohol *445* by *t*-butyl chromate oxidation. *445* is reacted with ethoxycarbonylmethylene-triphenylphosphorane *238* and subsequently hydrolyzed to *446* [239] (Scheme 78). ^{14}C-labelled 2-^{14}C-abscisic acid is also obtained by the use of phosphorus ylides [240]. Corey and Hamanaka [241] succeeded in synthesizing α-humulene *451* by Wittig reaction of the aldehyde *447* with the ylide *448*. Cleavage of the two protective groups and treatment of the resulting diol with phosphorus tribromide yields the *trans,cis,trans*-dibromo compound *449*. The latter is cyclized (nickel carbonyl) to the cis-isomer *450* of α-humulene. Photochemical isomerization in the presence of diphenyl disulfide leads to the formation of humulene *451* (Scheme 78) [241].

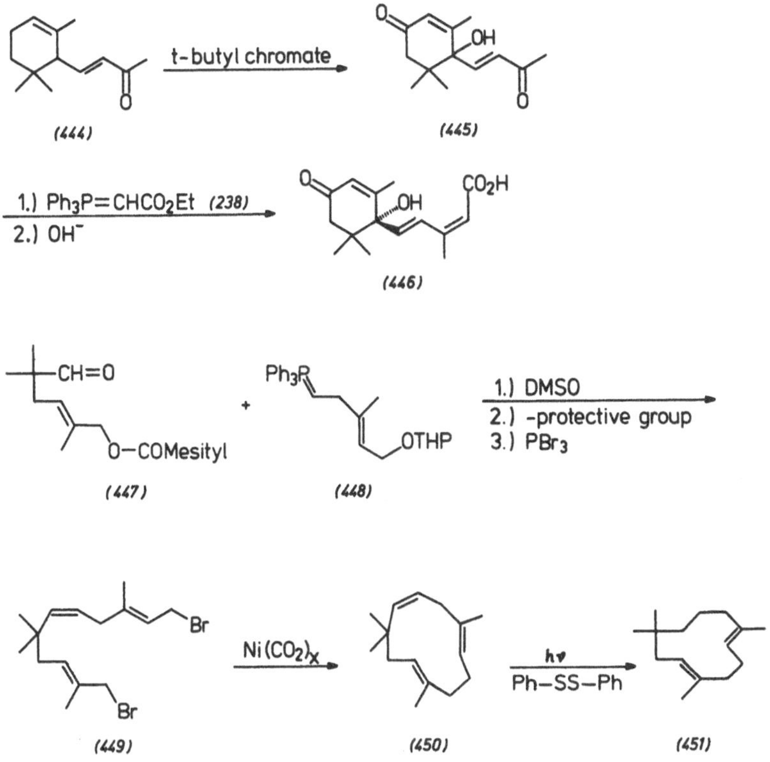

(444)   t-butyl chromate   (445)

1.) $Ph_3P=CHCO_2Et$ *(238)*
2.) OH^-

(446)

(447)   + Ph_3P (448)   1.) DMSO   2.) –protective group   3.) PBr_3

(449) $Ni(CO_2)_x$ (450) $\frac{h\nu}{Ph-SS-Ph}$ (451)

Scheme 78

Tanaka et al. [242] synthesized β-chamigrane *456*, a bicyclic sesquiterpene, starting from the 2-methylene-cyclohexanone derivative *452*. Diels Alder reaction of *452* with 2-ethoxy-1,3-butadiene *453* and hydrolysis of the resulting enol ether give the spiroketone *454*. *454* is subjected to a mono-Grignard reaction and subsequently dehydrated to ketone *455*. The latter is converted *via* a Wittig reaction with *205* into β-chamigrane *456* [242] (Scheme 79). β-Selinene *460*, with its trans-decalin skeleton belongig to the eudesmane group, is synthesized starting with octalinone *457* [243] which is converted via six steps into trans-decalone *458* from which *via* the ester *459a* and the nitril *459b*, followed by treatment with methyllithium, subsequent epimirization and Wittig reaction the sesquiterpene *460* can be obtained [243] (Scheme 79).

141

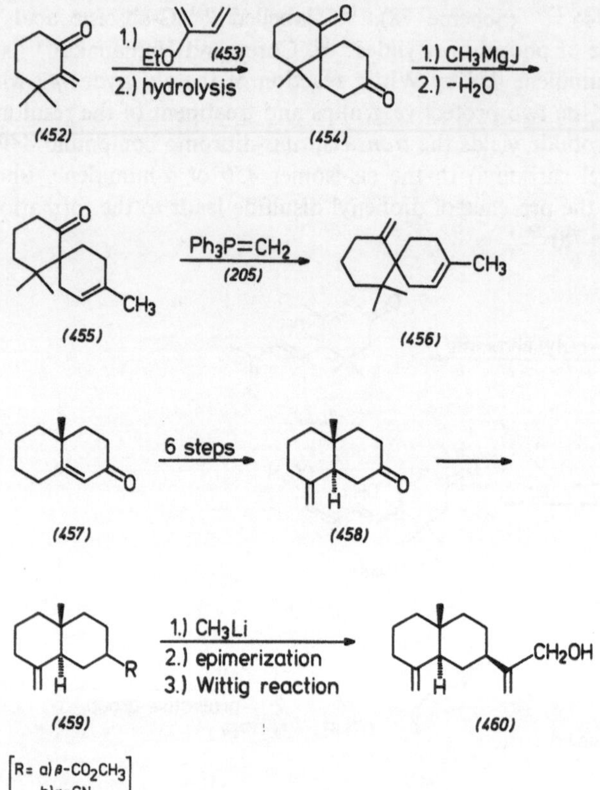

Scheme 79

Using methylene-triphenylphosphorane *205*, in many cases carbonyl functions have been converted into exomethylene groups in cyclic terpene chemistry. This technique was used, among others in the preparation of (±)-steviol methyl ether [244], dihydro-5,6-norcaryophyllene [245], (±)-nootkatone [246], (−)-phyllocladene [247], and (+)-ε-cadinene [248]. In the synthesis of the latter the trans-decalin derivative *462* was formed from the cis-decalone *461* the epimerization of which proceeded *via* the enol form during the methylenation [248] (Scheme 80).

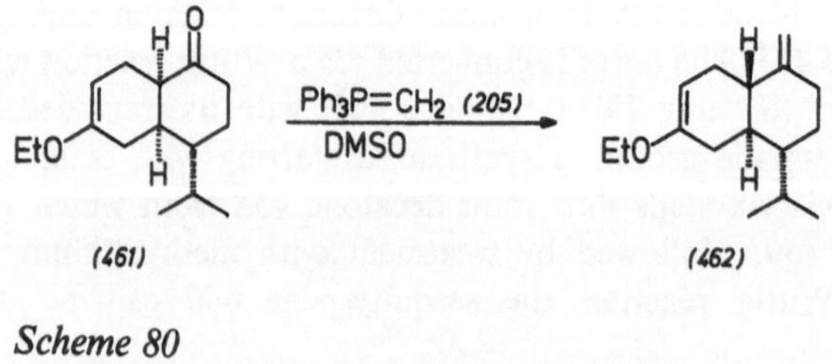

Scheme 80

142

(±)-Dihydro-β-santalol *464* is obtained in 90% yield when the hydroxy group of *463* is protected as borate ester prior to methylation [249]. The reaction of *463* with an unprotected OH-group with phosphorane *205* gives the isomeric compound *465* as the main product [249] (Scheme 81). In a similar way, the olefin *467* is formed by olefination of the hydroxy ketone *466* with *205*, whereas comparable rearrangements do not occur when using isopropylidene- or ethoxycarbonylmethylenephosphorane[249] as the ylide (Scheme 81).

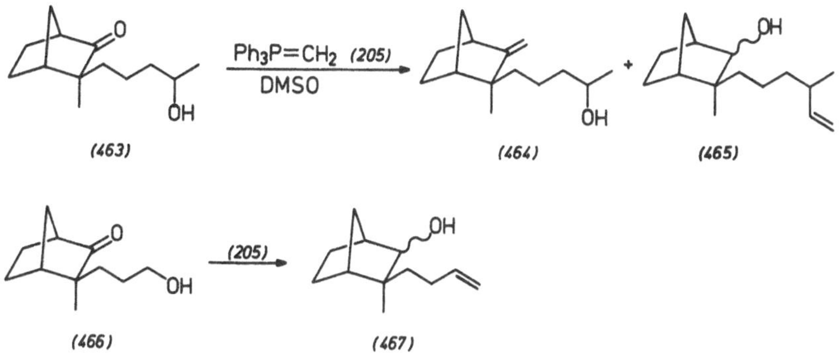

Scheme 81

Numerous tricyclic sesquiterpenes were isolated from different vetiver oils, possessing an exomethylene group, and associated to the tricyclovetivone group because of their carbon skeleton. As an example of these substances the synthesis of epizizanoic acid *473*, reported by Kido et al. [250], is briefly outlined. (+)-Methyl camphene-carboxylate *468* is converted into the corresponding aldehyde the aldol reaction of which with acetone gives enone *469* which is reacted to the cyano ketone *470*. From *470* the tricyclic ester *471* can be obtained the hydroxylation of which affords a glycol. The corresponding monomesylate is rearranged to the two C-4 epimeric ketones *472*. The Wittig reaction of the 4βH-epimer with *205* yields the tricyclic sesquiterpenecarboxylic acid *473* [250] (Scheme 82). In the last step of the preparation of an enantiomer of aromadendrene *479* the oxo group of the tricyclic ketone *478* is converted into an exomethylene group [251]. Perilla aldehyde *474* serves as the starting material and is reacted *via* the diene *475* and a Diels Alder reaction to give *476*. The monotosylate *477* of the corresponding glycol is rearranged to *478*. A final Wittig reaction of *478* with *205* yields the sesquiterpene hydrocarbon *479* with the exomethylene function [251] (Scheme 82).

The introduction of exomethylene groups *via* Wittig olefination of alicyclic ketones with Ph₃P=CH₂ *205*, as it was demonstrated in the previous examples, is probably the only carbonyl olefination which can be considered as a standard method for the preparation of terpenoid structures. Thus, the latter reaction occurs in some syntheses of sesquiterpenes belonging to the eudesmane and eremophilane series. In a series of papers dealing with the functionalization of *trans*-decalin Torii and Inokuchi [252] described the synthesis of racemic β-costol *483*, arctiol *485* and related eudesmane-type compounds starting from *trans*-decalin derivatives. The Wittig reaction of *480* with methylene-triphenylphosphorane *205* gives the exomethylene

143

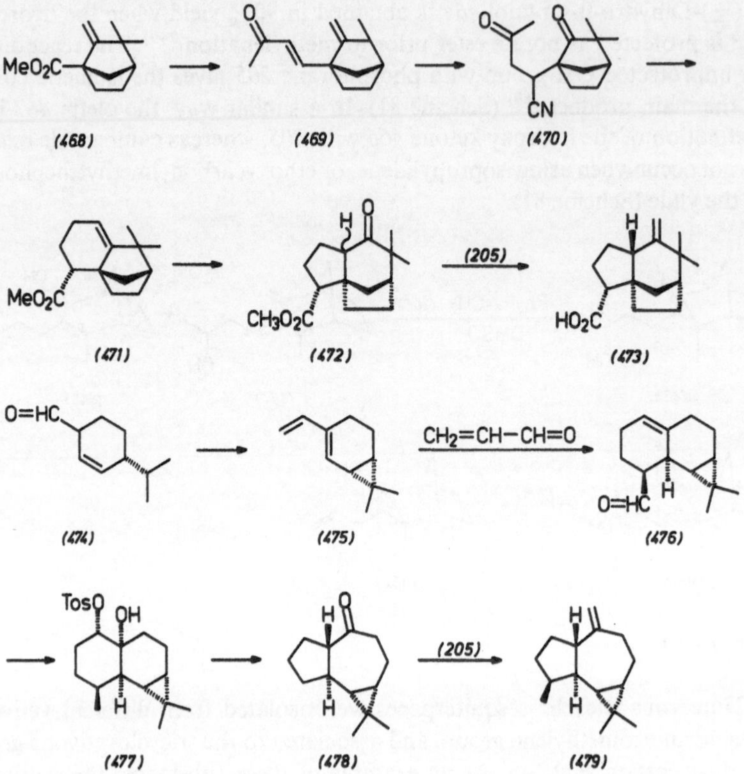

Scheme 82

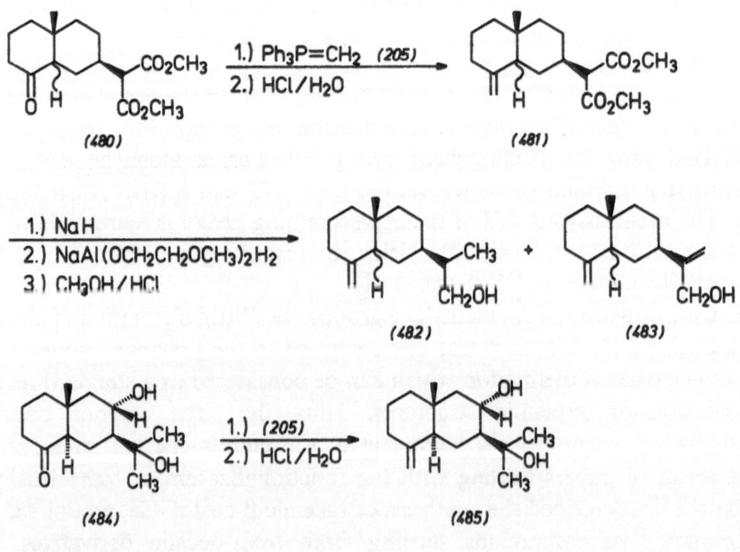

Scheme 83

compound *481* which is converted in two reduction steps into *482* and racemic β-costol *483*. The same methylenation of the decalone *484* leads to arctiol *485*, a further eudesmane sesquiterpene [252] (Scheme 83).

In a joint work coworkers of Firmenich SA (Genf) and of the MIT (Cambridge, USA) published the syntheses of the eight stereoisomeric sesquirose oxides starting from the corresponding optically active rose oxides *486–489* [253]. Ozonolysis of the latter compounds of which *486* and *487* are found in rose and geranium oil leads to the tetrahydropyrane carbaldehydes *490–493*. Olefination of these carbonyl

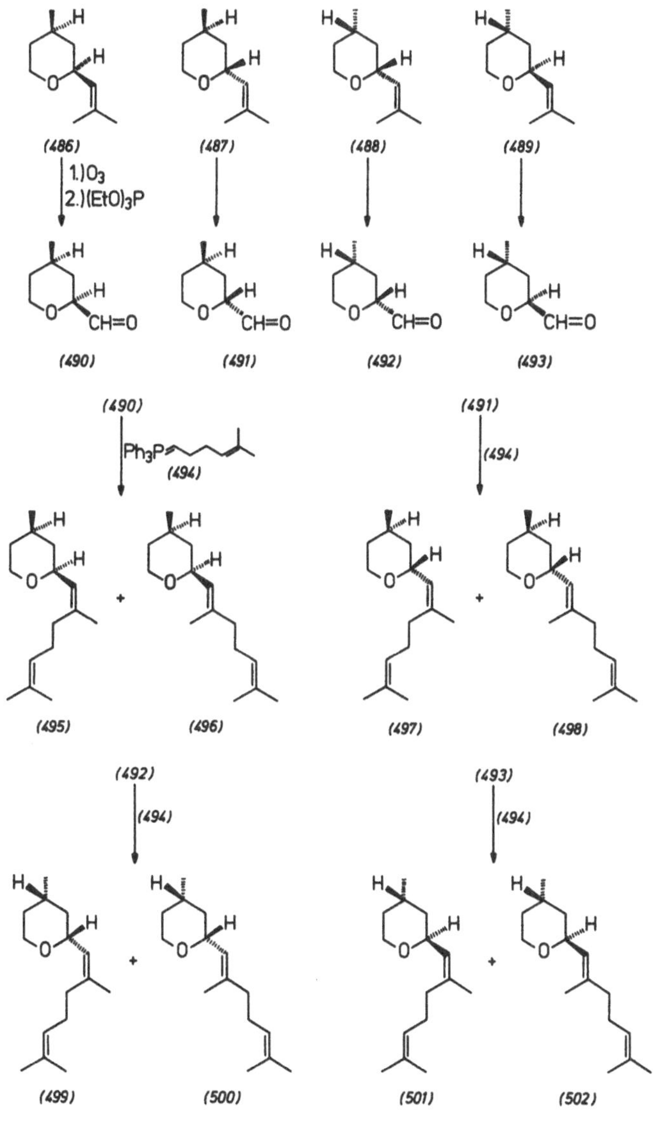

Scheme 84

145

compounds with the isoprenoid alkylidenephosphorane *494* yields a *Z/E*-mixture of each of the sesquirose oxides *495–502*, which have not been found in nature so far. It is of interest that from the *cis*-aldehydes *490* and *492* a 1:1-mixture of the stereo-isomers *495/496* and *499/500* is formed whereas the *trans* carbonyl compounds *491* and *493* are converted stereoselectively into the (*E*)-isomers *498* and *502* each with 32% of the (*Z*)-isomer (*497* and *501*) [253] (Scheme 84).

9 Retenoids, Carotenoids, Isoprenoids, and Polyenes

Among carotenoids, fat-soluble plant pigments, generally classified as tetraterpenes as well as among other isoprenoids and polyenes, isolated from plants or animals, many substances are found for the synthesis of which the Wittig reaction is of paramount importance. Quite often, these compounds consist of two symmetrically linked molecular halves and it might be sufficient to prepare one molecular half only which can subsequently be "dimerized".

The synthesis of vitamin A was certainly a pioneering work in the industrial application of the Wittig reaction [6]. The decisive step in this synthesis performed by the BASF, which had already established a plant for the production of vitamin A in 1971 [254], is the Wittig olefination of vinyl-β-ionol *503* with γ-formylcrotyl acetate *507* to vitamin A acetate *508*. The phosphonium salt *505* is obtained by reaction of the alcohol *503* with triphenylphosphine hydrobromide *504* [255] (Scheme 85).

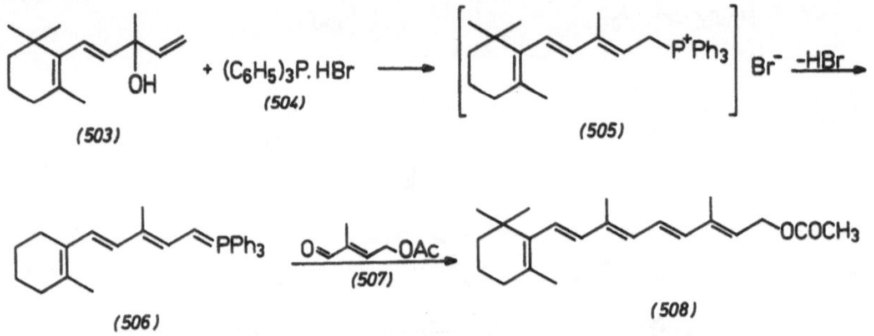

Scheme 85

Vitamin A acid, especially used in the treatment of acne for some years, can also be obtained by Wittig reaction. Thus, olefination of phosphorane *506*, with γ-formyl-crotonic ester *509* as well as the "reversed" Wittig olefination, i.e. the conversion of aldehyde *511* and ylide *512* yields the ester *510* of vitamin A acid [255], according to Scheme 86.

Cis-13-Vitamin A acid was synthesized by Bestmann and Ermann [256] starting from the aldehyde synthon *514*, one endocyclic double bond of which already exhibits *cis*-configuration. Olefination of *514* with the ylide *513* affords the 2*H*-pyran-2-one derivative *515*, which is reduced and hydrolyzed to the *cis*-13-vitamin A acid *516*. Applying the same synthetic technique also the aromatic analogs *517*, *518* and *519* are obtained [256] (Scheme 87).

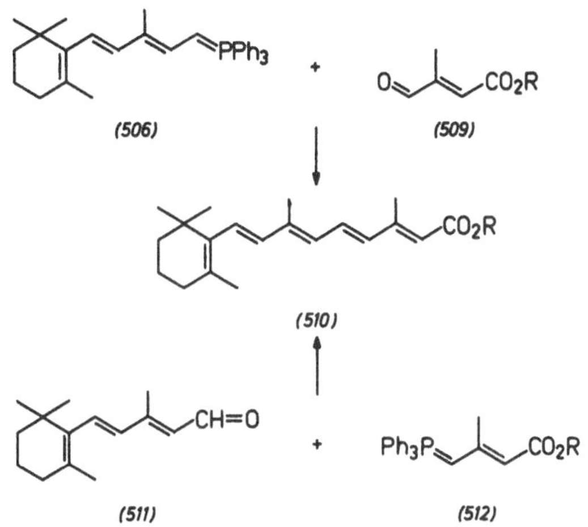

Scheme 86

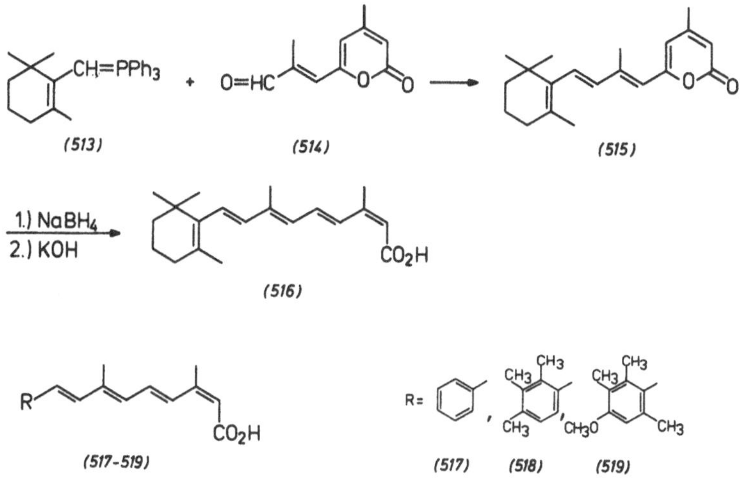

Scheme 87

To study the binding conditions of retinal in bacteriorhodopsin, a pigment of the purple membrane of *halobacterium halobium*, Nakanishi et al. [257] synthesized some retinal derivatives using the Wittig carbonyl olefination. The ylide of *505* is reacted with the silylated aldehyde *520* forming *521*. Cleavage of the protective group and subsequent oxidation afford *522*, which is subjected to bromination/dehydrobromination leading to 13-bromoretinal *523* (Scheme 88). The 9-bromo derivative *527*, corresponding to *523*, is obtained by the reaction of *524* with ethyl 3-bromo-3-formylacrylate *525*. The resulting 3-bromoester *526* is converted into the 9-bromoretinal *527* by the Emmons reaction, Dibal reduction and manganese dioxide oxidation [257] (Scheme 88).

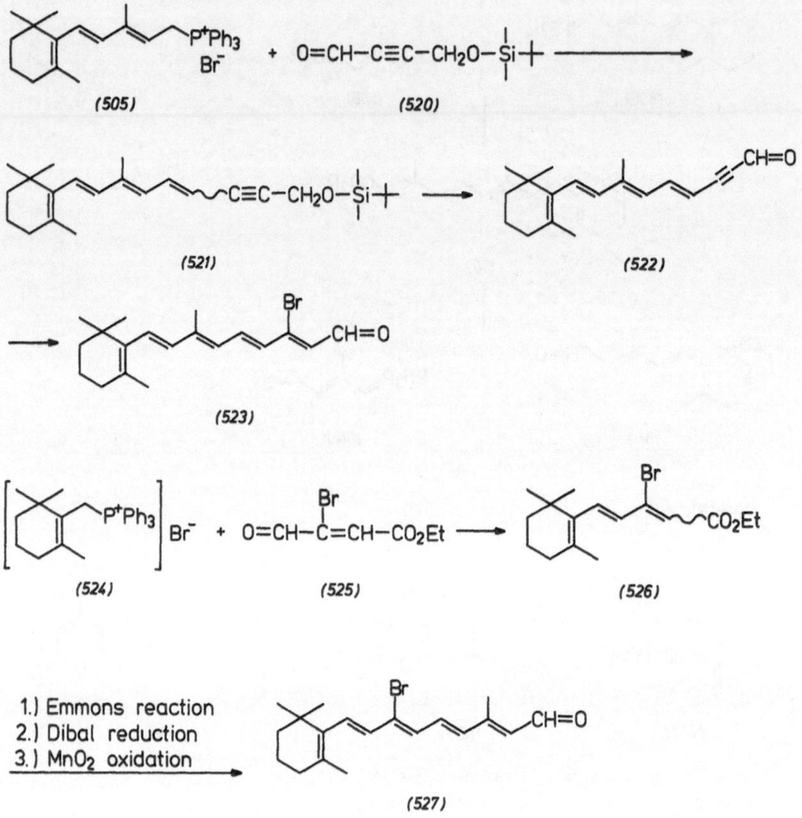

(505)   +   O=CH–C≡C–CH₂O–Si⟨   ⟶

(520)

(521)   ⟶   (522)

(523)

(524)   +   O=CH–C=CH–CO₂Et   ⟶   (526)

(525)

1.) Emmons reaction
2.) Dibal reduction
3.) MnO₂ oxidation   ⟶

(527)

Scheme 88

Oxidative coupling of two alkylidene groups, as it occurs in the autoxidation of alkylidenephosphoranes, [258,259] allows the synthesis of symmetrical carotenoids such as β-carotene, starting from a vitamin A synthon [260]. Using the adduct from ozone and triphenyl phosphite as the oxidizing reagent, high yields of β-carotene are obtained. More practicable from the technical point of view is the use of hydrogen peroxide because the reaction can be carried out in aqueous solution [261] (Scheme 89).

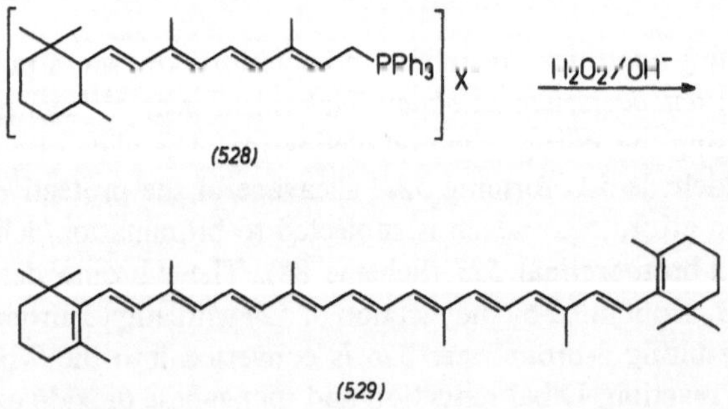

(528) H_2O_2/OH^- ⟶

(529)

Scheme 89

148

The oxidation of ylides carrying a hydrogen atom at the carbanionic centre gives primarily aldehydes. Since the rate of the Wittig reaction is faster than that of autoxidation [259], an oxaphosphetane *532* is formed immediately from the aldehyde generated and from still unreacted ylide *530*. This could be proved by ^{31}P-NMR analysis [259]. The recognition of the course of the Wittig reaction, discussed at the beginning of this chapter, allows a tritium atom to be attached to the central double bond of β-carotene during the oxidative coupling of the axerophthylidene-phosphorane *530* (Scheme 90).

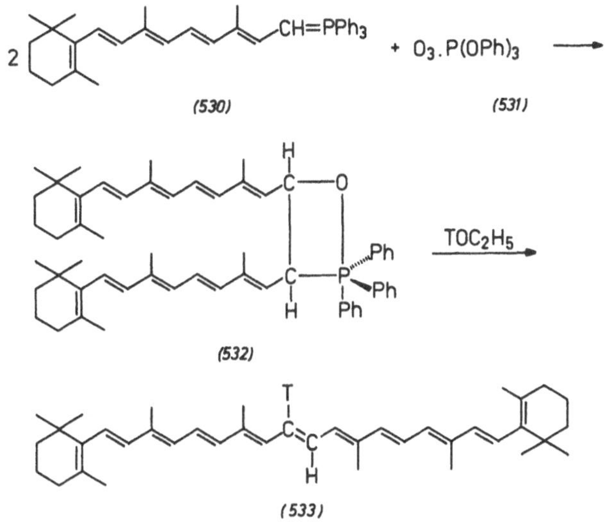

Scheme 90

The oxidation of *530* with the ozone adduct *531* is carried out at −78 °C, giving the oxaphosphetane *532*. Then, tritiated ethanol is added and the reaction mixture heated to room temperature. The rate of tritium incorporation thus obtained is very high [262] (for an explanation of the mechanism see Chapter 2).

Among the carotenoids with an aromatic end group like those isolated from the sea sponge *Reniera japonica*, the pigment renierapurpurin *538* with a symmetrical structure is found. This symmetrically built molecule can also be prepared by oxidation of an ylide [261b]. Thus, the Wittig reaction of phosphorane *534* with the ω-(2-tetrahydropyranyloxy)-alkatrienal *535* yields the polyenyl-tetrahydropyranyl ether *536*. Conversion into the phosphonium salt and subsequent treatment of the latter with Na_2CO_3 gives the ylide *537* which is oxidized with a 50 % solution of hydrogen peroxide to the carotenoid *538* in 73 % yield [261b] (Scheme 91).

Citranaxanthene *543* is found in citrus fruits and is used as a food dye like β-carotene. The same phosphonium salt synthon *505* as used for the vitamin A synthesis is monoolefinated with the polyene dialdehyde *539*. The Wittig reaction of the resulting *540* with phosphorane *541* followed by aldol condensation of the obtained *542* with acetone gives citranaxanthene *543* [255, 263] (Scheme 92). In the preparation of the polyenedial *539* 1,4-dibromo-2-butene *544* is reacted with trimethyl

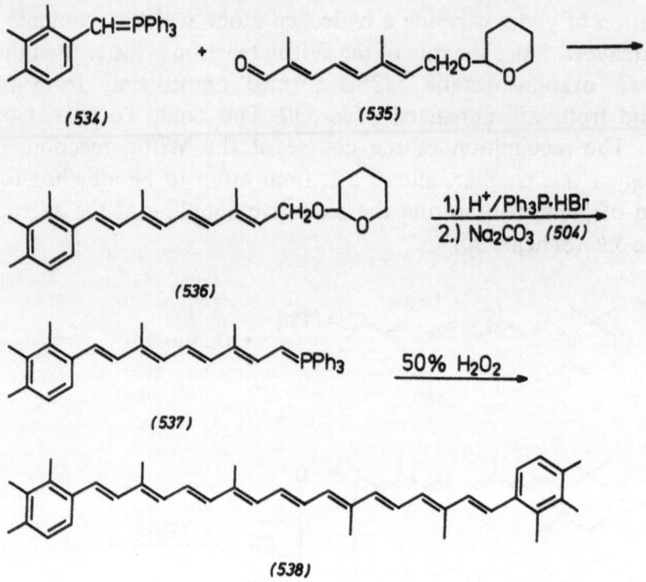
CH=PPh3
+
O
CH2O
O
(534)
(535)
1.) H+/Ph3P·HBr
2.) Na2CO3 (504)
CH2O
O
(536)
PPh3
50% H2O2
(537)
(538)

Scheme 91

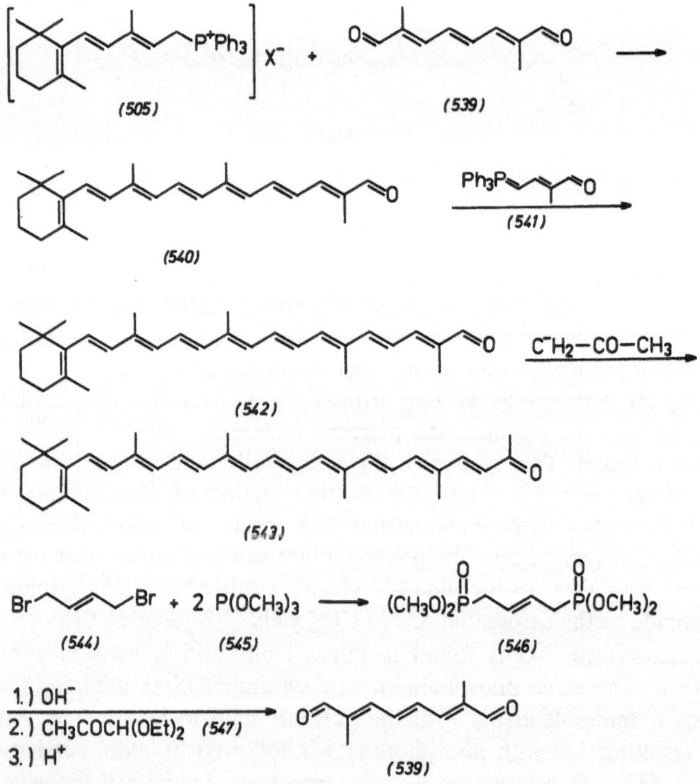
P+Ph3
X−
+
O
O
(505)
(539)
O
Ph3P
O
(541)
(540)
O
C−H2—CO—CH3
(542)
O
(543)
Br
Br
+ 2 P(OCH3)3
(CH3O)2P
O
O
P(OCH3)2
(544)
(545)
(546)
1.) OH−
2.) CH3COCH(OEt)2 (547)
3.) H+
O
O
(539)

Scheme 92

phosphite *545* to the bisphosphonate *546*. The double anion of the latter is converted with the methylglyoxal diethyl acetal *547* into the bisacetal of the C_{10}-dialdehyde *539* [264] (Scheme 92).

The synthetic principle of olefinating an α,ω-bisfunctional synthon *A*, representing the later molecular centre of the carotenoid, with two reaction partners *B*, the future molecular ends, can easily be described by the equation $A + 2B \rightarrow B—A—B$. This method was used by Mayer et al. [265] in the synthesis of rhodoxanthin *550*, a ketonic carotenoid, which was isolated for the first time from the red berrys of the yew-tree *Taxus baccata*. The authors describe the conversion of two equivalents of the stable phosphorane *548* with one equivalent of the C_{12}-dialdehyde *549* [265] (Scheme 93). The analogous reaction of crocetindial *552* with substituted benzylidenephosphoranes such as *551* gives carotenoids such as isorenieratene *553* [266,267] and related compounds [266] (Scheme 93). Condensation of the phosphoranes *554* and *555* with the C_{10}-dialdehyde *539* yields carotenoids of the dihydro type *556* [268] and such like *557* [269], respectively (Scheme 93).

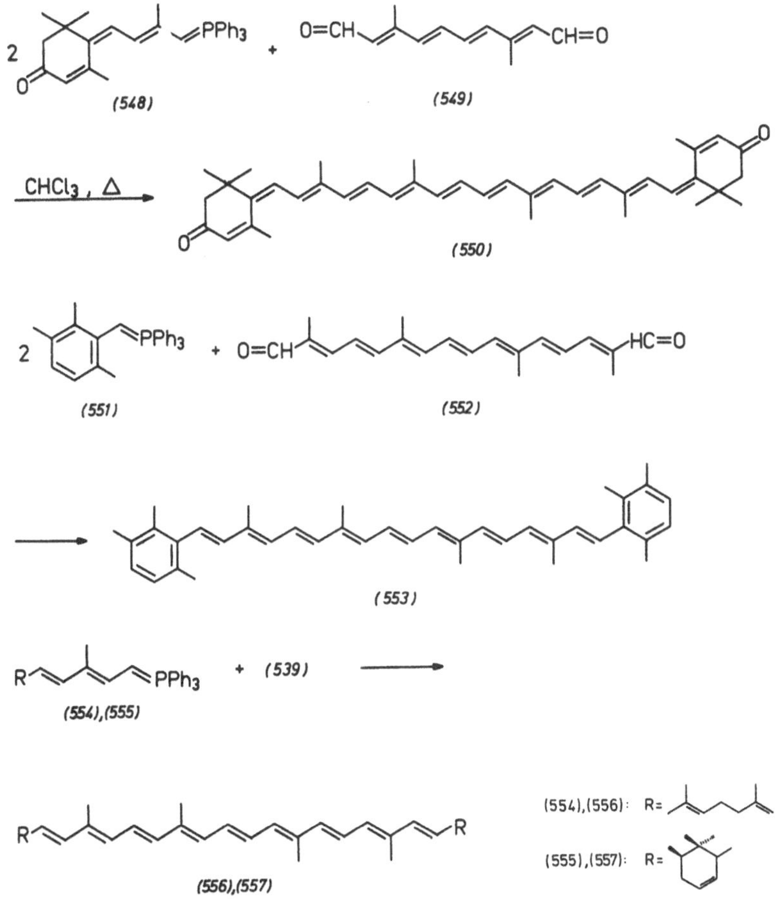

Scheme 93

The C_{10}-dialdehyde *539* serves as the central olefination synthon in the preparation of the symmetrical carotenoid zeaxanthin *560* [270]. In this synthesis the phosphonium salt *559* which can easily be prepared from the vinyl hydroxyionol *558* and triphenylphosphine hydrobromide, *504*, is reacted with *539* in 1,2-epoxybutane, a solvent which seems to be especially suitable for Wittig reactions with polyene dialdehydes [270]. The same phosphonium salt *559* was used in the synthesis of β-cryptoxanthin and zeinoxanthin.

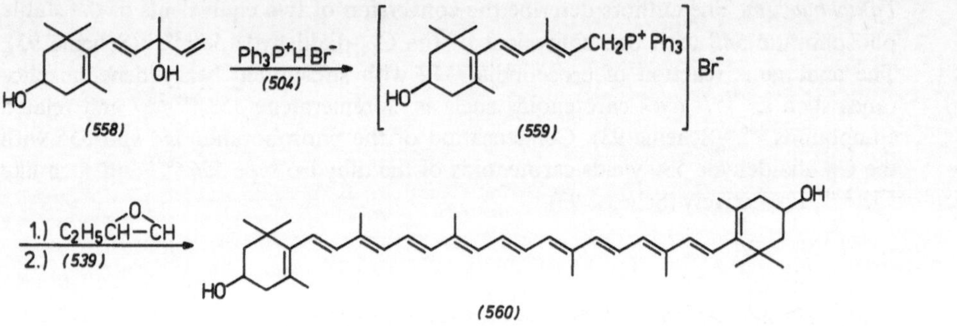

Scheme 94

Scheme 95 describes in principle the same synthesis technique using a bisphosphonium salt *562* as the central molecular part of the molecules *563* and *565*. Olefination of the ylide generated from *562* with two equivalents of the ketoaldehyde *561* yields ketocarotenoid *563* [271]. Unsymmetrical ketocarotenoid *565* has been synthesized by the Wittig reaction of *562* with a mixture of aldehydes *564* and *511* and subsequent hydrolysis of the acetal protective group [271] (Scheme 95).

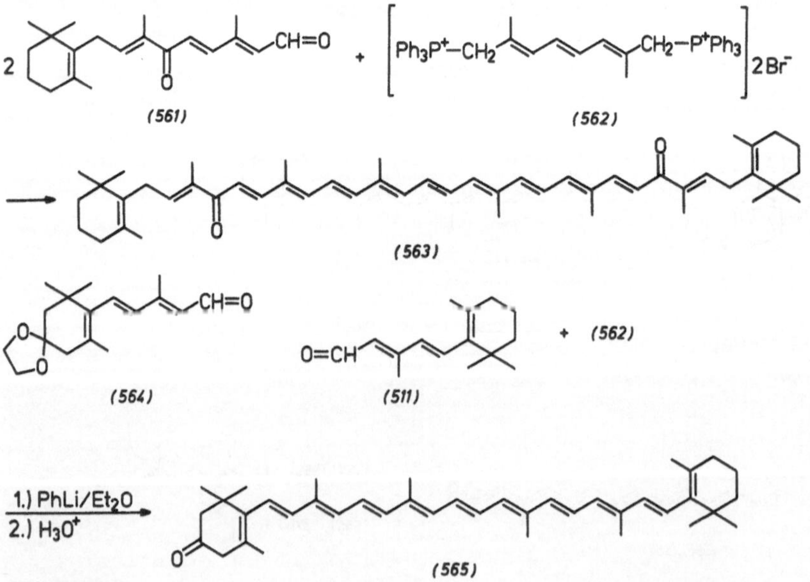

Scheme 95

Similar reactions lead to dehydroflexixanthin *566* [272] and β,β-carotene-2,2'-diol *567* [273]. Polyenes like *570* used in the combat against tumor, can be obtained from the phosphonium salt *568* and the formylpolyene ester *569* [274]. Applying the Wittig reaction, also branched unsaturated cross-conjugated carotenals *571* [275] and the polyene derivative *572* [276] can be prepared (Scheme 96).

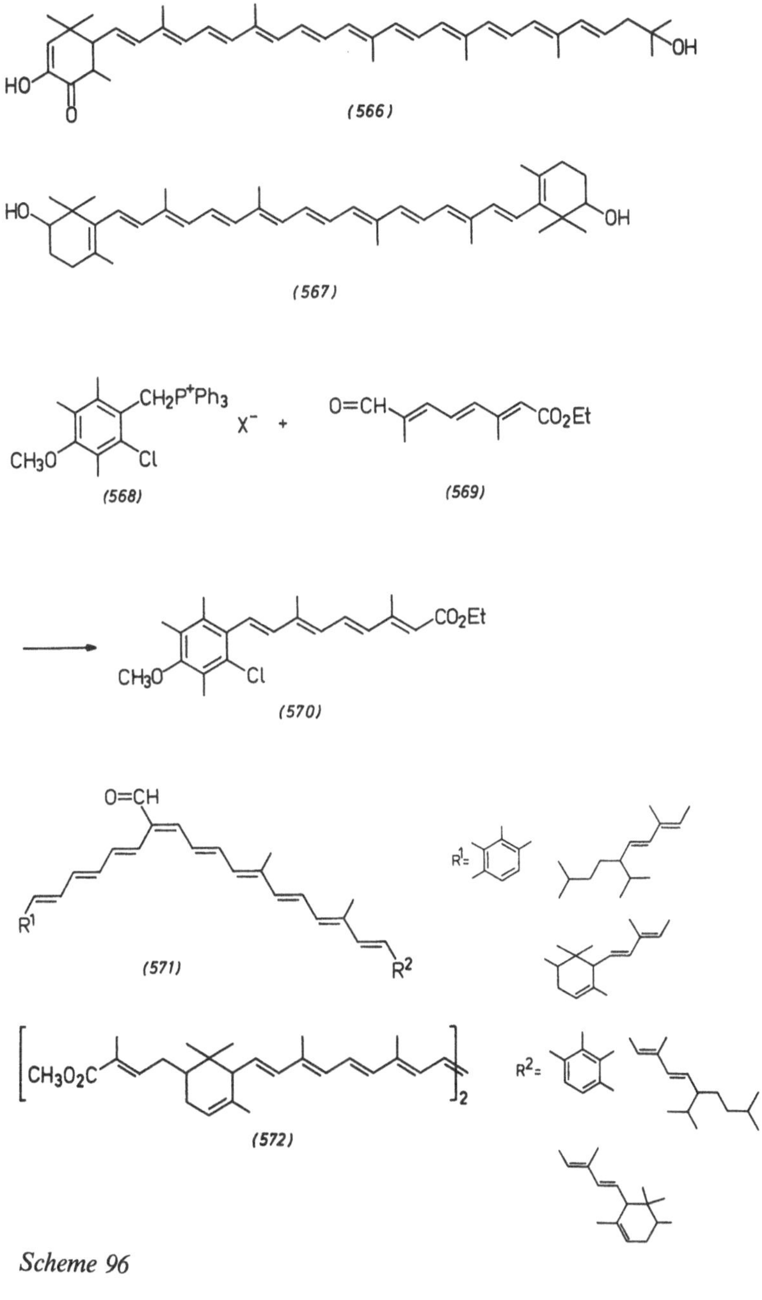

(*566*)

(*567*)

(*568*)   (*569*)

(*570*)

(*571*)

(*572*)

Scheme 96

153

The total synthesis of the symmetrical carotenoid (3S,3'S)-astaxanthin *576*, described by Kienzle and Mayer [277] also uses an α,ω-bifunctional synthon as the molecular center. These authors olefinated the dialdehyde *539* with the ylide generated from the optically active phosphonium salt *575*. Under the reaction conditions chosen cleavage of the phenoxyacetate group occurs simultanously with the formation of the desired polyene *576* [277] (Scheme 97). The chiral (4R,6R)-4-hydroxy--2,2,6-trimethylcyclohexanone *573*, which could be obtained from isophorone, is used as the starting material. Isomerization of the C-7 double bond of *574* is of interest in this synthesis. This bond, originally exhibiting cis-geometry due to partial hydrogenation of an alkyne intermediate by means of Lindlar catalyst, changes its configuration during treatment of the halogenide *574* with triphenylphosphine *27*, forming the trans-phosphonium salt *575*. Thus, the all-*trans* compound *576* is obtained [277] (Scheme 97).

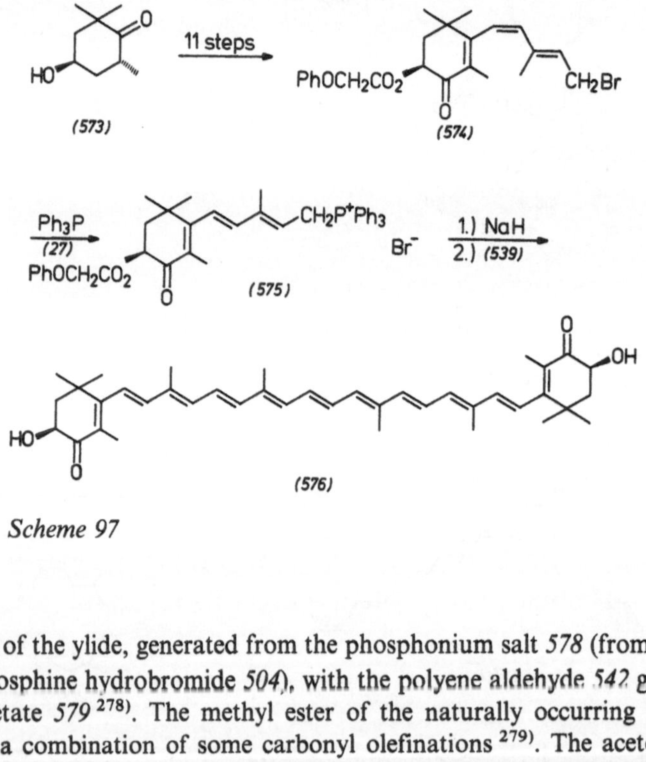

Scheme 97

Reaction of the ylide, generated from the phosphonium salt *578* (from the alcohol *577* and phosphine hydrobromide *504*), with the polyene aldehyde *542* gives aleuriaxanthin acetate *579* [278]. The methyl ester of the naturally occurring bixin *586* is formed by a combination of some carbonyl olefinations [279]. The acetoxyaldehyde *580* is olefinated with methoxycarbonylmethylene-triphenylphosphorane *67* to the (E)-unsaturated ester *581*. The latter is converted into the phosphonium salt *582* upon treatment with triphenylphosphine hydrobromide *504*. The corresponding ylide of *582* is reacted with the dialdehyde *539* to the polyene aldehyde ester *583*. The latter is reduced and converted into phosphonium salt *584*. The corresponding ylide is now reacted in a third carbonyl olefination with *585* to give the methyl ester *586* [279] (Scheme 98).

154

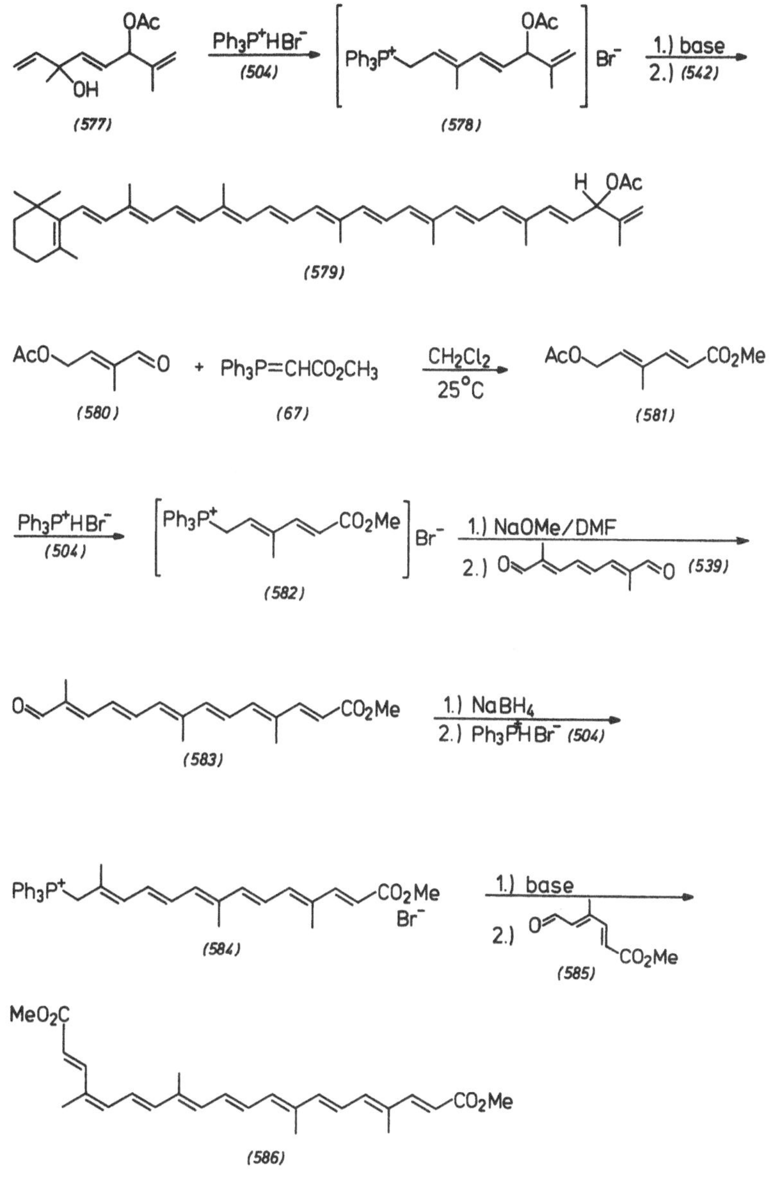

Scheme 98

The synthesis of tedanin *589*, an isoprenoid pigment of the marine sponge *Tedanis digitata*, starts with the Wittig reaction of phosphorane *506* with the carbonyl compound *587*, giving *588*. Bromination of the latter followed by treatment with KOH/methanol and twofold oxidation yield tedanin *589* [280] (Scheme 99).

Numerous other syntheses of carotenoids making use of the Wittig reaction have yielded β,γ-carotene, optically active γ,γ-carotene [281], (+)-α-carotene [282], and some

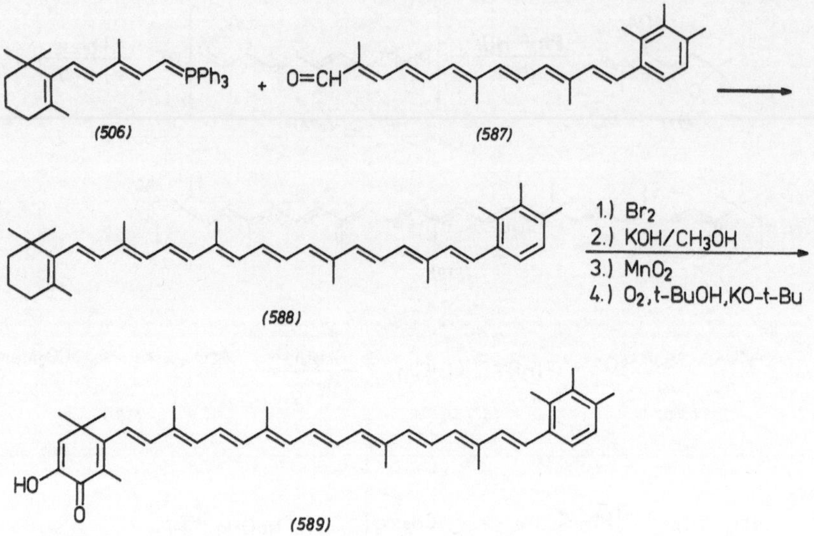

Scheme 99

specifically deuterated carotenes [283,284] such as 11,11-dideutero-ε-carotene and (19,19′-$^{2}H_6$)-β-carotene [283]. Furthermore, ylides were used for the synthesis of 3,3′,4,4′-bisdehydro-β-carotene [285] and (2*R*,2′*R*)-2,2′-dimethyl-β,β′-carotene [286]. The Wittig reaction was also applied in the synthesis of 7,8-didehydroisorenieratene [287], 7,8-didehydrorenieratene [287], and 3,3-dihydroxyisorenieratene [288]. Using ylides, ^{14}C-labelled (2-^{14}C)-abscisic acid [240], methyl *trans*-(10-^{14}C)-retenoate [289] and *trans*-(10-^{14}C)-retinol were prepared [289]. The higher methoxylated aromatic carotinoids [290], lycopen-20-al and rhodopen-20-al [291] can be obtained *via* ylide reactions.

Wittig reactions also were used for the synthesis of polyenic, nonisoprenoid ether-pigments like e.g. *590* [292]. Olefination was the key reaction in the preparation of the polyacetylenes of the fungus *Fistulina pallida* [293] and *Trachelium caeruleum* [294] as well as in the synthesis of the geometrical isomers of the alkatriene $C_5H_{11}\,(CH{=}CH)_3H$ isolated from the essential oil of *Galbanum* and the Hawaian seaweed *Dictyopteris* [295] and of the related polyene *591* isolated from the alga *Fucus vesiculosus* [296] (Scheme 100).

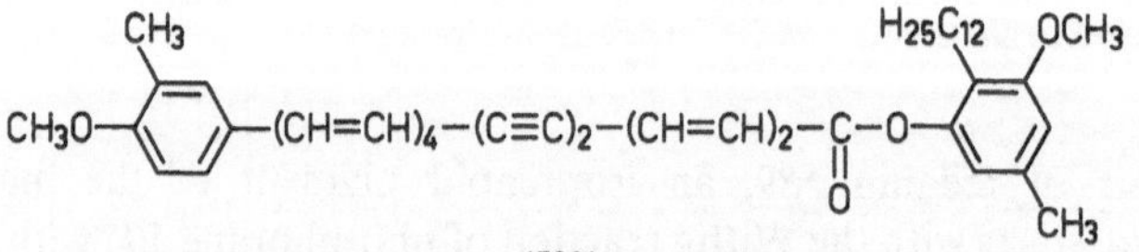

(590)

$$CH_3-(CH_2)_3-(CH_2-CH{=}CH)_4-(CH_2)_3-CH{=}CH_2$$

(591)

Scheme 100

10 Concluding Remarks

This article dealing with selected topics only and not claiming to be complete should demonstrate, on the occasion of *Prof. Dr. G. Wittig's* 85th birthday, the importance of the discovery of the reaction, named after the celebrant, for the development of the synthesis of natural products. Of further importance seems to us the possibility of varying the structure of biologically active compounds by means of the *Wittig* olefination. This enables systematic studies to be made on structure-activity relationships and may thus contribute to the elucidation of the mechanism of molecular interactions between substrate and receptor systems. Today, modern chemistry of natural products cannot be limited to the isolation and synthesis of the latter but should also be concerned with questions of the operational mechanisms of biologically active natural products.

11 References

1. G. Wittig, G. Geißler: Liebigs Ann. Chem. *580*, 44 (1953)
2. G. Wittig, U. Schöllkopf: Chem. Ber. *87*, 1318 (1954)
3. H. Pommer: Angew. Chem. *72*, 910 (1960)
4. H. J. Bestmann: Pure Appl. Chem. *51*, 515 (1979)
5. O. Isler: Carotenoids, Basel, Birkhäuser Verlag 1971
6. H. Pommer: Angew. Chem. *89*, 437 (1977); Angew. Chem. Intern. Ed. Engl. *16*, 423 (1977)
7. H. J. Bestmann: Pure Appl. Chem. *52*, 77 (1980)
8. H. J. Bestmann, et al.: Angew. Chem. *91*, 945 (1979); Angew. Chem. Intern. Ed. Engl. *18*, 876 (1979)
9. H. J. Bestmann, et al.: in preparation
10. M. Schlosser, in: Topics in Stereochemistry (eds.) E. L. Eliel, N. L. Allinger, Vol. 5, 13, New York, Intersci. Publ. 1970
11. M. Schlosser, K. F. Christmann: Liebigs Ann. Chem. *708*, 1 (1967)
12. E. Vedejs, K. A. J. Snoble: J. Am. Chem. Soc. *95*, 5778 (1973)
13. H. J. Bestmann, W. Downey, K. Geibel: in preparation
14. M. Schlosser, et al.: Chimia *29*, 341 (1975)
15. M. Schlosser, H. Ba Tuong: Angew. Chem. *91*, 675 (1979); Angew. Chem. Intern. Ed. Engl. *18*, 633 (1979)
16. F. Ramirez, C. P. Smith, J. F. Pilot: J. Am. Chem. Soc. *90*, 6726 (1968)
17. F. Ramirez: Bull. Soc. Chim. France *1970*, 3491
18. P. Gillespie, et al.: Angew. Chem. *83*, 691 (1971); Angew. Chem. Intern. Ed. Engl. *10*, 687 (1971)
19. D. Marquarding, et al.: Angew. Chem. *85*, 99 (1973); Angew. Chem. Intern. Ed. Engl. *12*, 91 (1973)
20. F. H. Westheimer: Accounts Chem. Res. *1*, 70 (1968)
21. H. J. Bestmann, et al.: J. Chem. Soc., Chem. Commun. *1980*, 978
22. R. Höller, H. Lischka: J. Am. Chem. Soc. *102*, 4632 (1980)
23. E. Vedejs, G. P. Meier, K. A. J. Snoble: J. Am. Chem. Soc. *103*, 2823 (1981)
24. H. B. Bürgi, et al.: Tetrahedron *30*, 1563 (1974)
25. H. J. Bestmann, et al.: unpublished results
26. M. Schlosser, in: Methodicum Chimicum, Vol. 7, 529, Stuttgart, Georg Thieme Verlag 1976
27. H. J. Bestmann, O. Kratzer: Chem. Ber. *95*, 1894 (1962)
28. L. D. Bergelson, L. J. Barsukow, M. M. Shemyakin: Tetrahedron Lett. *1967*, 2709
29. D. J. Burton, P. E. Greenlimb: J. Org. Chem. *40*, 2796 (1975)
30. H. O. House, V. K. Jones, G. A. Frank: J. Org. Chem. *29*, 3327 (1964)
31. G. Wittig, H. Eggers, P. Duffner: Liebings Ann. Chem. *619*, 10 (1958)

32. H. J. Bestmann: Angew. Chem. 77, 609, 651, 850 (1965); Angew. Chem. Intern. Ed. Engl. 4, 583, 645, 830 (1965); together edited in: Neuere Methoden der präparativen organischen Chemie (ed.) W. Foerst, Vol. 5, Weinheim, Verlag Chemie 1967
33. H. J. Bestmann, W. Stransky, O. Vostrowsky: Chem. Ber. 109, 1694 (1976)
34. H. J. Bestmann, W. Stransky: Synthesis 1974, 798
35. H. J. Bestmann, et al.: Chem. Ber. 108, 3582 (1975)
36. M. E. Jones, S. Tripett: J. Chem. Soc. (C) 1966, 1090
37. B. Giese, J. Schoch, C. Rüchard: Chem. Ber. 111, 1395 (1978)
38. H. J. Bestmann, P. Rösel: unpublished results; P. Rösel: thesis, University of Erlangen—Nürnberg 1978
39. H. J. Bestmann, A. Bomhard: unpublished results; A. Bomhard: thesis University of Erlangen—Nürnberg 1982
40. W. P. Schneider: J. Chem. Soc., Chem. Commun. 1969, 785
41. U. Axen, F. H. Lincoln, J. L. Thompson: J. Chem. Soc., Chem. Commun. 1969, 303
42. M. Schlosser, K. F. Christmann: Angew. Chem. 78, 115 (1966); Angew. Chem. Intern. Ed. Engl. 5, 126 (1966)
43. M. Schlosser, K. F. Christmann, A. Piskala: Chem. Ber. 103, 2814 (1970)
44. E. J. Corey, H. Yamamoto: J. Am. Chem. Soc. 92, 226 (1970)
45. E. J. Corey, P. Ulrich, A. Venkateswarbu: Tetrahedron Lett. 1977, 3231
46. M. Schlosser, K. F. Christmann: Synthesis 1969, 38
47. L. Horner, H. Hoffmann, H. G. Wippel: Chem. Ber. 91, 61 (1958)
48. W. S. Wadsworth, jr., W. D. Emmons: J. Am. Chem. Soc. 83, 1733 (1961); Org. Synthesis 45, 44 (1965)
49. L. D. Bergelson, M. M. Shemyakin: Angew. Chem. 76, 113 (1964); Angew. Chem. Intern. Ed. Engl. 3, 250 (1964)
50. L. D. Bergelson, M. M. Shemyakin, in: The Chemistry of Carboxylic Acids and Esters, p. 295, London, Intersci. Publ. 1969
51. A. S. Kovaleva, et al.: Zh. Org. Khim. 10, 696 (1974); C.A. 81, 37206 (1974)
52. G. Pattenden, B. C. L. Weedon: J. Chem. Soc. (C) 1968, 1984
53. D. L. Bergelson, et al.: Zh. Obshch. Khim. 32, 1802 (1962); C.A. 58, 4415 (1963)
54. L. D. Bergelson, et al.: Izv. Akad. Nauk SSSR, Ser. Khim. 59, 1417 (1963); C.A. 59, 15176 (1963)
55. J. M. Osbond: Progr. Chem. Fats Lipids 9, 121 (1966)
56. W. H. Kunau: Chem. Phys. Lipids 11, 254 (1973)
57. H. J. Bestmann, et al.: Liebigs Ann. Chem. 1981, 1705
58. H. J. Bestmann, O. Vostrowsky: Chem. Phys. Lipids 24, 335 (1979)
59. H. J. Bestmann, K. H. Koschatzky, O. Vostrowsky: Chem. Ber. 112, 1923 (1979)
60. H. J. Bestmann, P. Rösel, O. Vostrowsky: Liebigs Ann. Chem. 1979, 1189
61. H. J. Bestmann, et al.: Chem. Ber. 111, 248 (1978)
62. H. J. Bestmann, R. Wax, O. Vostrowsky: Chem. Ber. 112, 3740 (1979)
63. For the Hofmann degradation of phosphonium salts see H. J. Bestmann, H. Häberlein, I. Pils: Tetrahedron 20, 2079 (1964)
64. H. S. Corey, jr., J. R. D. McCormick, W. E. Swensen: J. Am. Chem. Soc. 86, 1884 (1964)
65. A. Jurasek, et al.: Z. Pr. Chemickotechnol. Fak. SVST 1972 (publ. 1974), 67; C.A. 82, 169954 (1975)
66. L. D. Bergelson, et al.: Zh. Obshch. Khim. 32, 1807 (1962); C.A. 58, 4416 (1963)
67. E. Ucciani, Y. Bensimon, P. Ranguis: Actes Congr. Mond. — Soc. Int. Etude Corps Gras (eds.) M. Nandet, M. Ucciani, A. Uzzan, 13e, Sect. C, 43, ITERG, Paris (1976)
68. L. D. Bergelson, V. D. Solodovnik, M. M. Shemyakin: Izv. Akad. Nauk SSSR, Ser. Khim. 1967, 843; C.A. 67, 99692 (1967)
69. H. J. Bestmann, et al.: Tetrahedron Lett. 1977, 121
70. H. J. Bestmann, J. Süß, O. Vostrowsky: Liebigs Ann. Chem. 1981, 2117
71. R. W. Bradshaw, et al.: J. Chem. Soc. (C) 1971, 1156
72. L. D. Bergelson, V. A. Vaver, M. M. Shemyakin: Izv. Akad. Nauk. SSSR, Otd. Khim. Nauk. 1962, 1894; C.A. 58, 4416 (1963)
73. R. Klok, et al.: J, Royal Netherlands Chem. Soc. 99, 132 (1980)
74. J. G. Gleason, D. Boles Bryan, C. M. Kinzig: Tetrahedron Lett. 21, 1129 (1980)

75. E. J. Corey, et al.: J. Am. Chem. Soc. *102*, 1436 (1980)
76. E. J. Corey, et al.: J. Am. Chem. Soc. *102*, 7984 (1980)
77. J. Rokach, et al.: Tetrahedron Lett. *21*, 1485 (1980)
78. B. Samuelsson, G. Ställberg: Acta Chem. Scand. *17*, 810 (1963)
79. J. F. Bagli, et al.: Tetrahedron Lett. *1966*, 465
80. Beal III P. F., J. C. Babcock, F. H. Lincoln: J. Am. Chem. Soc. *88*, 3131 (1966)
81. Beal III P. F., J. C. Babcock, F. H. Lincoln: Proc. 2nd Nobel Symp. (eds.) S. Bergström, B. Samuelsson, The Prostaglandins, Stockholm, Almquist and Wicksell 1967
82. G. Just, C. Simonovitch: Tetrahedron Lett. *1967*, 2093
83. K. G. Holden, et al.: Tetrahedron Lett. *1968*, 1569
84. G. Just, et al.: J. Am. Chem. Soc. *91*, 5364 (1969)
85. J. E. Pike, F. H. Lincoln, W. P. Schneider: J. Org. Chem. *34*, 3552 (1969)
86. U. Axen, F. H. Lincoln, J. L. Thompson: J. Chem. Soc., Chem. Commun. *1969*, 303
87. W. P. Schneider, et al.: J. Am. Chem. Soc. *90*, 5895 (1968)
88. G. Bundy, et al., in: Prostaglandins (eds.) P. Ramwell, J. E. Shaw, Ann. N.Y. Acad. Sci. *180*, 76 (1971)
89. E. J. Corey: ibid. *180*, 24 (1971)
90. E. J. Corey, et al.: J. Am. Chem. Soc. *90*, 3247 (1968)
91. E. J. Corey, et al.: J. Am. Chem. Soc. *91*, 5675 (1969)
92. E. J. Corey, T. Ravindranathan, S. Tereshima: J. Am. Chem. Soc. *93*, 4326 (1971)
93. E. J. Corey, R. Noyori, T. K. Schaaf: J. Am. Chem. Soc. *92*, 2586 (1970)
94. J. S. Bindra, R. Bindra: Prostaglandin Synthesis, New York, Academic Press 1977
95. A. Mitra: The Synthesis of Prostaglandins, New York, Wiley 1977
96. P. H. Bentley: Chem. Soc. Rev. *2*, 29 (1973)
97. M. Hayashi, H. Miyake: Japan Kokai 74 116,068; C.A. *82*, 170200 (1975)
98. P. Bollinger: Ger. Offen. 2,431,930; C.A. *83*, 58248 (1975)
99. H. J. E. Hess, L. J. Czuba, T. K. Schaaf: U.S. Patent 3,928,391; C.A. *84*, 121292 (1976)
100. Y. Iguchi, S. Kori, M. Hayashi: J. Org. Chem. *40*, 521 (1975); M. Hayashi, S. Kori, Y. Iguchi: Japan Kokai 75 35,133; C.A. *84*, 43430 (1976)
101. N. A. Nelson, R. W. Jackson: Tetrahedron Lett. *1976*, 3275
102. E. J. Corey et al.: Tetrahedron Lett. *1977*, 785; E. J. Corey, M. Shibasaki, J. Knolle: Tetrahedron Lett. *1977*, 1625
103. S. Hanessian, P. Lavallee: Canad. J. Chem. *55*, 562 (1977)
104. M. Brawner Floyd: Synth. Commun. *4*, 317 (1974)
105. E. J. Corey, G. Moinet: J. Am. Chem. Soc. *95*, 6831 (1973)
106. E. J. Corey, G. Moinet: J. Am. Chem. Soc. *95*, 7185 (1973); P. Crabbé, A. Guzman, M. Vera: Tetrahedron Lett. *1973*, 3021
107. W. Bartmann, G. Beck, U. Lerch: Tetrahedron Lett. *1974*, 2441
108. C.-L. J. Wang, P. A. Grieco, F. J. Okuniewicz: J. Chem. Soc., Chem. Commun. *1976*, 468, 639
109. C. Gandolfi, et al.: Farmaco Ed. Sci. *31*, 763 (1976)
110. J. Fried, et al., in: Advances in Prostaglandin and Thromboxane Research (eds.) B. Samuelsson, R. Paoletti, Vol. 1, p. 173, New York, Raven 1976
111. H. Miyake, et al.: Chem. Lett. *1976*, 211
112. G. J. Lourens, J. M. Koekemoer: Ger. Offen. 2,618,861; C.A. *86*, 89589 (1977), cf. Tetrahedron Lett. *1975*, 3719
113. Hoechst AG, Neth. Appl. 7512,794; C.A. *86*, 89596 (1977)
114. A. Sugie, et al.: Tetrahedron Lett. *1977*, 2759
115. K. Kojima, K. Sakai: Tetrahedron Lett. *1976*, 101
116. E. J. Corey, K. Narasaka, M. Shibasaki: J. Am. Chem. Soc. *98*, 6417 (1976)
117. N. Nakamura, K. Sakai: Tetrahedron Lett. *1976*, 2049
118. M. Hayashi, S. Kori, H. Miyake: Japan Kokai 75 49,259; C.A. *83*, 205824 (1975)
119. P. A. Grieco, C. S. Pogonowski, M. Miyashita: J. Chem. Soc., Chem. Commun. *1975*, 592
120. M. J. Dimsdale, et al.: J. Chem. Soc., Chem. Commun. *1977*, 716
121. A. E. Greene, et al.: Tetrahedron Lett. *1976*, 3755
122. E. J. Corey, et al.: Tetrahedron Lett. *1976*, 737
123. N. S. Crossley: Tetrahedron Lett. *1971*, 3327
124. N. Hamanaka, H. Nakai, M. Kuruno: Bull. Chem. Soc. Japan *53*, 2327 (1980)

125. K. C. Nicolaou, R. L. Magolda, D. A. Claremon: J. Am. Chem. Soc. *102*, 1404 (1980)
126. P. Barraclough: Tetrahedron Lett. *21*, 1897 (1980)
127. N. Finch, J. J. Fitt: Tetrahedron Lett. *1969*, 4639
128. R. B. Morin, et al.: Tetrahedron Lett. *1968*, 6023
129. M. Miyano, C. R. Dorn: Tetrahedron Lett. *1969*, 1615
130. M. Miyano, et al.: J. Chem. Soc., Chem. Commun. *1971*, 425; cf. also M. Miyano, M. A. Stealy: J. Org. Chem. *40*, 1748 (1975)
131. G. Ambrus, et al.: Hung. Teljes *11*, 745 (1977); C.A. *86*, 16352 (1977)
132. J. Himuzu, et al.: Japan Kokai 76 01461; C.A. *85*, 123751 (1976)
133. F. Kienzle, R. E. Minder: Helv. Chim. Acta *63*, 1425 (1980)
134. M. Hayashi, et al.: Japan Kokai 75 137,961; C.A. *84*, 164263 (1976)
135. E. J. Corey, et al.: J. Am. Chem. Soc. *93*, 1490 (1971)
136. J. Ernest: Angew. Chem. *88*, 244 (1976); Angew. Chem. Intern. Ed. Engl. *15*, 207 (1976); W. Bartmann: Angew. Chem. *87*, 143 (1975); Angew. Chem. Intern. Ed. Engl. *14*, 337 (1975)
137. H. Röller, et al.: J. Insect Physiol. *15*, 379 (1969)
138. H. Röller, et al.: Angew. Chem. *79*, 190 (1967); Angew. Chem. Intern. Ed. Engl. *6*, 179 (1967)
139. A. S. Meyer, et al.: Arch. Biochem. Biophys. *137*, 190 (1970)
140. K. J. Judy, et al.: Proc. Nat. Acad. Sci. USA *70*, 1509 (1973)
141. C. M. Williams: Scientific American *217*, July 13, 1967
142. K. H. Dahm, B. M. Trost, H. J. Röller: J. Am. Chem. Soc. *89*, 5292 (1967)
143. For the reaction of phosphorus ylides and halogen compounds cf. ref. [32]
144. B. H. Braun, et al.: J. Econ. Entomol. *61*, 866 (1968)
145. G. W. K. Cavill, D. G. Laing, P. J. Williams: Austral. J. Chem. *22*, 2145 (1969)
146. W. Hoffmann, H. Pasedach, H. Pommer: Liebigs Ann. Chem. *729*, 52 (1969)
147. H. Schulz, I. Sprung: Angew. Chem. *81*, 258 (1969); Angew. Chem. Intern. Ed. Engl. *8*, 271 (1969)
148. J. A. Findlay, W. D. MacKay: J. Chem. Soc., Chem. Commun. *1969*, 733
149. J. A. Findlay, W. D. MacKay: J. Chem. Soc. (C) *1970*, 2631
150. J. S. Cochrane, J. R. Hanson: J. Chem. Soc., Perkin I, *1972*, 361
151. E. J. Corey, et al.: J. Am. Chem. Soc. *90*, 5618 (1968)
152. E. J. Corey, H. Yamamoto: J. Am. Chem. Soc. *92*, 226, 3523 (1970)
153. E. J. Corey, J. I. Shulman, H. Yamamoto: Tetrahedron Lett. *1970*, 447
154. E. J. Corey, H. Yamamoto: J. Am. Chem. Soc. *92*, 6636 (1970)
155. E. J. Corey, H. Yamamoto: J. Am. Chem. Soc. *92*, 6637 (1970)
156. C. A. Henrick, F. Schaub, J. B. Siddall: J. Am. Chem. Soc. *94*, 5374 (1972)
157. R. J. Anderson, et al.: J. Am. Chem. Soc. *94*, 5374 (1972)
158. G. Ohloff: Fortschr. Chem. org. Naturst. *35*, 431 (1978)
159. H. J. Bestmann, O. Vostrowsky: unpublished results
160. H. J. Bestmann: Chem. Ber. *95*, 58 (1962)
161. D. E. Heinz, W. G. Jennings: J. Food Sci. *31*, 69 (1966)
162. G. Ohloff, M. Pawlak: Helv. Chim. Acta *56*, 1176 (1973)
163. M. Baumann, W. Hoffmann: DOS 2534859, BASF (1975); M. Baumann, W. Hoffmann: Synthesis *1977*, 681
164. M. J. Devos, et al.: Tetrahedron Lett. *1977*, 3911
165. H. J. Bestmann, J. Süß: Liebigs Ann. Chem. *1982*, 363
166. H. J. Bestmann, K. Roth, M. Ettlinger: Angew. Chem. *91*, 748 (1979); Angew. Chem. Intern. Ed. Engl. *18*, 687 (1979)
167. H. J. Bestmann, K. Roth, M. Ettlinger: Chem. Ber. *115*, 161 (1982)
168. H. J. Bestmann, O. Vostrowsky, in: Chemie der Pflanzenschutz- und Schädlingsbekämpfungsmittel (ed.) R. Wegler, Vol. 6, p. 29, Berlin, Springer-Verlag 1980
169. N. Petragnani, G. Schill: Chem. Ber. *97*, 3293 (1964): L. B. Hendry, et al.: Experientia *30*, 886 (1974)
170. A. S. Kovaleva et al.: Zh. Org. Chem. *8*, 2613 (1972)
171. O. Vostrowsky, H. J. Bestmann: Mitt. dtsch. Ges. allg. angew. Entomol. *2*, 252 (1981)
172. O. Vostrowsky, H. J. Bestmann, E. Priesner: Nachr. Chem. Techn. *21*, 501 (1973)

173. E. Priesner, M. Jacobson, H. J. Bestmann: Z. Naturforsch. *30c*, 283 (1975)
174. E. Priesner, et al.: Z. Naturforsch. *32c*, 979 (1977)
175. B. A. Bierl, M. Beroza, C. W. Collier: Science *170*, 87 (1970)
176. a) R. T. Cardé, W. L. Roelofs, C. C. Doane: Nature *241*, 474 (1973);
     b) B. A. Bierl, M. Beroza, W. C. Collier: J. Econ. Entomol. *65*, 659 (1972)
177. H. J. Bestmann, O. Vostrowsky: Tetrahedron Lett. *1974*, 207; H. J. Bestmann, O. Vostrowsky,
     W. Stransky: Chem. Ber. *109*, 3375 (1976)
178. H. J. Bestmann, et al.: Liebigs Ann. Chem. *1982*, 536
179. H. J. Bestmann, K. H. Koschatzky, O. Vostrowsky: Liebigs Ann. Chem. *1982*, 1478
180. A. Butenandt, et al.: Z. Naturforsch. *14b*, 283 (1959)
181. G. Kasang, et al.: Angew. Chem. *90*, 74 (1978); Angew. Chem. Int. Ed. Engl. *17*, 60
     (1978)
182. B. F. Nesbitt, et al.: Nature, New Biol. *244*, 208 (1973)
183. Y. Tamaki, H. Noguchi, T. Yushima: Appl. Entomol. Zool. *8*, 200 (1973)
184. a) W. L. Roelofs, et al.: Mitt. Schweiz. Entomol. Ges. *46*, 71 (1973);
     b) R. Buser, H. Arn: J. Chromatogr. *106*, 83 (1975)
185. a) W. L. Roelofs, et al.: Science *174*, 297 (1971);
     b) H. Beroza, B. A. Bierl, H. R. Moffitt: Science *183*, 89 (1974)
186. A. Butenandt, E. Hecker: Angew. Chem. *73*, 349 (1961)
187. A. Butenandt, et al.: Liebigs Ann. Chem. *658*, 39 (1962)
188. H. J. Bestmann, J. Süß, O. Vostrowsky: Tetrahedron Lett. *1979*, 2467
189. H. J. Bestmann, J. Süß, O. Vostrowsky: Tetrahedron Lett. *1978*, 3329
190. H. J. Bestmann, et al.: unpublished results
191. C. A. Henrick, M. A. Geigel, W. E. Willy: unpublished results, cited in C. A. Henrick: Tetra-
     hedron *33*, 1845 (1977)
192. D. R. Hall, et al.: Chem. Ind. (London) *1975*, 216
193. G. Goto, et al.: Chem. Lett. *1975*, 103
194. C. Descoins, C. A. Henrick: unpublished results, cited in ref. [191]
195. B. F. Nesbitt, et al.: Tetrahedron Lett. *1973*, 4669
196. R. E. Doolittle, et al.: J. Chem. Ecol. *2*, 399 (1976)
197. M. D. Chisholm, et al.: Chem. Ecol. *7*, 159 (1981)
198. H. J. Bestmann, et al.: Liebigs Ann. Chem. *1982*, 1356
199. M. Julia, S. Julia, R. Guegan: Bull. Soc. Chim. France *1960*, 1072
200. H. J. Bestmann, A. Plenchette, O. Vostrowsky: Tetrahedron Lett. *1974*, 779
201. J. S. Read, P. S. Beevor: J. Stored Prod. Res. *12*, 55 (1976)
202. E. Sonnet: J. Org. Chem. *39*, 3793 (1974)
203. H. J. Bestmann, et al.: Tetrahedron Lett. *1976*, 353
204. R. J. Anderson, C. A. Henrick: J. Am. Chem. Soc. *97*, 4327 (1975)
205. A. Hammond, C. Descoins: Bull. Soc. Chim. France *1978*, 299
206. H. J. Bestmann, et al.: Tetrahedron Lett. *23*, 4007 (1982)
207. H. J. Bestmann, et al.: in preparation
208. K. Mori: Tetrahedron Lett. *1975*, 2187
209. K. Mori: Tetrahedron *32*, 1101 (1976)
210. K. Mori: Tetrahedron Lett. *1976*, 1609
211. K. Mori: Tetrahedron *32*, 1979 (1976)
212. K. Mori: Tetrahedron *31*, 3011 (1975)
213. P. D. Hobbs, P. D. Magnus: J. Chem. Soc., Chem. Commun. *1974*, 856
214. P. D. Hobbs, P. D. Magnus: J. Am. Chem. Soc. *98*, 4594 (1976)
215. R. Zurflüh, et al.: J. Am. Chem. Soc. *92*, 425 (1970)
216. H. Kosugi, et al.: Bull. Chem. Soc. Japan *49*, 520 (1976)
217. G. Stork, J. F. Cohen: J. Am. Chem. Soc. *96*, 5270 (1974)
218. J. H. Tumlinson, et al.: Science *197*, 789 (1977)
219. D. A. Carlson, et al.: Science *174*, 76 (1971)
220. D. A. Carlson, et al.: J. Agric. Food Chem. *22*, 194 (1974)
221. H. J. Bestmann, O. Vostrowsky, H. Platz: Chem.-Ztg. *98*, 161 (1974)
222. H. J. Bestmann, R. Kunstmann, H. Schulz: Liebigs Ann. Chem. *699*, 33 (1966)
223. G. Magnusson: Tetrahedron Lett. *1977*, 2713

224. R. Nishida, H. Fukami, S. Ishii: Experientia *30*, 978 (1974)

225. R. Nishida, H. Fukami, S. Ishii: Appl. Entomol. *10*, 10 (1975)

226. M. Schwarz, J. E. Oliver, P. E. Sonnet: J. Org. Chem. *40*, 2410 (1975)

227. L. D. Rosenblum, R. J. Anderson, C. A. Henrick: Tetrahedron Lett. *1976*, 419

228. A. W. Burgstahler, et al.: J. Org. Chem. *42*, 566 (1977)

229. S. B. Vinson, et al.: Entomol. Exp. Appl. *18*, 443 (1975)

230. L. Ruzicka: Experientia *9*, 357 (1953)

231. K. Sato, S. Inoue, S. Ota: J. Org. Chem. *35*, 565 (1970)

232. W. E. Bondinell, C. D. Snyder, H. Rapoport: J. Am. Chem. Soc. *91*, 6889 (1969)

233. E. J. Corey, H. Yamamoto: J. Am. Chem. Soc. *92*, 6637 (1970)

234. Organophosphorus Chemistry (ed.), S. Tripett, Specialist Periodical Reports, London, Vol. 2, p. 165, The Chemical Society 1971

235. E. J. Corey, J. I. Shulman, H. Yamamoto: J. Am. Chem. Soc. *92*, 226 (1970)

236. O. P. Vig, J. C. Kapur, C. K. Khurana: J. Indian Chem. Soc. *46*, 505 (1969)

237. Swiss Patent 493451; C.A. *73*, 120775 (1970)

238. M. Baumann, W. Hoffmann, H. Pommer: Liebigs Ann. Chem. *1976*, 1626

239. D. L. Roberts, et al.: J. Org. Chem. *33*, 3566 (1968)

240. H. Lehmann, et al.: Z. Chem. *13*, 255 (1973)

241. E. J. Corey, E. Hamanaka: J. Am. Chem. Soc. *89*, 2758 (1967)

242. A. Tanaka, H. Uda, A. Yoshikoshi: J. Chem. Soc., Chem. Commun. *1967*, 188

243. J. A. Marshall, M. T. Pike, R. D. Carroll: J. Org. Chem. *31*, 2933 (1966)

244. K. Mori, Y. Nakahara, M. Matsui: Tetrahedron Lett. *1970*, 2411

245. J. L. Gras, R. Maurin, M. Bertrand: Tetrahedron Lett. *1969*, 3533

246. J. A. Marshall, R. A. Ruden: Tetrahedron Lett. *1970*, 1239

247. R. A. Appleton, P. A. Gunn, R. McCrindle: J. Chem. Soc. (C) *1970*, 1148

248. M. D. Soffer, L. A. Burk: Tetrahedron Lett. *1970*, 211

249. W. I. Fanta, W. F. Erman: J. Org. Chem. *37*, 1624 (1972)

250. F. Kido, H. Uda, A. Yoshikoshi: J. Chem. Soc., Chem. Commun. *1969*, 1335

251. G. Buchi, W. Hofheinz, J. V. Paukstelis: J. Am. Chem. Soc. *91*, 6473 (1969)

252. S. Torii, T. Inokuchi: Bull. Chem. Soc. Japan *53*, 2642 (1980)

253. G. Ohloff, et al.: Helv. Chim. Acta *63*, 1589 (1980)

254. W. Reif, H. Grassner: Chem.-Ing. Techn. *45*, 646 (1973)

255. H. Pommer: Angew. Chem. *72*, 811, 911 (1960); H. Pommer, A. Nürrenbach: Pure Appl. Chem. *43*, 527 (1975)

256. H. J. Bestmann, P. Ermann: unpublished results

257. M. G. Motto, et al.: J. Am. Chem. Soc. *102*, 7947 (1980)

258. H. J. Bestmann, O. Kratzer: Chem. Ber. *96*, 1899 (1963)

259. H. J. Bestmann, L. Kisielowski, W. Distler: Angew. Chem. *88*, 297 (1976); Angew. Chem. Intern. Ed. Engl. *15*, 298 (1976)

260. H. J. Bestmann, et al.: Liebigs Ann. Chem. *1973*, 760

261. a) B. Schulz, J. Paust, J. Schneider: DOS 2 505 869 (1976), BASF;
b) A. Nürrenbach, et al.: Liebigs Ann. Chem. *1977*, 1146

262. H. J. Bestmann, W. Rieck: unpublished results

263. H. Freyschlag, et al.: DBP 1 210 780 (1963); Brit. Patent 1 137 429 (1966); Swiss Patent 506 513 (1967); all BASF

264. W. Stilz, H. Pommer: DBP 1092 472 (1958); W. Stilz: Diplomarbeit, Universität Tübingen 1954

265. H. Mayer, et al.: Helv. Chim. Acta *50*, 1606 (1967)

266. R. D. G. Cooper, J. B. Davis, B. C. L. Weedon: J. Chem. Soc. *1963*, 5637

267. Hamasaki, et al.: Bull. Chem. Soc. Japan *46*, 1553 (1973)

268. A. Eidem, et al.: Acta Chem. Scand. (B), *29*, 1015 (1975)

269. A. G. Andrewes, G. Borch, S. Liaaen-Jensen: Acta Chem. Scand. (B) *30*, 214 (1976)

270. D. E. Loeber, et al.: J. Chem. Soc. (C) *1971*, 404

271. J. D. Surmatis, et al.: Helv. Chim. Acta *53*, 974 (1970)

272. R. E. Coman, B. C. L. Weedon: J. Chem. Soc., Perkin I, *1975*, 2529

273. K. Tsukida, et al.: J. Nutr. Sci. Vitaminol. *21*, 147 (1975); C.A. *83*, 131784 (1975)

274. W. Bollag, R. Rueegg, G. Ryser: Ger. Offen. 2 542 612; C.A. *85*, 32639 (1976)

275. J. E. Johansen, S. Liaaen-Jensen: Tetrahedron *33*, 381 (1977)
276. A. K. Chopra, et al.: J. Chem. Soc., Chem. Commun. *1977*, 357
277. F. Kienzle, H. J. Mayer: Helv. Chim. Acta *61*, 2609 (1978)
278. H. Kjøsen, S. Liaanen-Jensen: Acta Chem. Scand. *27*, 2495 (1973)
279. G. Pattenden, J. E. Wray, B. C. L. Weedon: J. Chem. Soc. (C), *1970*, 235
280. M. Yasuhara, et al.: Bull. Chem. Soc. Japan *53*, 1629 (1980)
281. A. G. Andrewes, S. Liaanen-Jensen: Acta Chem. Scand. *27*, 1401 (1973)
282. L. Bartlett, et al.: J. Chem. Soc. (C) *1969*, 2527
283. A. Eidem, S. Liaanen-Jensen: Acta Chem. Scand. *B28*, 273 (1974); J. E. Johansen, S. Liaanen-Jensen: ibid. *B28*, 301 (1974)
284. H. Brzezinka, B. Johannes, H. Budzikiewicz: Z. Naturforsch. *29B*, 429 (1974)
285. J. D. Surmatis, et al.: J. Org. Chem. *35*, 1053 (1970)
286. A. G. Andrewes, S. Liaanen-Jensen, G. Borch: Acta Chem. Scand. *28B*, 737 (1974)
287. T. Ike, et al.: Bull. Chem. Soc. Japan *47*, 350 (1974)
288. N. Okukado, T. Kimura, M. Yamaguchi: Mem. Fac. Sci. Kyushu Univ. *9* (C) 139 (1974); C.A. *81*, 152447 (1974)
289. J. D. Bu'Lock, S. A. Quarrie, D. A. Taylor: J. Labelled Compounds *9*, 311 (1973)
290. N. Okukado: Bull. Chem. Soc. Japan *47*, 2345 (1974)
291. O. Puntervold, S. Liaanen-Jensen: Acta Chem. Scand. *28B*, 1096 (1974)
292. H. Achenbach, J. Witzke: Angew. Chem. *89*, 198 (1977); Angew. Chem. Intern. Ed. Engl. *16*, 191 (1977)
293. M. Ahmed, et al.: J. Chem. Soc., Perkin I, *1974*, 1981
294. R. K. Bentley, et al.: J. Chem. Soc., Perkin I *1974*, 1987
295. F. Näf, et al.: Helv. Chim. Acta *58*, 1016 (1975)
296. T. G. Halsall, J. R. Hills: J. Chem. Soc., Chem. Commun. *1971*, 448

Industrial Applications of the Wittig Reaction

Horst Pommer and Peter C. Thieme

BASF Aktiengesellschaft, D-6700 Ludwigshafen/Rh., FRG

Table of Contents

1 Introduction 166
   1.1 Discovery, Structure and First Syntheses of Vitamin A and β-Carotene . 166
   1.2 Commercial Significance of the Carotenoid Terpenes 167

2 The Wittig Reaction in the BASF Vitamin A Synthesis 167
   2.1 Building Block Principle 167
   2.2 Linking Reactions 168
   2.3 The Wittig Reaction in the Industrial Vitamin A Synthesis 169
      2.3.1 Retinoic Acid as a $C_{15+}C_5$ Linkage 169
      2.3.2 Adaptation of the Wittig Reaction to Industrial Conditions . . . 170
         2.3.2.1 Simplification of the Reaction Conditions 170
         2.3.2.2 Preparation and Regeneration of Triphenylphosphine . . 171
         2.3.2.3 Development of the Alternative Phosphonate Method . . 172
         2.3.2.4 Cis-trans Isomerization of Vitamin A Acetate 174
      2.3.3 The Industrial Vitamin A Process 174
      2.3.4 Synthesis of the C_5 Building Blocks 175
      2.3.5 Synthesis of the C_{15} Building Blocks 177

3 The Wittig Reaction in Carotenoid Syntheses 179
   3.1 β-Carotene 180
   3.2 Apo-Carotenals 181
   3.3 Citranaxanthine 183
   3.4 Canthaxanthine 183
   3.5 Astaxanthine 184

4 Prospects 185

5 References 185

1 Introduction

The reaction, discovered by G. Wittig and named after him, for converting a carbonyl into an olefine was introduced into industry at an unusually early stage. It became known at a time when new possibilities for *industrial syntheses of Vitamin A and carotenoids* were sought in the chemical industry. The core of many of these syntheses was the linking of individual molecular building blocks, with the formation of an olefinic double bond. The Wittig reaction opened up for this synthesis a completely new path, the efficiency of which was to be quickly established in the course of investigations on the industrial vitamin A synthesis. Both developments have complemented and mutually furthered each other. In the development of the industrial vitamin A synthesis an important hurdle was mounted and the propagation of the Wittig synthesis was given an additional impulse. Under the necessity of establishing an economical process for the manufacture of vitamin A, it was important to adapt the reaction originally conceived by G. Wittig to industrial conditions. Proton acceptors and solvents which were industrially readily accessible and industrially easy to handle, as well as a simple synthesis for the auxiliary reagent triphenylphosphine, had to be found. The article which follows is a summarizing report on the application of the Wittig reaction in industrial syntheses of vitamin A and its derivatives and of carotenoids.

1.1 Discovery, Structure and First Syntheses of Vitamin A and β-Carotene

Vitamin A (2) was first described as a fat-soluble growth factor by W. Stepp in 1909, and was later investigated in more detail by US research groups. Its structure was solved by P. Karrer and was published in 1931 [1].

Vitamin A (2) belongs to the class of the polyunsaturated diterpenes with 20 C atoms and 5 conjugated double bonds. It is formally produced by the hydrolytic cleavage of β-carotene (1), which constantly occurs in nature in association with chlorophyll.

The first synthesis of a vitamin A derivative was described by H. H. Inhoffen, D. A. von Dorp and J. F. Arens in 1944 for the vitamin A acid (retinoic acid 5) [2]. The first syntheses [3] of vitamin A (2), discovered independently of one another, became known in 1947, and were followed as early as 1948 by an industrial synthesis by O. Isler [4] at Hoffmann-La Roche. This long interval between the discovery and the first laboratory synthesis of the crystalline compound can be explained in the case of vitamin A (2) by the fact that the structure of the molecule was complicated for the synthetic capabilities of that time.

The first synthesis of a C_{40} carotenoid, which was identical in structure to the dihydro-β-carotene obtainable from β-carotene (1), was successfully carried out by H. H. Inhoffen, H. Pommer and E. G. Meth in 1950 [5]. In the same year, the research groups of H. H. Inhoffen and P. Karrer [6] also published the first laboratory syntheses of β-carotene (1).

A comprehensive review on the synthesis of carotenoid terpenes from various building blocks is contained in the monograph "Carotenoids" by O. Isler [7].

1.2 Commercial Significance of the Carotenoid Terpenes

Vitamin A participates in important biological processes.

The differentiation of epithelial tissue, growth, reproduction and the process of sight are dependent, in mammals and in humans, on an adequate provision of vitamin A. Ensuring this provision is therefore of great importance for nutrition and health in man. Via the vitaminization of animal feedstuff, which ensures healthy animal stocks, the vitamins additionally contribute to the provision of man with adequate and high quality foodstuffs. The vitamin demand arising therefrom could only be covered by synthetically produced products identical to the natural vitamin, a fact which gave a commercial stimulus to the development of industrial production processes.

Carotenoids are used to an increasing extent for the coloration of foodstuffs and animal feedstuffs, since, as natural constituents of many foodstuffs, they have been part of the human diet for thousands of years and are thus toxicologically acceptable [8]. The precursors from the vitamin A and carotenoid syntheses also include some of the important fragrances which are used in the perfume industry. In addition, this chemistry led to the synthesis of a number of other interesting flavors and fragrances.

2 The Wittig Reaction in the BASF Vitamin A Synthesis

2.1 Building Block Principle

At the beginning of the 50s, work which was aimed at providing an industrial synthesis of vitamin A was begun in the BASF research [9].

The work is based on the idea of W. Reppe of applying the acetylene chemistry developed by him to the synthesis of terpenes. The 2-methylbutyn-2-ol (40, see page 14) formed by the addition of acetylene to acetone (39) was intended, as the C_5 building block, as the starting point for terpene syntheses. A C_5 building block appeared to be small enough to ensure the required flexibility for terpenoid vitamins and carotenoids and the extensive area of terpenoid flavors and fragrances.

On the other hand, it was sufficiently large for many individual steps of the linking reactions to be avoided. The latter consideration was particularly important for the synthesis of triterpenes and tetraterpenes, such as apocarotenals and carotenoids, the industrial production of which was included in the concept from the beginning.

2.2 Linking Reactions

The simplest possible reactions which, when repeated, lead to identical structural elements should be used as linking reactions. The combination

Ethynylation and partical hydrogenation (C_2 addition)

and

Carroll reaction (C_3 addition)

proved to be suitable for this purpose. As already mentioned, the ethynylation of acetone yields 2-methylbutyn-2-ol (40). C_3 fragments can also be introduced into the molecule in the form of acetoacetate, with CO_2 and ethanol being split off, instead of being introduced via the C_3 building block acetone. This is a reaction discovered by M. F. Carroll [10] in 1940, which extends tertiary allyl alcohols by 3 carbon atoms, with the formation of α, β-unsaturated ketones.

The following reaction steps were thus possible:

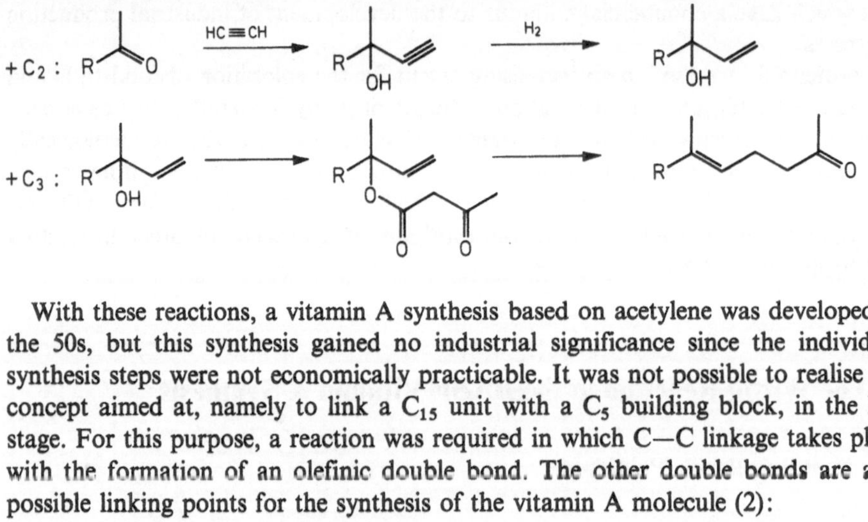

$+ C_2$:

$+ C_3$:

With these reactions, a vitamin A synthesis based on acetylene was developed in the 50s, but this synthesis gained no industrial significance since the individual synthesis steps were not economically practicable. It was not possible to realise the concept aimed at, namely to link a C_{15} unit with a C_5 building block, in the last stage. For this purpose, a reaction was required in which C—C linkage takes place with the formation of an olefinic double bond. The other double bonds are also possible linking points for the synthesis of the vitamin A molecule (2):

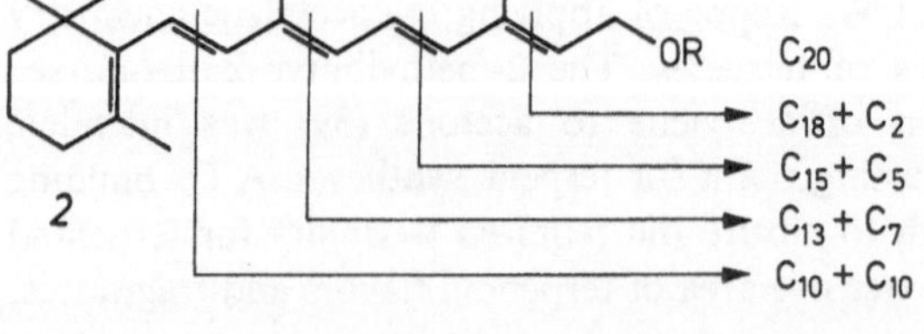

OR $\quad C_{20}$

$\longrightarrow C_{18} + C_2$

$\longrightarrow C_{15} + C_5$

$\longrightarrow C_{13} + C_7$

$\longrightarrow C_{10} + C_{10}$

2

At the beginning of the 50s, however, a suitable reaction for the desired C—C linking with double bond formation was lacking. Instead, the detour via the addition of an organometallic compound onto a carbonyl function, with subsequent splitting off of water, was customary [4].

2.3 The Wittig Reaction in the Industrial Vitamin A Synthesis

2.3.1 Retinoic Acid as a $C_{15} + C_5$ Linkage

In 1953, G. Wittig and his co-worker G. Geissler [11] found, in the reaction of phosphorylides with carbonyl compounds, a reaction which produced C—C linkages with the formation of olefinic double bonds. Owing to the traditionally close contact between university teachers of chemistry and industrial chemists in Germany, the reaction was known in the BASF laboratories at an early stage. G. Wittig introduced his new reaction in a discussion with W. Reppe and H. Pommer, who immediately recognized its significance for syntheses of the vitamin A type. Only a few days after the discussion, the synthesis of retinoic acid (5) was successfully carried out with the aid of the new reaction. From the work with vitamin A building blocks, suitable precursors, such as β-ionylideneacetaldehyde (3) and β-formylcrotonates (7), were available, with which it was possible to accomplish the synthesis of retinoic acid (5), according to the principle of a $C_{15} + C_5$ linkage [12].

Retinoic acid prepared according to this process is employed in pharmaceutical preparations as an active ingredient against acne [52].

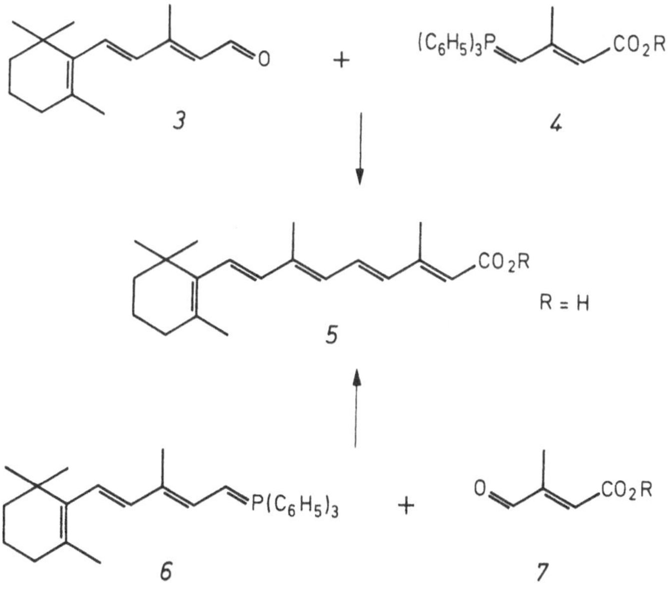

In experiments on animals, it was possible to show that all-trans retinoic acid and 13-cis retinoic acid, and derivatives of these acids, can prevent the development of tumors from pre-cancerous conditions [53]. Should intensive research in this area lead to a marketable drug, the Wittig reaction will again be a preferred method for its preparation.

2.3.2 Adaptation of the Wittig Reaction to Industrial Conditions

After it had been proved that the Wittig reaction was suitable, in principle, for polyene syntheses, the linking of a C_5 building block with a C_{15} ylide (6) was chosen for the synthesis of vitamin A acetate (9). Suitable C_5 building blocks, such as, for example, β-formylcrotyl acetate (8), were, of course, still unknown at that time.

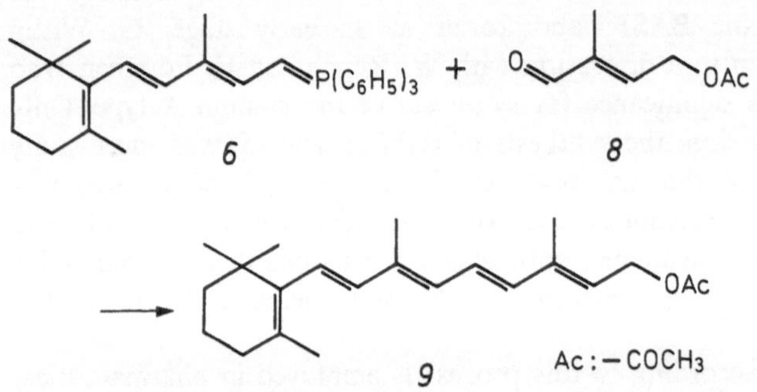

In order to work up the synthesis into an industrial process, the following questions had to be answered:

How can the reaction conditions of the Wittig reaction be realised industrially?

Which phosphine shows the best ylide reaction and is the cheapest to prepare?

Are other building blocks possibly more economical? Each of the non-cyclic double bonds in vitamin A could be formed as the last one, and each building block could be a carbonyl as well as a halide function.

Firstly, it was necessary to develop the reaction sequence of the Wittig reaction — synthesis of the phosphonium salt, formation of the ylide and reaction with a carbonyl compound to give the olefin — into an industrial process, under the stringent criteria of safety and cost-efficiency.

The organo-lithium compounds for the production of the phosphorus ylide had to be replaced by other proton acceptors, an industrially suitable solvent had to be found, and the triphenylphosphine reagent used in equimolar amounts had to be manufactured economically on a scale of tonnes.

2.3.2.1 Simplification of the Reaction Conditions

It was soon found that it was possible to form ylides of the C_{15} ylide type (6), which were resonance-stabilized by conjugated double bonds, from the phosphonium salts, using alcoholates, also in protic solvents, such as alcohols. In addition, they are so retarded in their reactivity that they are no longer hydrolyzed by water below 5 °C, whereby the stringent exclusion of moisture required, for example, for the reactive triphenylmethylenephosphorane can be dispensed with. Methanol, as the solvent, and sodium methylate, as the proton acceptor, where suitable industrially. The Wittig reaction in an aqueous medium [13] and in a two-phase system similar to that under the conditions of phase transfer catalysis [14] has also been used with success for the synthesis of vitamin A and carotenoids.

The considerable heat of reaction of 60 ± 2 kcal/mole of the vitamin A acetate formation proved to be problematic, since the by-products formed in the reaction, namely sodium chloride and triphenylphosphine oxide, made it difficult to conduct away the heat. The problem was solved by carrying out the reaction continuously in a small reaction volume and conducting away the heat adiabatically [15].

A further difficulty was the instability of the unsaturated terpenoid halogen compounds, which reacted with tertiary phosphines to phosphonium salts only in unsatisfactory yields. A new method for the preparation of phosphonium salts, in which alcohols (for example 10) are reacted with a tertiary phosphine in the presence of an acid [16 a], brought decisive advantages. Allyl alcohols (11) [16 b], as produced, for example, in the C_2 addition (ethynylation with subsequent partial hydrogenation) to β-ionone (35), proved to be particularly suitable. The allyl rearrangement takes place simultaneously with the formation of the phosphonium salts (13). The hydrocarbon (12) formed by "retro-rearrangement" can also be converted into the phosphonium salt (13), using phosphine in the presence of an acid [16 c].

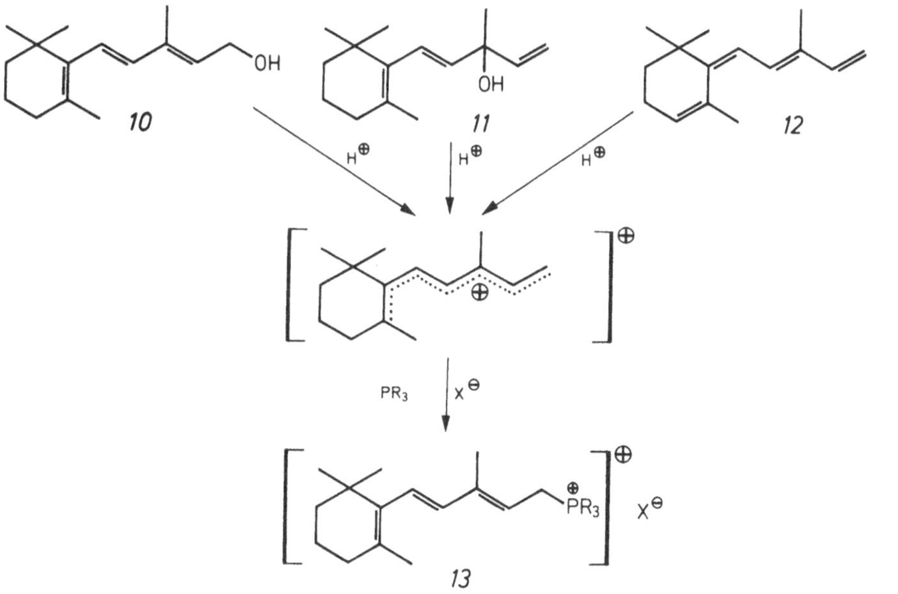

2.3.2.2 Preparation and Regeneration of Triphenylphosphine (15)

The use of triphenylphosphine (15) as an auxiliary reagent for the Wittig reaction posed a further economical and technical problem, since triphenylphosphine (15) was not obtainable in industrial quantities at that time. A long-established synthetic procedure, based on phosphorus trichloride, chlorobenzene (14) and metallic sodium [17], served as the basis for the development of an industrial process. The process was at first carried out batchwise, but increasing demand and greater process safety led to a continuous process being preferred [34 a].

$$3\,C_6H_5Cl + 6\,Na + PCl_3 \rightarrow (C_6H_5)_3P + 6\,NaCl$$
$$14 \qquad\qquad\qquad\qquad 15$$

The use of molar quantities of triphenylphosphine (15) in the Wittig reaction gave a stimulus to the development of processes by which it is possible to regenerate triphenylphosphine (15) from the waste product triphenylphosphine oxide (16).

$$(C_6H_5)_3PO + COCl_2 \rightarrow (C_6H_5)_3PCl_2 + CO_2$$
$$16 \qquad\qquad\qquad\qquad 17$$

$$3\,(C_6H_5)_3PCl_2 + 2\,P \rightarrow 3\,(C_6H_5)_3P + 2\,PCl_3$$
$$17 \qquad\qquad\qquad\qquad 15$$

In one of the processes, triphenylphosphine oxide (16) is reacted with phosgene to give triphenylphosphine dichloride (17) [18], which in turn disproportionates with red phosphorus to give triphenylphosphine and phosphorus trichloride [19]. The phosphorus trichloride formed thereby can be recycled into the triphenylphosphine synthesis. Processes have also been described for the reduction of the dichloride (17) to give triphenylphosphine (15), for example using hydrogen with noble metal catalysis [20], or under pressure at elevated temperature [21]. Other processes use aluminum hydrides [22], phosphites [23] or silanes [24] as the reducing agent for triphenylphosphine oxide (16).

2.3.2.3. Development of the Alternative Phosphonate Method

In the course of the work aimed at simplifying the Wittig reaction, other phosphorus derivatives were also included in the investigations at an early stage, in addition to phosphonium salts. In these investigations, particular significance was attached to phosphonates, which are formed in the reaction of phosphites with alkyl halides, by the Michaelis-Arbusow rearrangement. Whilst G. Wittig employed triaryl phosphites [25], H. Pommer and W. Stilz [26] investigated the dialkylphosphonates obtainable from trialkyl phosphites and found that they can also be subjected to olefine formation. L. Horner and co-workers [27] independently discovered the reaction for phosphine oxides. It is now known as the PO-activated olefination or the Wittig-Horner reaction [28].

The phosphonates stabilized by an adjacent electron-withdrawing substituent, such as the carboxylate group or nitrile group, are particularly suitable. The phosphonate carbanions are very readily formed, even with aqueous alkali, and, in contrast to the triphenylalkylenephosphoranes stabilized by carbonyl groups, react not only with aldehydes but also with ketones.

Differences are also found in the reaction behavior of the potential dicarbanions, the preliminary stages of which are the corresponding bistriphenylphosphonium salts

(18) or vinylogous α,β-bisphosphonates (19). Whilst the bistriphenylphosphonium salts (18) fragment [29] in the presence of alkali to give triphenylphosphine (15), triphenylphosphine oxide (16) and olefine (20), the analogous bisphosphonates (19) can readily be subjected to double olefine formation.

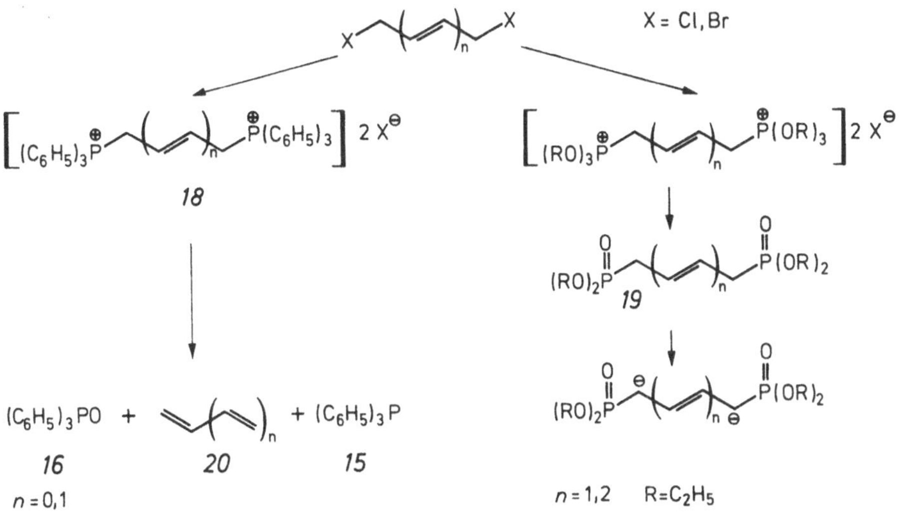

By this route, the C_{10} dialdehyde, 2,7-dimethyl-2,4,6-octatriene-1,8-dial (22), is readily obtainable [26 a)] from 2-butene-1,4-bis(dimethylphosphonate) (19, n = 1) and methylglyoxaldimethylacetal (21). It is employed as the C_{10} building block in an industrially used β-carotene synthesis (see page 16).

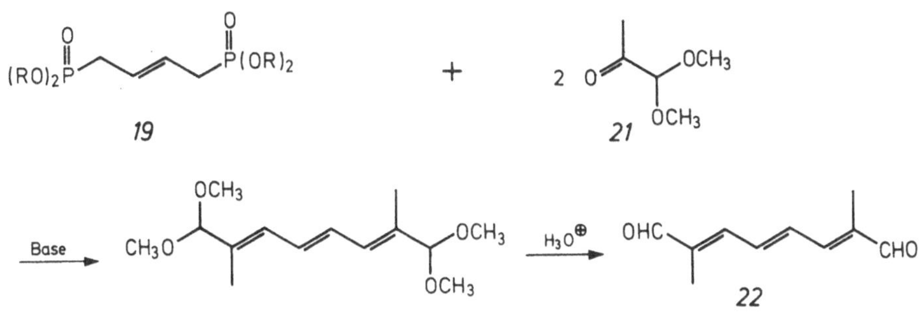

The PO-activated olefination was also used in industrial practice, outside the area of terpene chemistry. The bisolefine product 2,2'-(p-phenylenedivinylene)-dibenzo-nitrile (25) is obtained, in a smooth reaction, from o-cyanobenzyl phosphonate (23) and terephthalaldehyde (24). It is used as an optical brightener for polyester and polyamide fibers [30], under the name Palanilbrillantweiß R®.

173

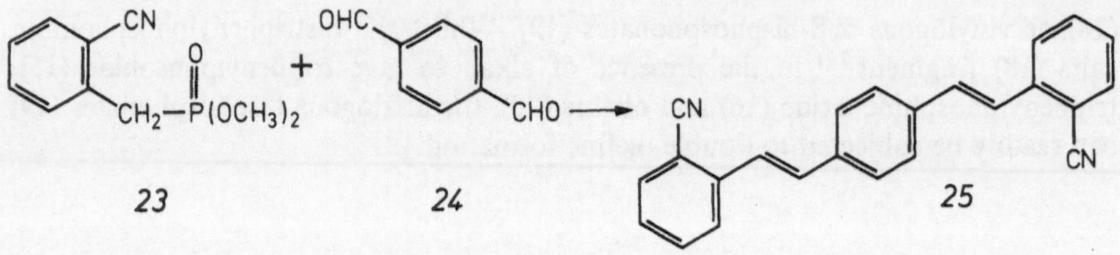

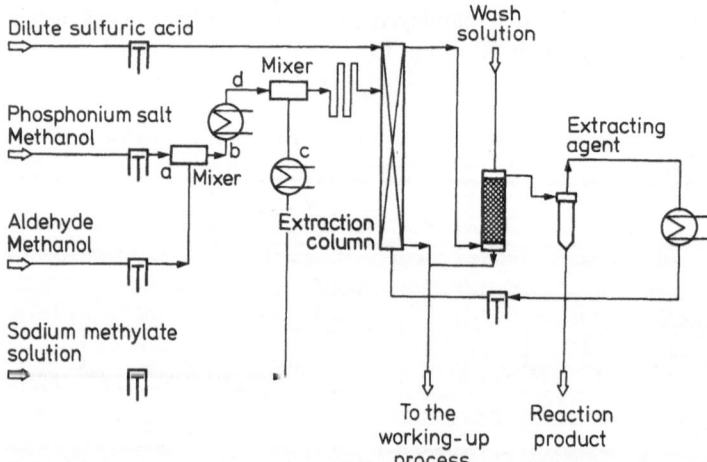

23   24   25

2.3.2.4 Cis-trans Isomerization of Vitamin A Acetate

A further limitation of the Wittig reaction is its inadequate stereospecificity [11 e)], which frequently leads to mixtures of cis-trans isomers. In the industrial Wittig synthesis of vitamin A acetate (9), the natural all-trans isomer is formed to approx. 70% and is separated off by crystallization from heptane. In order to avoid losses, the 9-cis or 11-cis isomer can be rearranged into the all-trans form using cis-trans isomerization catalysts, such as iodine [15)], or palladium complexes [31)], and by photochemical isomerization with UV light [32)]. The photochemical isomerization, sensitized with dyes [33)], is also used industrially.

2.3.3 The Industrial Vitamin A Process

The industrial development of the vitamin A synthesis from the C_{15} and C_5 building blocks was characterized by the special features of the Wittig reaction and the sensitivity of the starting materials as well as of the end products to chemical and thermal effects.

The figure shows a reaction scheme for the continuous Wittig synthesis. The solution of the C_{15} triphenylphosphonium salt (13) in methanol is continuously mixed, in a mixing reactor, a, with a methanolic solution of β-formylcrotyl acetate (8), and the mixture is cooled to −30 °C in a pre-cooler, b. A proton acceptor is simultaneously cooled in a further pre-cooler, c. Both solutions are then mixed intensively, by spraying, in a further mixing reactor, d, and are brought to reaction. The reaction begins immediately and proceeds to completion, after a very short residence time, with an increase in temperature. The yield is 98 %, relative to C_{15} salt. The reaction temperature is taken up by the reaction mixture (adiabatic heat conduction) and is controlled by the pre-cooling of the reaction solutions. In this way, the C_{15} phosphorane (6) is maintained below its decomposition threshold. By choosing the arrangement of the jets, the shape of the jets, the flow rate and the concentration, the course of the reaction can be exactly controlled [15].

The mixture of reaction product and triphenylphosphine oxide (16), which mixture is gelatinous owing to the alkali metal salt formed, enters an extraction column e, in which the desired reaction product, after acidification with sulfuric acid, is extracted, in counter-current, by hydrocarbons. In a second column, the hydrocarbon extract is washed, also continuously, with aqueous alcohol in order to remove the last residues of triphenylphosphine oxide (16). After concentration, under mild conditions, of the hydrocarbon extract, the all-trans vitamin A acetate (9) formed in the reaction crystallizes out and is separated off. The cis isomer, which does not crystallize, remains in the mother liquor and can be converted, according to one of the above-mentioned techniques, into the all-trans compound.

The Wittig reaction was first introduced into industry, in the BASF Aktiengesellschaft, according to this process. The product plant capacity was designed for 600 tonnes of vitamin A per annum.

2.3.4 Synthesis of the C_5 Building Blocks

The linking of two building blocks to form a double bond under the conditions of the Wittig reaction required that one of the building blocks has phosphorus ylide functionality and that the other has carbonyl functionality. The appropriate C_5 building block for vitamin A can thus be either 1-triphenyl-2-methyl-4-hydroxy-2-butenylidenephosphorane (26) [9] or β-formylcrotyl alcohol (2-methyl-4-hydroxy-2-butenal) (8a).

$(C_6H_5)_3P$ ... OR   26   O ... OR   8

a) R = H   b) R = Ac

β-Formylcrotyl alcohol is employed in the form of its acetate (8b) (C_5 acetate) in the industrial vitamin A synthesis. Two routes have essentially proved suitable for its preparation. One uses methylglyoxalacetal (21) as the starting material and thus again realises the $C_3 + C_2$ principle, whilst the other uses the C_4 building block butenediol diacetate (28).

Methylglyoxalacetal (21) is obtained from acetone (39) by oxidation with methyl nitrite in the presence of methanol and an acid catalyst. Ethynylation and partial hydrogenation yield 2-hydroxy-2-methyl-but-3-enal-dimethylacetal (27), which, after acetylation, yields the β-formylcrotyl acetate (8b) [34] via a copper-catalyzed allyl rearrangement with subsequent hydrolysis.

Starting from butenediol acetate (28), a further carbon atom is introduced by rhodium-catalyzed hydroformylation, likewise after a copper-catalyzed allyl re-arrangement to give vinylglycol diacetate. Splitting of an acetyl group leads to the β-formylcrotyl acetate (8b) (C_5 acetate) [34a)].

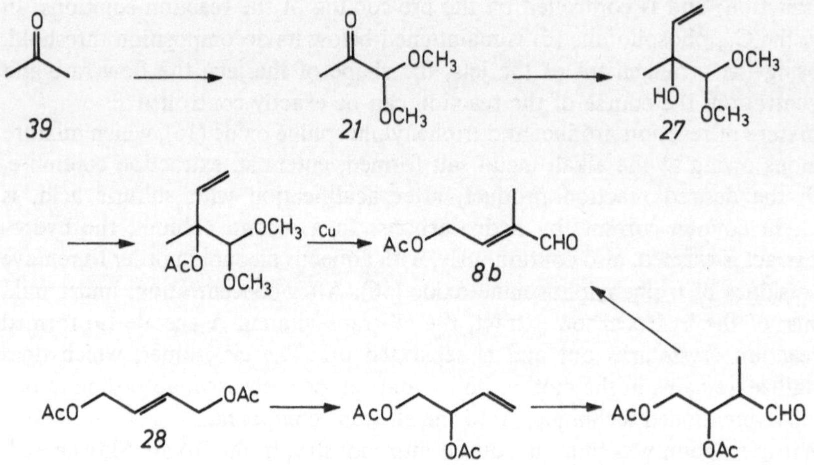
39
21
27
8b
28
OCH3
HO
AcO
Cu
CHO
OAc

For syntheses of vitamin A aldehyde (retinal) and higher terpene aldehydes, appropriate 2-methyl-2-butene-1,4-dials (32) with a protected aldehyde function were used as the C_5 building blocks. They are accessible, in turn, via the C_5 building block β-formylcrotyl acetate (8b) in the form of its dimethyl acetal, or via β-formylcrotyl methyl ether (29) [34b)]:

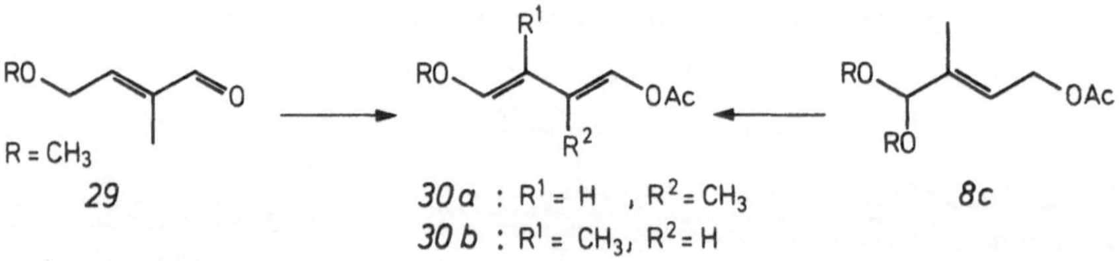
R = CH3
29
30a : R¹ = H , R² = CH3
30b : R¹ = CH3, R² = H
8c
RO
OAc

The oxidation of the dienes 30a and 30b is carried out in a simple manner, via the intermediate (31), by the addition of molar quantities of halogen and subsequent alcoholate treatment.

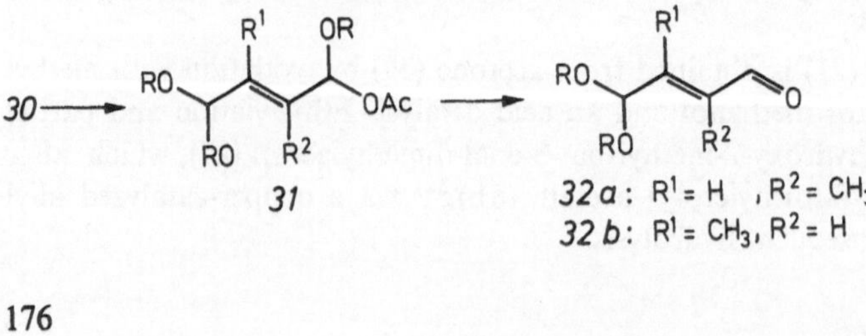
30
31
32a : R¹ = H , R² = CH3
32b : R¹ = CH3, R² = H
RO
OR
OAc

The aldehyde group in the end product (32) is formed, by trans-esterification, from the methoxy-acetoxy group in the intermediate (31).

A C_5 synthetic unit which has ylide functionality and which is likewise accessible from 2-hydroxy-2-methyl-but-3-enal-dimethylacetal (27) has proved suitable particularly for the synthesis of apocarotenals. The copper-catalyzed reaction of (27) with triphenylphosphine (15) in the presence of aqueous acid leads to 4-triphenyl-phosphonium-2-methyl-buten-2-al (33). The bifunctional C_5 ylenal (34), which is important for carotenoid syntheses, is formed therefrom with proton acceptors.

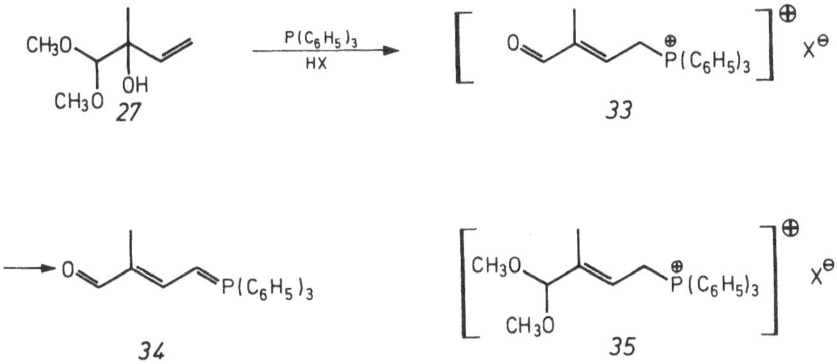

To avoid side-reactions, the appropriate dimethylacetal (35) (1,1-dimethoxy-2-methyl-2-butenyl-4-triphenylphosphonium salt) is customarily employed with a protected aldehyde function [34 c)].

2.3.5 Synthesis of the C_{15} Building Blocks

All industrial vitamin A syntheses use β-ionone as the starting compound (36, see page 14) [35]. This monocyclic C_{13} ketone can be obtained either completely synthetically from acetone and acetylene by consequent use of the C_2 and C_3 addition reaction, or via citral (59, see page 14) obtainable from natural sources (lemongrass oil).

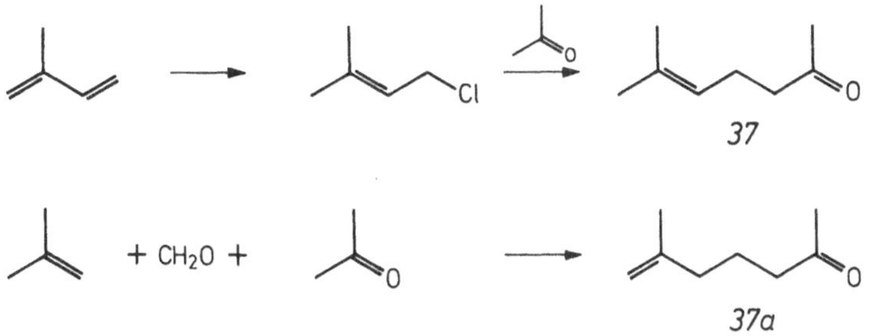

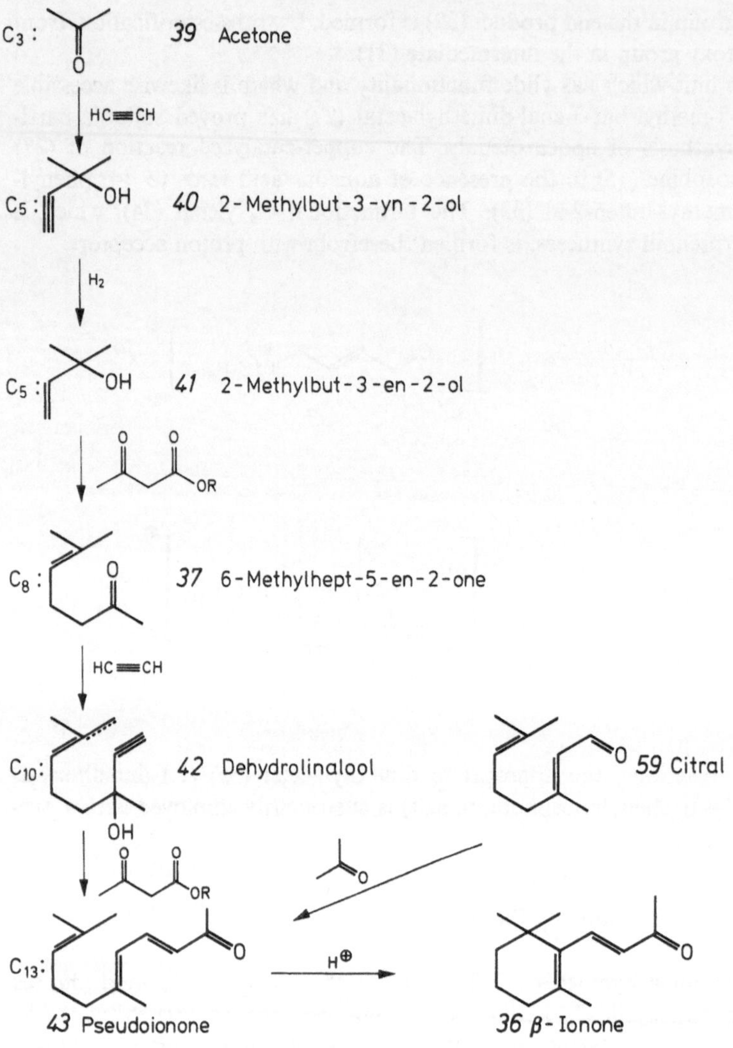

C_3 :　　　　　　39　Acetone

C_5 :　　OH　　40　2-Methylbut-3-yn-2-ol

C_5 :　　OH　　41　2-Methylbut-3-en-2-ol

C_8 :　　　37　6-Methylhept-5-en-2-one

C_{10}:　　42　Dehydrolinalool　　　　59　Citral

C_{13}:　43　Pseudoionone　　　　36　β-Ionone

An important intermediate for synthetic β-ionone (36) is the C_8 building block methyl heptenone (37). In addition to the synthesis shown above, two further processes are known for its industrial production. In the process of Rhodia INC [36], the starting material is isoprene, and methyl heptenone (37) is obtained via prenyl chloride. At BASF, methyl heptenone (37) is produced, for economic reasons, in the form of its double bond isomer (37a) by thermal condensation of isobutylene, formaldehyde and acetone [37] (see page 13). By suitable choice of the reaction conditions, various side-reactions, such as the Cannizzaro reaction of formaldehyde, the oligomerization of isobutene and aldol condensation between formaldehyde and acetone, can largely be suppressed.

This reaction based on the petrochemical crude material isobutylene makes the synthetic route to β-ionone (36) substantially shorter and cheaper, especially since the isomeric double bond proves to be advantageous in the subsequent reactions. In addition, i-methylheptenone (37a) can be converted into methylheptenone (37) by noble metal-catalyzed isomerization. The reaction steps ethynylation (C_2 addition), Carroll reaction (C_3 addition), ethynylation and partial hydrogenation (C_2 addition) lead from methylheptenone (37) via dehydrolinalool (42), pseudoionone (43) and β-ionone (36) to the C_{15} alcohol β-vinylionol (44). With triphenylphosphine (15), the desired C_{15} phosphonium salt (13), which is the second important synthetic building block for vitamin A and carotenoids [16], is obtained directly from β-vinylionol, by allyl rearrangement.

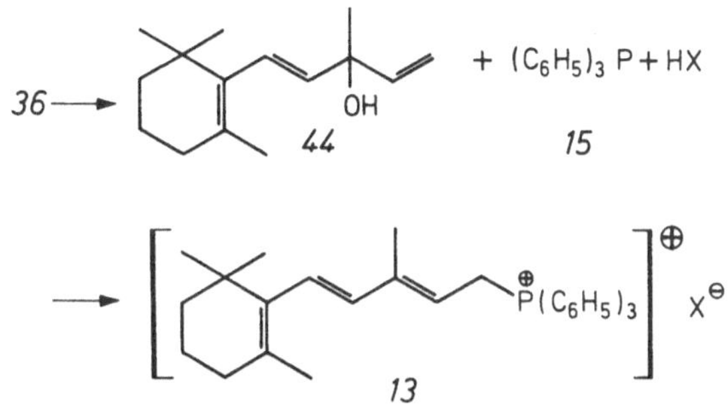

The individual stages for the preparation of C_5 acetate and C_{15} salt are well-tried processes on the industrial scale and are carried out continuously or by a program-controlled process [15].

3 The Wittig Reaction in Carotenoid Syntheses

Owing to their extensive distribution in the plant and animal world [38], the carotenoids stand out amongst the natural pigments. They have the important function of affording protection against light and oxidation, and prevent, for example in plant cells, the light-induced destruction of chlorophyll by oxygen [39]. Their commercial significance lies in their use as foodstuff colorants having a natural structure [8]. In addition, β-carotene could be of interest for inhibiting the uncontrolled growth of potential cancer cells, and thus as a prophylactic against certain tumors [54].

In their industrial production, the Wittig reaction is again a preferred process step for linking selected building blocks which are identical to those for the vitamin A synthesis or are logically derived from these. The advantages of the building block principle become particularly clear in the carotenoid syntheses [8b, 9c, 35].

3.1 β-Carotene (C_{40})

An industrial synthesis for β-carotene (1), the orange-red colorant of carrots and the important provitamin A, is based on the linking of two C_{15} phosphonium salts (13) with a C_{10} dialdehyde (22), in a double Wittig reaction [40]:

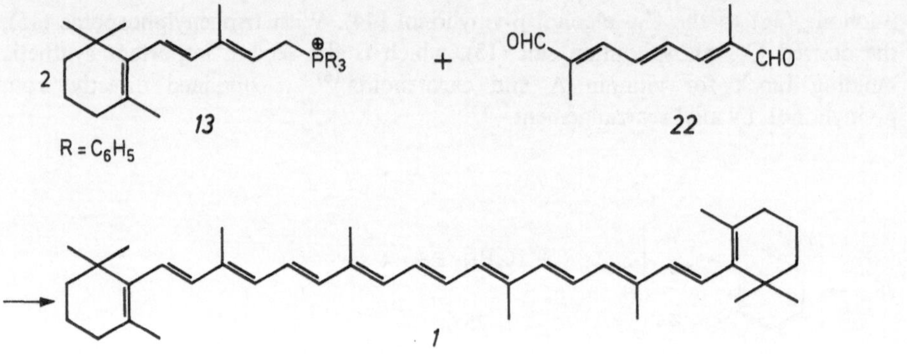

$$C_{15} + C_{10} + C_{15} = \text{β-carotene (1)}$$

In other syntheses, vitamin A (2) is used as the starting compound and is formally dimerized to give the C_{40} framework of β-carotene. By converting vitamin A acetate (9) into the C_{20} retinyltriphenylphosphonium salt (45), which is carried out using triphenylphosphine in the presence of an acid, a further interesting building block for carotenoid syntheses [41] is obtained. Its reaction with the vitamin A aldehyde retinal (46) under the conditions of the Wittig reaction yields β-carotene (1), according to the equation:

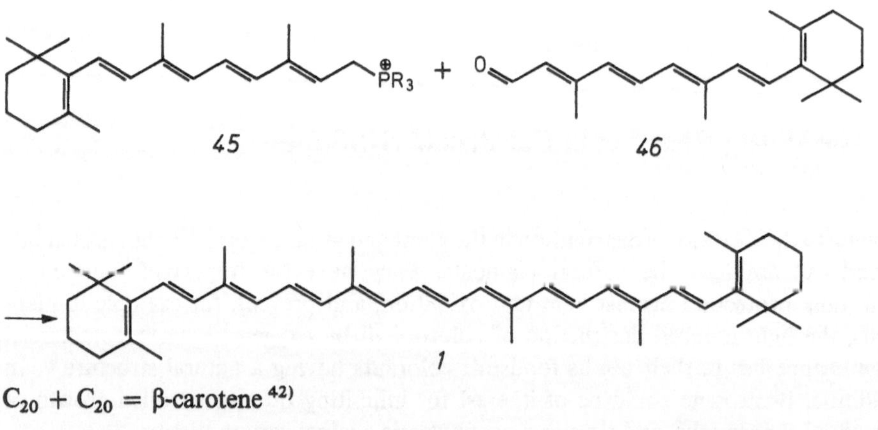

$$C_{20} + C_{20} = \text{β-carotene } [42]$$

A further simplification is obtained by the oxidative dimerization of the C_{20} phosphonium salt (45). The oxidation with molecular oxygen and with phosphite-ozone adducts was described by H. J. Bestmann and co-workers [43]. The use of hydrogen peroxide [44] brought industrially useful yields and a simple operation of the process, the symmetrical olefine being formed, with the loss of triphenylphosphine oxide (16). Any 20-cis constituents formed can be converted into the all-trans β-carotene (1) by simple thermal treatment.

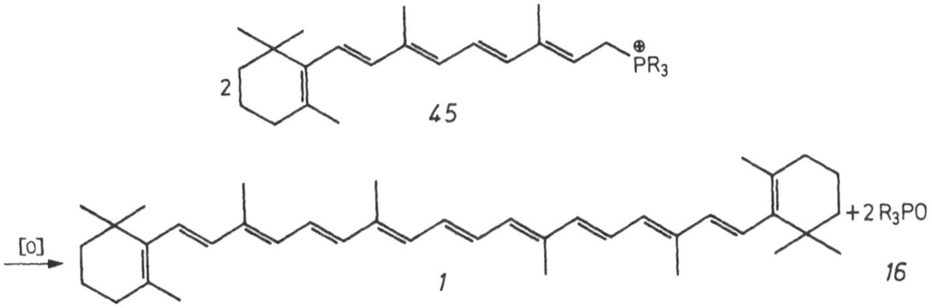

45

1   16   +2 R₃PO

[o]

3.2 Apo-Carotenals (C₂₅–C₃₅)

Apocarotenals and their corresponding carboxylates are natural colorants which occur in citrus fruits, spinach and other plants and which have vitamin A activity. They are used as fat-soluble foodstuff colorants. For their synthesis, specifically selected building blocks can be linked with one another under the conditions of the Wittig reaction:

β-Apo-12′-carotenal (47) (C₂₅):

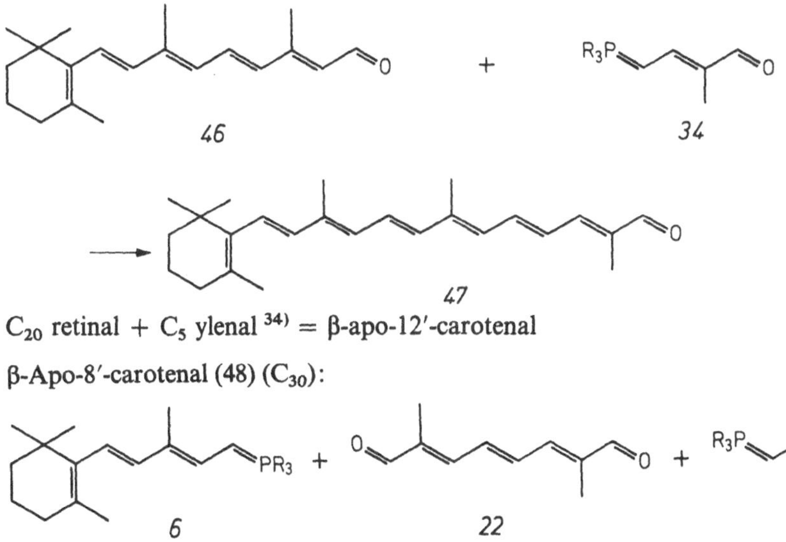

46   +   34

47

C_{20} retinal + C_5 ylenal [34] = β-apo-12′-carotenal

β-Apo-8′-carotenal (48) (C₃₀):

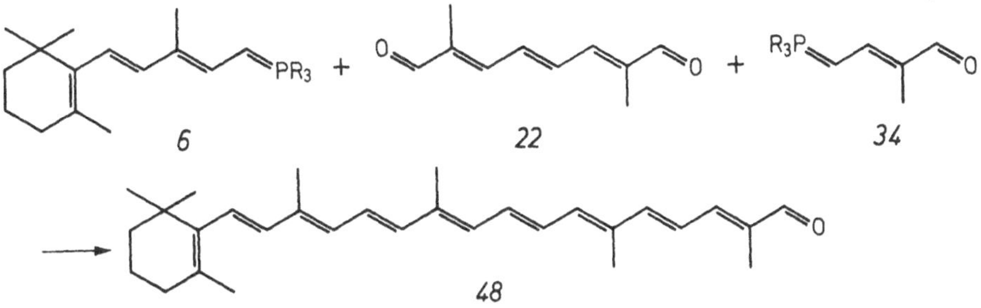

6   +   22   +   34

48

C_{15} ylide + C_{10} dialdehyde (2,7-dimethyl-) + C_5 ylenal [45] = β-apo-8′-carotenal

181

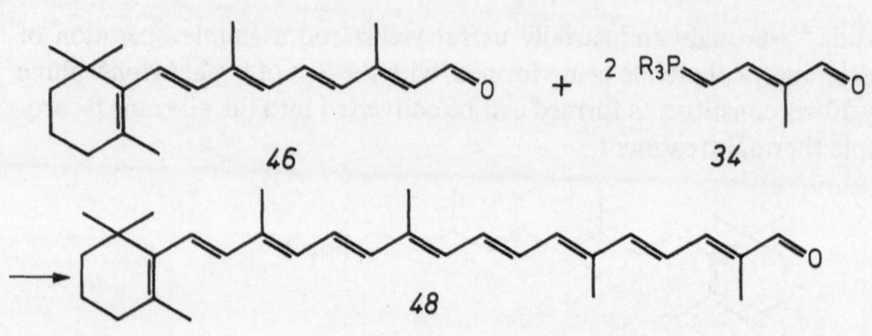

C_{20} retinal $+ 2 \times C_5$ ylenal [34] = β-apo-8'-carotenal

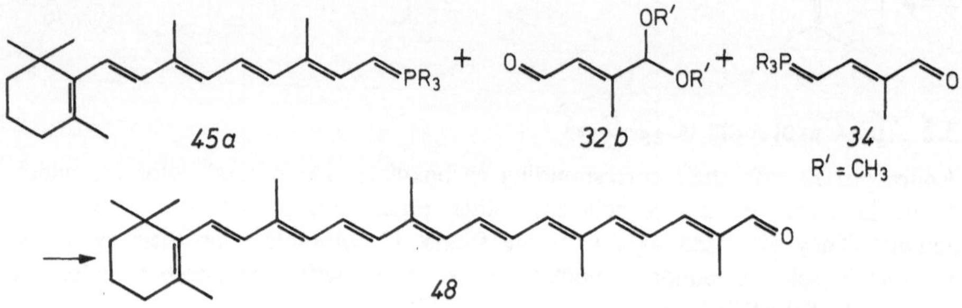

C_{20} ylide $+ C_5$ dialdehyde monoacetal $+ C_5$ ylenal [34] = β-apo-8'-carotenal

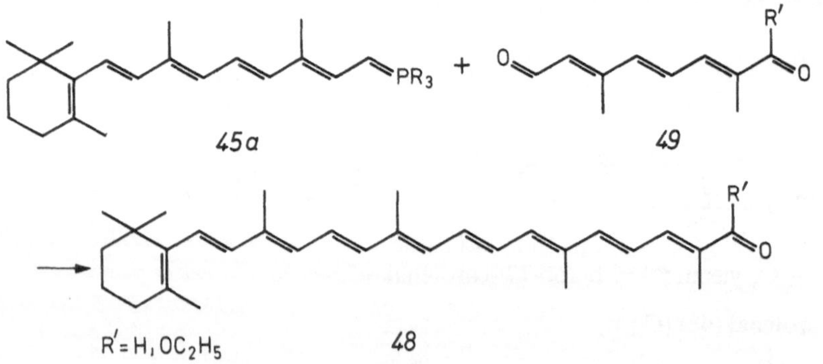

C_{20} ylide $+ C_{10}$ dialdehyde (2,6-dimethyl-) [46] = β-apo-8'-carotenal (R' = H)

The reaction of the C_{20} ylide (45a) with the unsymmetrical C_{10} dialdehyde, 2,6-dimethyl-2,4,6-octatriene-1,8-dial (49, R' = H), shows a remarkable regioselectivity. Only one of the two aldehyde groups participates in a Wittig olefination, leading uniformly to β-apo-8'-carotenal (48) [46]. The Wittig reaction between the C_{20} ylide (45a) and ethyl-2,6-dimethyl-8-al-octatrienoate (49, R' = OC_2H_5) correspondingly yields an ethyl-β-apo-8'-carotenate (48, R' = OC_2H_5) [58].

182

3.3 Citranaxanthine (C$_{33}$)

A simple aldol condensation with acetone leads from the C$_{30}$ aldehyde β-apo-8'-carotenal (48) to the C$_{33}$ ketone citranaxanthine (50), which is also used as a foodstuff colorant and feed additive [50]. It has a reddish shade, similar to that of canthaxanthine (53).

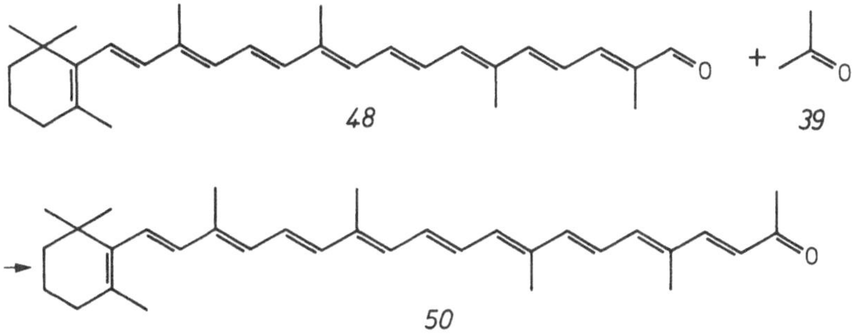

3.4 Canthaxanthine (C$_{40}$)

Canthaxanthine (53) is to be regarded as an oxidative metabolite of β-carotene (1). In its industrial production, 4,4'-diacetoxy-β-carotene (52) is used as a starting compound, being hydrolyzed and oxidized [51]. This compound, in turn, is obtained by the dimerization of 4-acetoxy-retinal or its phosphorus ylide (51), according to one of the methods described above [48 a]. The reaction of 4-oxo-C$_{15}$-phosphonium salt (54) with C$_{10}$ dialdehyde (22) likewise leads to canthaxanthine (53) [48 b]. In a further production process, β-carotene (1) is directly oxidized with chlorate, under catalysis with iodine [49].

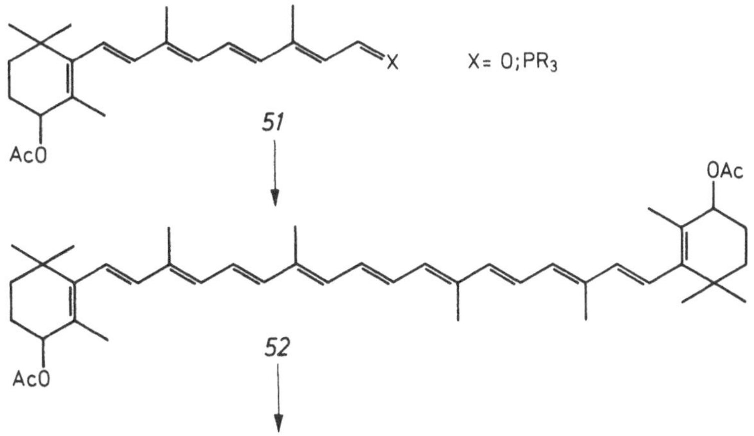

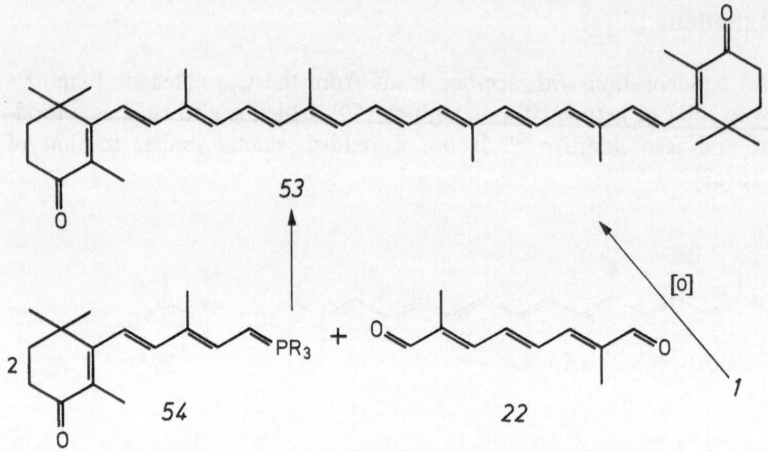

Canthaxanthine (53) has a reddish color, is somewhat more stable than β-carotene, and was introduced commercially as a feed additive in 1964.

In addition to the phosphorus ylide building blocks which are employed in industrial processes for the preparation of carotenoid terpenes, a large number of phosphorus ylides or phosphonium salts have been developed for laboratory syntheses of natural carotenoids. The monograph "Carotenoids" by O. Isler [7] contains an extensive summary of the examples described in the literature and in the patent literature.

3.5 Astaxanthine (C_{40})

According to the list of natural carotenoids by O. Straub [38], more than half of the over 400 natural carotenoids described are chiral. The asymmetric optically active terpene phosphonium salts which have recently become known, and which can be employed for the synthesis of chiral carotenoids, are contained in a review article by H. Mayer [47].

Astaxanthine (58) is such a chiral carotenoid and occurs in nature both in the optically active form and in the racemic form [55]. Astaxanthine is the red skin pigment of lobster, salmon and trout. It occurs, in addition, in certain green algae and yeasts. Its structure was determined as early as 1938 by R. Kuhn and N. A. Sörensen [56]. A synthesis uses oxo-isophorone (55), or 4-hydroxy-2,2,6-trimethylcyclohexanone (56) prepared therefrom, as the starting material, which can also be employed in the optically active form [57]. After conversion to the C_{15} phosphonium salt (57) with a protected hydroxyl group, this compound is reacted, according to the scheme $C_{15} + C_{10} + C_{15}$, with C_{10} dialdehyde (22), with simultaneous removal of the phenoxyacetyl protective group R', to give astaxanthine (58). If (4R, 6R)-4-hydroxy-2,2,6-trimethyl-cyclohexanone (56) is used as the starting compound, (3S, 3S)-astaxanthine (58) is obtained.

For racemic (3RS, 3'RS)-astaxanthine (58), an industrially practicable synthesis was developed which uses oxo-isophorone (55) as the starting compound, and yields

(3RS, 3'RS)-astaxanthine (58) via the unprotected C_{15} phosphonium salt (57, R' = H), in 7 simply practicable reaction steps [57 c].

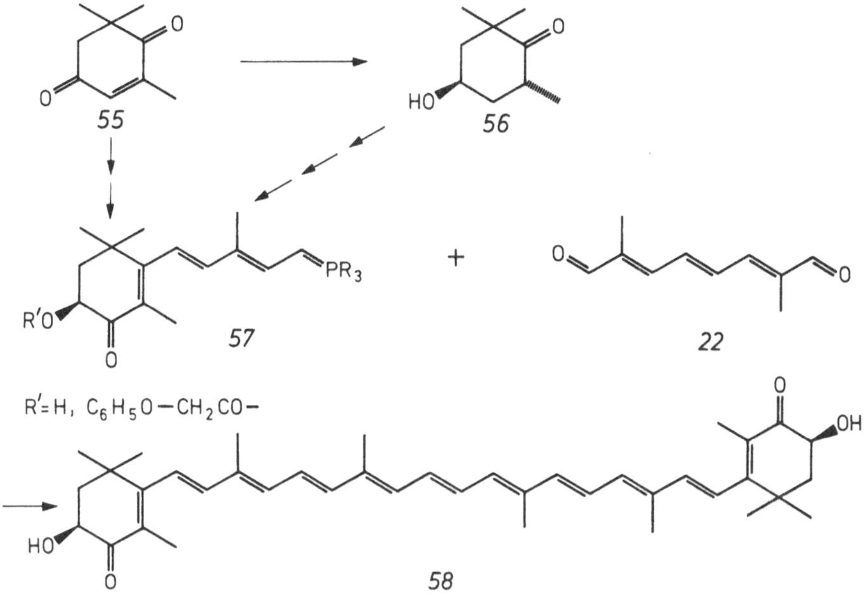

4 Prospects

In the industrial production of carotenoid polyenes, great significance has been attained by the Wittig reaction for linking synthetic building blocks with the formation of a carbon-carbon double bond, after it proved possible to reduce the originally high requirements on the reaction to simple industrially practicable conditions. Its advantages with respect to structural selectivity and high reactivity will also be utilized in the future in industrial syntheses, particularly of biologically active compounds and natural products with double bonds.

5 References

1. a) Stepp, W.: Biochem. Z. 22, 452 (1909)
   b) McCollum, E. V., Davis, M.: J. Biol. Chem. 15, 167; 19, 245 (1914); 23, 181 (1915)
   c) Karrer, P., Morf, R., Schöpp, K.: Helv. Chim. Acta 14, 1036, 1431 (1931)
2. a) Schering AG (Inventors H. H. Inhoffen, D. A. van Dorp, J. F. Arens), Germ. Pat. Appl., 729, 439 (1944)
   b) Van Dorp, D. A., Arens, J. F.: Rec. Trav. Chim. Pays-Bas 65, 338 (1946)

Horst Pommer and Peter C. Thieme

3. a) Isler, O. et al.: Helv. Chim. Acta 30, 1911 (1947)
   b) Van Dorp, D. A., Arens, J. F.: Nature 160, 189 (1947)
   c) Arens, J. F., van Dorp, D. A.: Nature 157, 190 (1946)
4. a) Isler, O. et al.: Helv. Chim. Acta 32, 489 (1949)
   b) Lindlar, H.: Helv. Chim. Acta 35, 446 (1952)
5. a) Inhoffen, H. H., Pommer, H., Meth, E. G.: Chemiker Z. 74, 211 (1950)
   b) Inhoffen, H. H., Pommer, H., Meth, E. G.: Liebigs Ann. Chem. 569, 74 (1950)
6. a) Inhoffen, H. H., Pommer, H., Bohlmann, F.: Chemiker Z. 74, 309 (1950); Liebigs Ann.
   Chem. 569, 237 (1950); Inhoffen, H. H. et al.: Liebigs Ann. Chem. 570, 54 (1950),
   Chemiker Z. 74, 285 (1950)
   b) Karrer, P., Eugster, C. H.: Helv. Chim. Acta 33, 1172 (1950)
   c) see also: Milas, N. A. et al.: J. Amer. Chem. Soc. 72, 4844 (1950)
7. a) Mayer, H., Isler, O., in O. Isler: Carotenoids, Birkhäuser-Verlag, Basel and Stuttgart 1971
   b) Reppe, K.: Houben Weyl, Vol. V/1d pages 88–137
8. a) Kläui, H., Isler, O.: Chemie in unserer Zeit 15, 1 (1981)
   b) Müller, R. K. et al.: Food Chem. 5, 15 (1980)
9. a) Pommer, H.: Angew. Chem. 72, 811 (1960)
   b) ibid. 72, 911 (1960)
   c) ibid. 89, 437 (1977), Angew. Chem. Int. Ed. Engl. 16, 423 (1977)
10. a) Hoffmann, W., Pasedach, H., Pommer, H.: Liebigs Ann. Chem. 729, 52 (1969)
    b) Carroll, M. F.: J. Chem. Soc. 1941, 507
11. a) Wittig, G., Geissler, G.: Liebigs Ann. Chem. 580, 44 (1953)
    Reviews on the Wittig reaction:
    b) Schöllkopf, U.: Angew. Chem. 71, 260 (1959)
    c) Maercker, A.: Org. Reactions 14, 270 (1965)
    d) Johnson, A. W.: Ylide Chemistry, Academic Press, New York, 1966
    e) Gosney, I., Rowley, A. G., in J. I. G. Cadogan: Organophosphorus Reagents in Organic
    Synthesis, Academic Press, New York, 1979
    for the mechanism of the Wittig reaction see:
    Bestmann, H. J.: Pure Appl. Chem. 52, 771 (1980); Giese, B., Schoch, J., Rüchardt, Ch.: Chem.
    Ber. 111, 1395 (1978)
12. BASF AG (Inventors G. Wittig, H. Pommer) Germ. Pat. 950,552 (1954)
13. a) BASF AG (Inventors W. Reif, B. Schulz, P. Grafen, H.-U. Scholz, H. Grassner), DAS
    2,729,974 (1977)
    b) Hoffmann-La Roche (Inventors K. Schleich, H. Stoller), DAS 2,733,231 (1977)
14. a) Hoffmann-La Roche (Inventors K. Schleich, H. Stoller), DAS 2,636,879 (1976)
    compare also Märkl, G., Merz, A.: Synthesis 1973, 295 and Boden, R. M.: Synthesis 1975,
    784
15. a) Zechmeister, L.: Cis-trans-isomeric Carotenoids, Vitamins A and Arylpolyenes, Springer-
    Verlag Vienna 1962, page 51
    b) Reif, W., Grassner, H.: Chem.-Ing.-Tech. 45, 646 (1973)
16. a) BASF AG (Inventors W. Sarnecki, H. Pommer) Germ. Pat. 1,046046 (1956)
    b) BASF AG (Inventors W. Sarnecki, H. Pommer) Germ. Pat. 1,060,386 (1957)
    c) Freyschlag, H. et al.: Angew. Chem. 77, 277 (1965), Angew. Chem. Int. Ed. Engl. 4, 287
    (1965)
17. a) Michaelis, A., Reese, A.: Chem. Ber. 15, 1610 (1882)
    b) Maier, L., in: Org. Phosphorus Compounds 1, 1 (1972)
18. BASF AG (Inventors R. Appel, W. Heinzelmann) Germ. Pat. 1,192,205 (1962)
19. a) BASF AG (Inventors G. Wunsch, K. Wintersberger, H. Geierhaas) Germ. Pat. 1,247,310
    (1966)
    b) Wunsch, G., Wintersberger, K., Geierhaas, H.: Z. anorg. allg. Chem. 369, 33 (1969)
20. Hoffmann-La Roche (Inventors E. A. Broger) Europ. Pat. 5,747 (1978)
21. a) UBE Industries, Ltd. JP 057009 (1979)
    b) Masaki, M., Kakeya, N.: Angew. Chem. 89, 558 (1977), Angew. Chem. Int. Ed. Engl. 16,
    552 (1977)
22. Texas Alkyls Inc. (Inventors D. B. Malpass, G. S. Yeargin), DAS 2,714,721 (1973)
23. Hoffmann-La Roche (Inventors J. N. Gardner, J. Kïchling), DAS 2,437,153 (1973)

24. a) M + T Chemicals Inc. (Inventors W. R. Davis, M. D. Gordon), DAS 2,828,604 (1978)
    b) Hoffmann-La Roche (Inventors J. M. Townsend, D. H. Valentine jun.), DAS 2,455,371 (1973)
25. Wittig, G.: personal communication (1957), see also under 9 b
26. BASF AG (Inventors W. Stilz, H. Pommer)
    a) Germ. Pat. 1,092,472 (1958)
    b) Germ. Pat. 1,109,671 (1958)
    c) Germ. Pat. 1,108,208 (1959)
27. a) Horner, L., Hoffmann, H., Wippel, H. G.: Chem. Ber. *91*, 61 (1958)
    b) Horner, L. et al.: Chem. Ber. *92*, 2499 (1959)
28. More recent reviews:
    a) Walker, B. J., in J. I. G. Cadogan: Organophosphorus Reagents in Organic Chemistry, Academic Press, New York, 1979
    b) Wadsworth Jr., W. S.: Organ. React. *25*, 73 (1978)
    c) Boutagy, J., Thomas, R.: Chem. Rev. *74*, 87 (1974)
29. Stilz, W.: Thesis, Univers. Tübingen 1954
30. a) BASF AG (Inventors W. Stilz, H. Pommer) Germ. Pat. 1,108,219 (1959)
    b) BASF AG (Inventors W. Stilz, H. Pommer, K.-H. König) Germ. Pat. 1,112,072 (1959)
31. Hoffmann-La Roche (Inventors H. Stoller, H. P. Wagner), DAS 2,439,860 (1973)
32. Hoffmann-La Roche (Inventors M. Lalonde, H. Stoller), DAS 2,548,883 (1975)
33. BASF AG (Inventors M. Fischer, W. Wiersdorf, A. Nürrenbach, D. Horn, F. Feichtmayer), DAS 2,210,800 (1972)
34. a) Pommer, H., Nürrenbach, A.: Pure Appl. Chem. *43*, 527 (1975)
    b) Paust, J., Reif, W., Schumacher, H.: Liebigs Ann. Chem. *1976*, 2194
    c) BASF AG (Inventors H. Freyschlag, W. Reif, A. Nürrenbach, H. Pommer), DAS 1,216,962 (1964), 1,211,616 (1963)
       BASF AG (Inventors H. Freyschlag, W. Reif, A. Nürrenbach, W. Sarnecki) Germ. Pat. 1,210,832 (1963)
    d) see reference 45 a)
35. a) Isler, O.: Pure Appl. Chem. *51*, 447 (1979)
    b) Kienzle, F.: Pure Appl. Chem. *47*, 183 (1976)
36. Meuly, W. C.: Riechstoffe, Aromen, Körperpflegemittel *22*, 191 (1972)
37. BASF AG (Inventors H. Pommer, H. Müller, H. Overwien) Germ. Pat. 1,259,876 (1966)
38. Straub, O.: Key to Carotenoids: List of Neutral Carotenoids, Birkhäuser-Verlag, Basel, Stuttgart, 1976
39. Thommen, H.: Pure Appl. Chem. *51*, 867 (1979)
40. BASF AG (Inventors H. Pommer, W. Sarnecki) Germ. Pat. 1,068,710 (1958) and 1,068,703 (1958)
41. a) BASF AG (Inventors W. Sarnecki, A. Nürrenbach, W. Reif) Germ. Pat. 1,155,126 (1963) and 1,158,505 (1963)
    b) Schwieter, U. et al.: Helv. Chim. Acta *49*, 369 (1966)
42. a) BASF AG (Inventors H. Pommer, W. Sarnecki) Germ. Pat. 1,068,709 (1958)
    b) see ref. 9 b
43. a) Bestmann, H. J., Kratzer, O.: Chem. Ber. *96*, 1899 (1963)
    b) Bestmann, H. J., Kisielowski, L., Distler, W.: Angew. Chem. *88*, 297 (1976), Angew. Chem. Int. Ed. Engl. *15*, 298 (1976)
44. a) BASF AG (Inventors B. Schulz, J. Paust, J. Schneider) Germ. Pat. 2,505,869 (1975)
    b) Nürrenbach, A. et al.: Liebigs Ann. Chem. *1977*, 1146
45. a) BASF AG (Inventors H. Freyschlag, W. Reif, A. Nürrenbach, H. Pommer), Germ. Pat. 1,210,780 (1963)
    b) see ref. 34 a
46. BASF AG (Inventor H. Jaedicke), DAS 2,851,051 (1978)
47. Mayer, H.: Pure Appl. Chem. *51*, 535 (1979)
48. a) Hoffmann-La Roche (Inventor J. D. Surmatis), US Pat. 3,311,656 (1964)
    b) Hoffmann-La Roche (Inventor M. Rosenberger), DAS 2,801,908 (1977)
49. BASF AG (Inventors H. Jaedicke, J. Paust, J. Schneider) Germ. Pat. 2,534,807 (1975), 2,534,805 (1975)

50. a) Yokoyama, H., White, M. J.: J. Org. Chem. *30*, 2481 (1965)
    b) see ref. 9c and 45a
51. a) Entschel, R., Karrer, P.: Helv. Chim. Acta *41*, 402,983 (1958)
    b) see ref. 35a
52. a) Stüttgen, G.: Dermatologica *124*, 65 (1962)
    b) König, H. et al.: Arzneimittel-Forsch. *24*, 1184 (1974), and subsequent papers
53. a) Sporn, M. B., Newton, D. L.: Fed. Proc. *38*, 2528 (1979)
    b) Newton, D. L., Henderson, W. R., Sporn, M. B.: Cancer Res. *40*, 3413 (1980)
54. Peto, R. et al.: Nature *290*, 201 (1981)
55. a) Ronneberg, H. et al.: Helv. Chim. Acta *63*, 711 (1980)
56. b) Schiedt, K., Leuenberger, F. J., Vecchi, M.: Helv. Chim. Acta *64*, 449 (1981), Kuhn, R., Sörensen, N. A.: Ber. deut. chem. Ges. *71*, 1879 (1938)
57. a) Kienzle, F., Mayer, H.: Helv. Chim. Acta *61*, 2609 (1978)
    b) Hoffmann-La Roche (Inventors E. A. Broger, Y. Crameri, H. G. W. Leuenberger, E. Widmer, R. Zell) Europ. Pat. 5,748 (1978)
    c) Widmer, E. et al.: Abstr. 6. Int. Symp. Carotenoids, July 1981, Liverpool
58. a) see ref. 41b
    b) Hoffmann-La Roche (Inventors U. Schwieter, R. Rüegg) Germ. Pat. 1,247,299 (1968)
    c) Isler, O. et al.: Helv. Chim. Acta *42*, 864 (1959)

Angle Strained Cycloalkynes [1]

Adolf Krebs and Jürgen Wilke

Institut für Organische Chemie und Biochemie, Universität Hamburg, Hamburg, FRG

Dedicated to Prof. Dr. Dr. h. c. mult. G. Wittig who Contributed so much to the Chemistry of Angle Strained Cycloalkynes.

Table of Contents

1 Introduction 191

2 Preparation and Generation of Angle Strained Cycloalkynes 191
  2.1 Isomerization 191
  2.2 Oligomerization 192
  2.3 Addition of Base 192
  2.4 Generation of Unstable Cycloalkynes in a Matrix. 193
  2.5 Generation of Unstable Cycloalkynes in Solution 193

3 Experimental Criteria for Angle Strain 202

4 Structural Effects on the Isolability of Angle Strained Cycloalkynes 202

5 Molecular Structure of Angle Strained Cycloalkynes 205
  5.1 X-Ray Analyses and Electron Diffraction 205
  5.2 Force Field and Quantum Mechanical Calculations 207

6 Thermochemical Data 208

7 Spectroscopic Properties 210
  7.1 Photoelectron (PE) and Electron Transmission (ET)-Spectroscopy . . 210
  7.2 ^{1}H- and ^{13}C-NMR-Spectroscopy 212
  7.3 IR- and Raman-Spectroscopy 215
  7.4 Electronic Absorption Spectra 217

8 Reactions 217
  8.1 General Remarks 217
  8.2 Isomerization 217
  8.3 Oligomerization 218
  8.4 Electrophilic Additions 219

8.5 Nucleophilic Additions 221
8.6 Radical Addition Reactions 221
8.7 Hydrogen Transfer Reactions 223
8.8 Cycloadditions 224
    8.8.1 [2+1]-Cycloadditions 224
    8.8.2 [2+2]-Cycloadditions 225
    8.8.3 [3+2]-Cycloadditions 226
    8.8.4 [4+2]-Cycloadditions 226
8.9 Reactions with Metal Salts or Metal Complexes 227

9 Conclusions and Outlook 228

10 References 228

1 Introduction

The smallest cycloalkanes and cis-cycloalkenes, cyclopropane (*1*) and cyclopropene (*2*) are isolable, although very reactive, compounds.

$$\triangle \qquad \underline{\triangle} \qquad \underline{\underline{\triangle}} \qquad (1)$$

1   *2*   *3*

In contrast, cyclopropyne (*3*) [3] could not even be generated and detected spectroscopically at very low temperatures. This is due to the enormous angle deformation in *3* of the normally linear C—C≡C—C arrangement in an open-chain acetylene.

For this reason the following three problems have been investigated for a long time:

1. Which is the smallest isolable [2] cycloalkyne?
2. How does ring strain in a cycloalkyne influence physical properties and chemical reactivity of these compounds?
3. Can one definitely prove the occurrence of non-isolable short-lived cycloalkynes?

Throughout this article the expression "angle strained cycloalkynes" is used. Therefore it is necessary to define the borderline between unstrained and strained systems. Since it is much easier to deform a C—C≡C-angle than a C—C=C or C—C—C-angle relatively large angle deformations are possible in cycloalkynes without significant changes in energy. Therefore, we deliberately consider here only cycloalkynes in which angle deformation of the C—C≡C-angles is larger than 10° for each angle. For example, within this definition cyclononyne (*6*) is qualified as an angle strained cycloalkyne, while cyclodecyne is not. In some borderline cases where neither experimental data nor calculations exist, models have been used to differentiate between strained and unstrained cycloalkynes according to the mentioned definition. 1,2-Didehydrobenzene (benzyne) and other five and six membered 1,2-didehydroaromatic systems, although they could qualify as angle strained cycloalkynes are not considered in this review.

2 Preparation and Generation of Angle Strained Cycloalkynes

The methods of preparation of cycloalkynes are essentially the same as those used for their open-chain analogs, which are discussed in detail elsewhere [1]. However, for special reasons connected with the nature of these compounds some methods have become more important in the preparation and generation of angle strained cycloalkynes. Some problems are:

2.1 Isomerization

In the presence of strong bases or metals isomerization to the corresponding allene and 1,3-diene can occur. Cyclic allenes with less than 11 carbon atoms are less strained than the isomeric acetylenes since an allene requires only three colinear carbon atoms; this is shown by equilibration experiments [4]. Therefore, in the synthesis of angle strained cycloalkynes strongly basic conditions should be avoided, if such an isomerization is possible. For example, 1-chloro-cyclononene (4) on treatment with sodium amide in liquid ammonia yields 1,2-cyclononadiene (5) and cyclononyne (6) in 6 to 1 ratio [5]. In this case (5) may be formed by isomerization of (6) or by direct HCl elimination from (4).

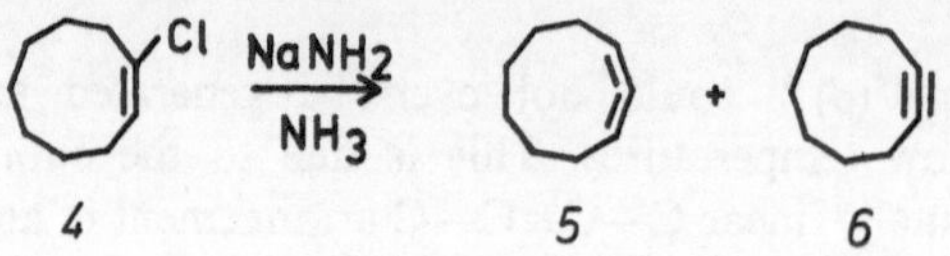

$$\tag{2}$$

Pure (6) can be prepared via the oxidative decomposition of the bishydrazone (7) with mercuric oxide [6] or thermal decomposition of the selenadiazole (8) [7].

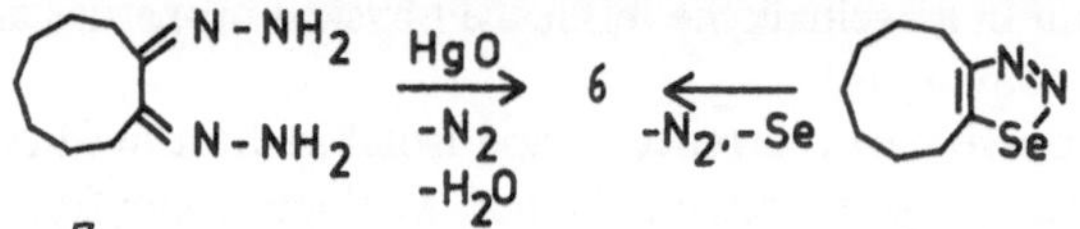

$$\tag{3}$$

2.2 Oligomerization

If a very reactive cycloalkyne is synthesized, the necessary fast reaction at low temperatures must be worked up in a short period. These conditions are met by the oxidative decomposition of bishydrazones with lead tetraacetate in dichloromethane. By this method the most strained, but still isolable cycloalkyne 3,3,7,7-tetramethylcycloheptyne (10) was prepared [8], which on standing at room temperature

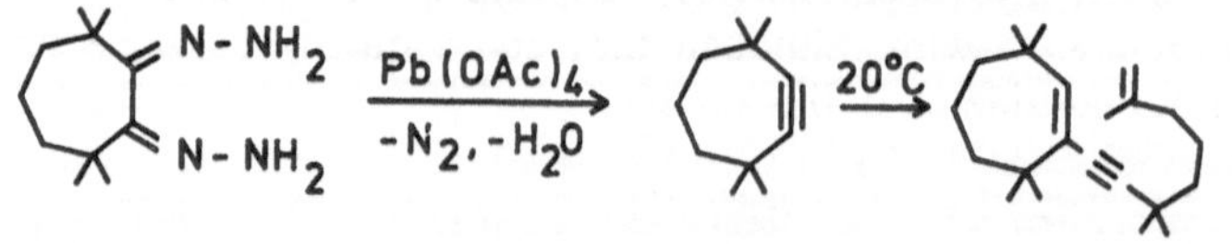

$$\tag{4}$$

gives almost quantitatively the dimer (11) [9]. On oxidation of (9) with mercuric oxide only the dimer is observed since the reaction is too slow to allow isolation of the intermediate (10).

2.3 Addition of Base

In the attempted preparation of cyclooctyne (14) [10] from 1-chlorocyclooctene (12) using lithium piperidide as a base only 1-(piperidyl)-cis-cyclooctene (13) was obtained [10]. This result was interpreted by assuming (14) as an intermediate which then adds lithium piperidide to yield (13).

However, with the proper starting material, base and conditions this elimination reaction can be used for an efficient synthesis of (*14*) in good yield by applying sterically hindered lithium diisopropylamide as a base and 1-bromo-cis-cyclooctene (*15*) in excess [11].

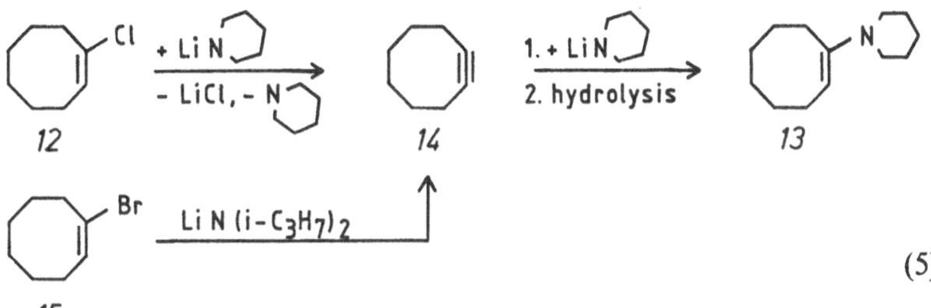

$$(5)$$

2.4 Generation of Unstable Cycloalkynes in a Matrix

Cyclooctyne (*14*) is the smallest isolable unsubstituted cycloalkyne. In order to obtain spectroscopic evidence for the occurrence of smaller cycloalkynes, such as cycloheptyne (*16*) [12], 3,3,6,6-tetramethylcyclohexyne (*17*) [12], cyclopentyne (*18*) [13] and acenaphthyne (*19*) [13] matrix photolysis of the corresponding cyclopropenones was used. For example, in the photolysis of (*21*) in an argon matrix at 20 K a band at 2108 cm^{-1} of medium intensity was observed in the IR spectrum which was assigned to the C≡C stretching vibration in (*17*) [12].

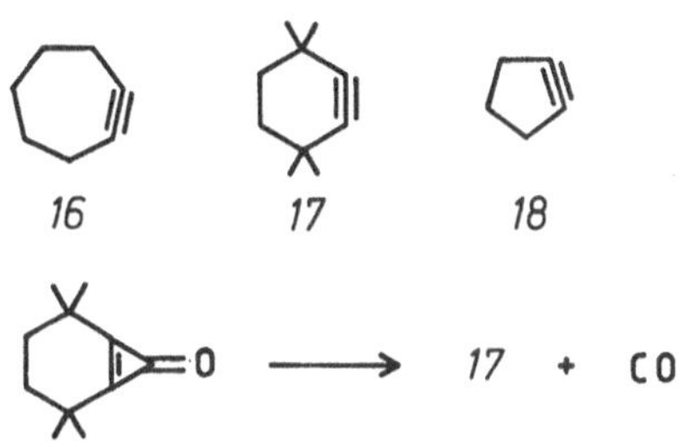

$$21 \longrightarrow 17 + CO \qquad (6)$$

2.5 Generation of Unstable Cycloalkynes in Solution

It is also possible to prove the occurrence of cycloheptyne (*16*) and cyclohexyne (*20*) in solution. Here, as in the isolation of very reactive cycloalkynes, it is essential to generate the intermediate in a fast reaction at low temperatures in the absence of a reactive reagent which could add to the cycloalkyne. Again, an oxidative decomposition with lead tetraacetate, here of the corresponding 1-aminotriazoles (*22*) resp (*27*), is used for the generation of (*16*) and (*20*) [14].

In these reactions, the kinetic stability of (*16*) and (*20*) was determined by trapping reactions with tetraphenylcyclopentadienone (TPCP). The half life of (*16*) in an 0.016 m CH_2Cl_2 solution at $-76\ °C$ is about one hour, while that of (20) under analogous conditions at even $-110\ °C$ is only a few seconds [14].

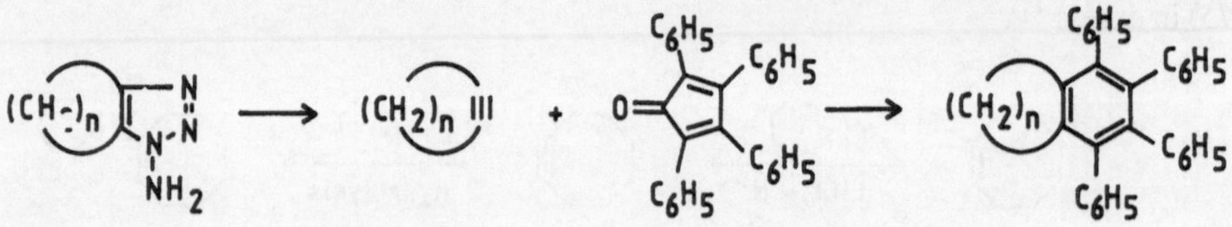

27: n = 4   20: n = 4
22: n = 5   16: n = 5   (7)

However, from a preparative point of view the situation is completely different, if high yields of adducts are desired. Then, a slow reaction, which leads to only low concentrations of the reactive intermediate, thereby preventing or at least reducing oligomerization reactions, offers an advantage over fast reactions in which relatively high concentrations of the reactive intermediate are generated. An

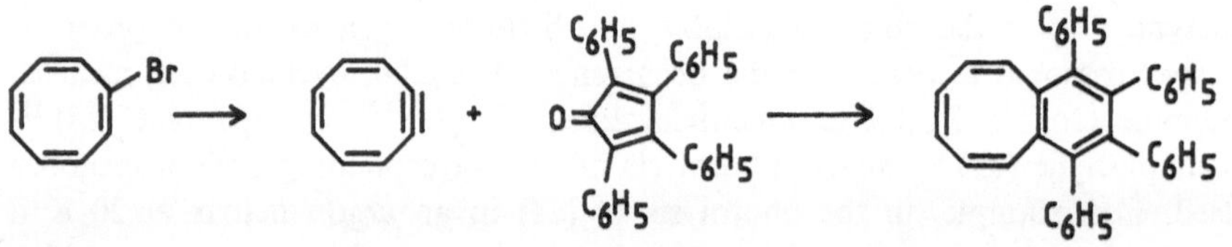

                                                                          (8)

24   23   25

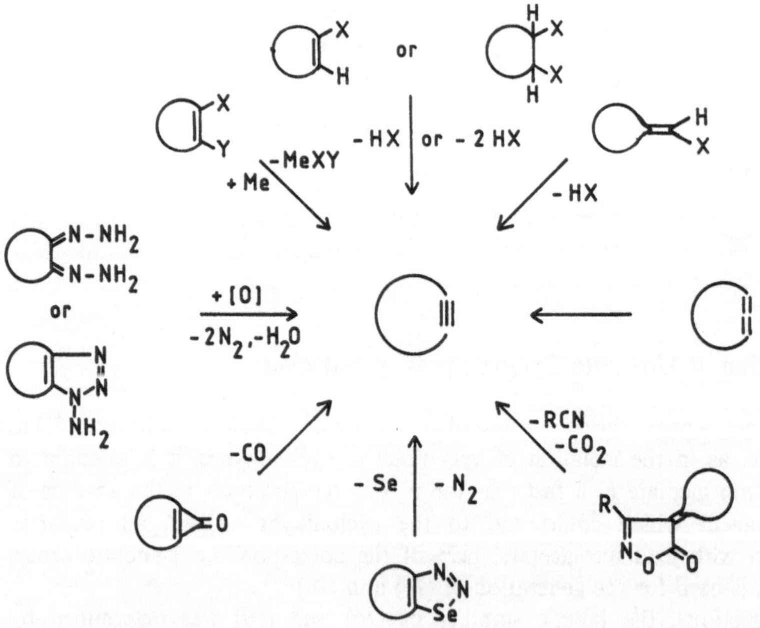

Fig. 1. General methods for the synthesis and generation of angle strained cycloalkynes

194

example is presented in the trapping of 1,2-didehydrocyclooctatetraene (*23*), which is generated from 1-bromo-cyclooctatetraene (*24*) with potassium-t-butoxide in 70 hours with TPCP [15]; yields of up to 91 % (*25*) are obtained in this reaction [16].

The scheme on p. 194 summarizes the most important methods for the synthesis and generation of angle strained cycloalkynes (Fig. 1).

In Tables 1 and 2 the methods of generation and experimental evidence for the intermediate occurrence of unstable cycloalkynes and methods of synthesis of isolable angle strained cycloalkynes, resp. are given.

Table 1. Unstable Angle Strained Cycloalkyne Intermediates

Ring Size	Structure	Methods of generation	Evidence based on	Ref.
5		a–f	L, M, T, K, O	10, 13, 17 – 24, 27 – 29, 53, 54)
		d	T, L, O	30 – 32)
	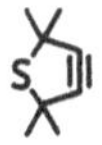	a	T	25)
	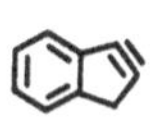	a	T	26)
		c, g	M, O	13, 33)
6		a, b, d, e, h, i	T, L, O	7, 14, 17 – 19, 23, 24, 34 – 36, 40 – 42, 46, 47, 50 – 54, 59, 61)
	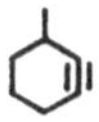	d	T, L	35, 37)
	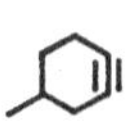	d	T, L	35)
	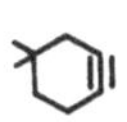	d	T, L	37)
		d	T, L	37)
		a, b, c	T, M, O	12, 38, 39)
		e	L, T	27, 43)
		d	L, T	44)

195

Table 1. (continued)

Ring Size	Structure	Methods of generation	Evidence based on	Ref.
	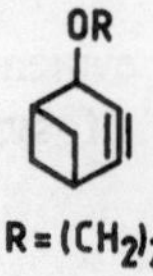 OR (structure) R = (CH₂)₂OH	d	L, T	45)
	 (structure)	d	L, T	48,49)
7	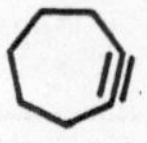 (structure)	a–d, g–i	T, K, L, M, O	7,12,14,17–19,46,47, 51–57,59,61)
	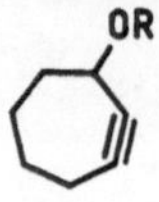 OR (structure) R = (CH₂)₂OH	d	T	45)
	 (structure)	i	T	105)
	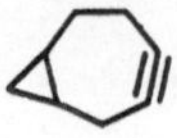 (structure)	d	T	55)
	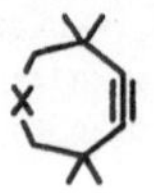 X (structure) X = O, N-CH₃	a	T, K	62,63)
	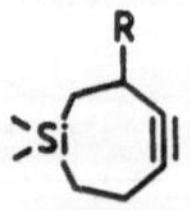 R Si (structure) R = H, CH₃	a	T	64,65)
	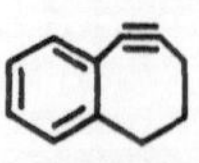 (structure)	a, l	T	26,58)
	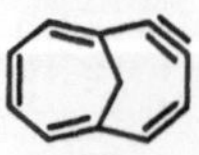 (structure)	d	T	82)
	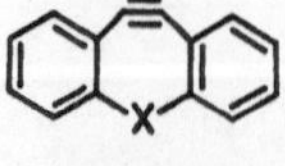 (structure) X = CH₂, CO C(OCH₃)₂, S SO₂, Si(CH₃)₂	d, i	T, K	66–70)

Table 1. (continued)

Ring Size	Structure	Methods of generation	Evidence based on	Ref.
		d	T	90 – 92)
8		d	T, O	15, 16, 71 – 74)
		i	T	111)
		i	T	112)
		d	T	80)
		d	T	75)
	Br	d	T	76)
	OtBu	d	T	76)
		a, d, i	T	77, 78)
10		k	T	81)
		d		129)

Table 1. (continued)

Ring Size	Structure	Methods of generation	Evidence based on	Ref.
11	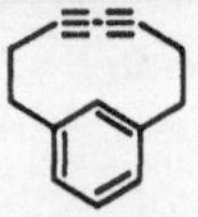	k		130)
12		k		131–133)
12	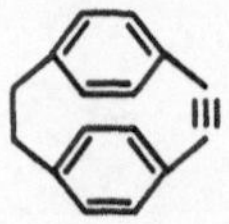	d	T, O	83)

a) Oxidative decomposition of 1,2-dihydrazones or 1-amino-1,2,3-triazoles.
b) Dehalogenation of 1,2-dihalocycloalkenes with metals or organometallic compounds.
c) Photolytic or thermal decarbonylation of cyclopropenones.
d) Dehydrohalogenation of 1-halocycloalkenes or 1,2-dihalocycloalkanes with strong bases.
e) Dehydrohalogenation of bromomethylenecycloalkanes with strong bases.
f) Debromination of dibromomethylenecycloalkanes with organometallic compounds.
g) Base catalyzed decomposition of 1,2-ditosylhydrazones or 1-tosylamino-1,2,3-triazoles.
h) Photolytic decomposition of alkali salts of 1-tosylamino-1,2,3-triazoles or 1,2-bis-tosylhydrazones.
i) Pyrolysis of 1,2,3-selenadiazoles or reaction of 1,2,3-selenadiazoles with n-butyllithium.
k) Oxidative coupling of α,ω-diynes or coupling of metal acetylides with haloorganic compounds.
l) Fragmentation of α,β-epoxi-keto-derivatives
T = trapping reactions
K = kinetic investigations
M = matrix spectroscopy
L = isotope or substituent labeling experiments
O = isolation of oligomerization products

Table 2. Isolable Angle Strained Cycloalkynes

Ring Size	Structure	Synthetic methods	Ref.
7	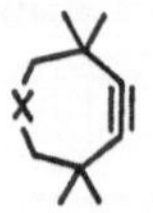 X=CH$_2$, S, SO, SO$_2$, Si(CH$_3$)$_2$, C(CH$_3$)$_2$	a	8, 9, 84–87)
	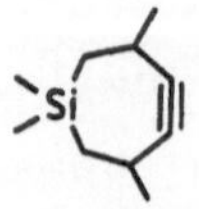	a	87)
	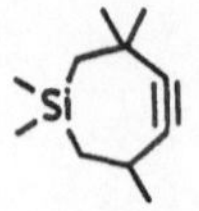	a	65)

Table 2 (continued)

Ring Size	Structure	Synthetic methods	Ref.
8		a, b, d, g–i	11, 18, 19, 59, 89, 93 – 97, 110)
	R_1 R_2 ... R_3 R_4	a, d	65, 98, 99)

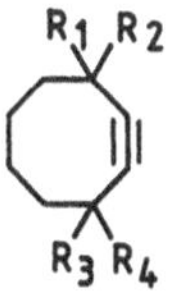

$R^1 = CH_3$
$R^2, R^3, R^4 = H$

$R^1, R^2 = CH_3$
$R^3, R^4 = H$

$R^1, R^3 = CH_3$
$R^2, R^4 = H$

$R^1, R^2, R^3 = CH_3$
$R^4 = H$

$R^1, R^2, R^3, R^4 = CH_3$

	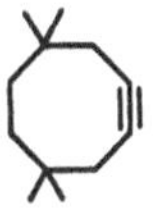	a	100)
	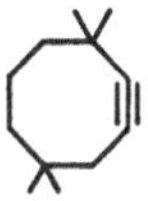	a, c	101, 102)
		1,4-dibromo-2-butyne + Zn	103, 104)
		i	106, 108)
	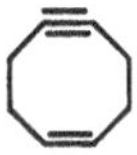	a, i	107, 108)
		i	108)
		i	109)
		a	112)

199

Table 2. (continued)

Ring Size	Structure	Synthetic methods	Ref.
8	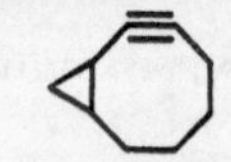	i	105)
	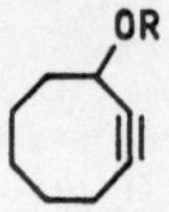 OR R = H, CH₃	d	113, 114)
	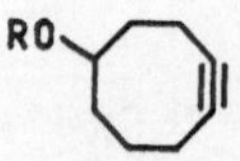 R = H, OCC₆H₄NO₂	i	115, 116)
	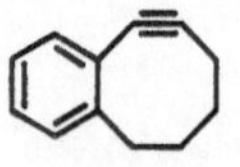	i	117)
	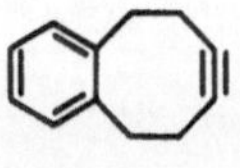	a	118)
	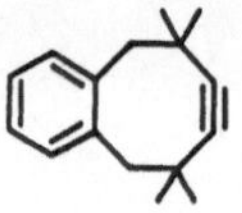	a	118)
	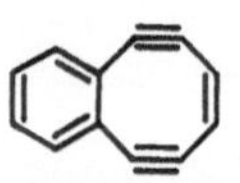	d	119)
	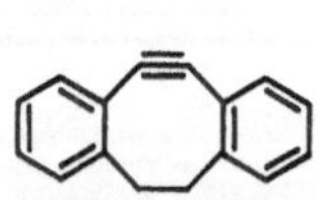	d, i	110, 120, 121)
	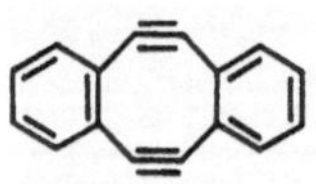	d	122, 123)
	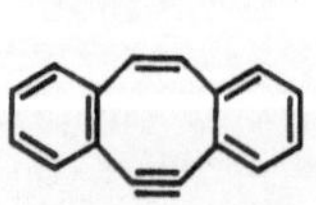	d	122, 123)
9	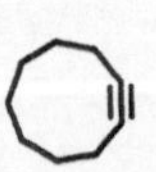	a, b, d, e, i	5–7, 27, 114, 124, 126)
	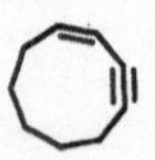	d	124)

Table 2. (continued)

Ring Size	Structure	Synthetic methods	Ref.
9		d	124)
	(structure with OCH₃)	d	113,114)
		l	142)
		l	125,116)
		a	127)
		a	128)
		a	128)
12		Cl — C≡CLi + CuCl	134,135)
		k	136–138)
		k	139)

For further angle strained cycloalkynes of larger ring size see Ref. [1c], p. 658–662. Some isolable ten- or eleven membered cycloalkynes having additional unsaturation in the ring are not considered here, since angle deformation at the triple bond, at least according to models, should be less than 10°.

3 Experimental Criteria for Angle Strain [143]

The most direct experimental evidence of angle strain in cycloalkynes is, of course, provided by structural analyses, such as X-ray analyses or electron diffraction (see Sect. 4). However, for most molecules discussed in this review such information is not available. In these cases for isolable molecules some spectroscopic data give at least an indication of angle strain. Particularly the ^{13}C-NMR-shifts of the $C \equiv C$-carbons [85, 86, 123, 144] and the positions of the $C \equiv C$-stretching vibration in the IR and Raman spectra [85, 86, 144] provide valuable information on angle strain, if related open-chain reference compounds are known. $^{13}C-^{13}C$ coupling constants, although only determined in a few cases, seem to depend on angle strain in cycloalkynes, too [145]. Since cis-bending of the $C-C \equiv C$ angles in cycloalkynes dramatically lowers the LUMO energy, electron transmission spectroscopy (ETS) is a very sensitive probe for ring strain in cycloalkynes [146, 147]. In contrast, the HOMO energy is not affected to such an extent by angle bending [146-149].

Heats of hydrogenation provide quantitative information about ring strain in cycloalkynes [150, 151]. Since in most cases ring strain leads to an increase in angle bending at the $C-C \equiv C$ bonds, a larger heat of hydrogenation ($-\Delta H$) reflects a larger angle deformation.

In most cases reactivity in addition reactions to the triple bond increases with increasing angle deformation. For transient cycloalkynes comparative studies of the reactivity, e.g. competition experiments, are one of the few experimental methods to assess ring strain in these reactive intermediates.

Last not least, force field [152-154] and semiempirical [106, 144] quantummechanical calculations have been used to calculate structural parameters and enthalpies of formation of angle strained cycloalkynes. These calculations are at present the only way to obtain quantitative estimates of ring strain in transient cycloalkynes, such as cycloheptyne [152].

If none of the aforementioned information is available, then construction and inspection of models can give a rough idea of the angle strain involved. For example, cyclodecyne can be built with stiff models without problems, but cyclononyne and cyclooctyne can only be constructed with a flexible $C-C \equiv C-C$ moiety.

4 Structural Effects on the Isolability of Angle Strained Cycloalkynes

The most important influence on the isolability is the ring size, i.e. the number of atoms in the ring. Cyclononyne (6) and cyclooctyne (14) are isolable compounds, but

$$\text{(9)}$$

202

cycloheptyne (*16*), cyclohexyne (*20*) and cyclopentyne (*18*) cannot be isolated. However, the intermediate occurrence of (*16*), (*20*) and (*18*) can be proven unequivocally by several methods. Experimental evidence for cyclobutyne (*26*) and cyclopropyne (*3*) has not been published yet.

In the absence of trapping reagents the main stabilization reactions of (*16*) and (*20*) are oligomerizations. Depending on starting material and conditions the products arising from (*16*) are two hexamers and the trimer of (*16*) [14, 56, 57], whereas (*20*) yields tetramers and the trimer [41, 42].

The oligomerization can be prevented in some cycloheptyne derivatives by introduction of four methyl groups in the α-positions. Thus, in contrast to (*16*) 3,3,7,7-tetramethylcycloheptyne (*10*) is isolable. However, 3,3,6,6-tetramethylcyclohexyne (*17*) cannot be isolated; apparently in the six membered cycloalkynes the four methyl groups cannot block the dimerization of (*17*) [38, 39].

The influence of α-methyl substituents on isolability has been investigated in detail in the silacycloheptyne system. Here the unsubstituted (*28*) [64] and the α-methyl substituted (*29*) [65] compounds are not isolable, while the silacycloheptynes having two or more methyl groups in the α-position are isolable [65, 87]. Thus, an increasing number of methyl groups in the α-position to the triple bond leads, as expected, to increasing kinetic stability.

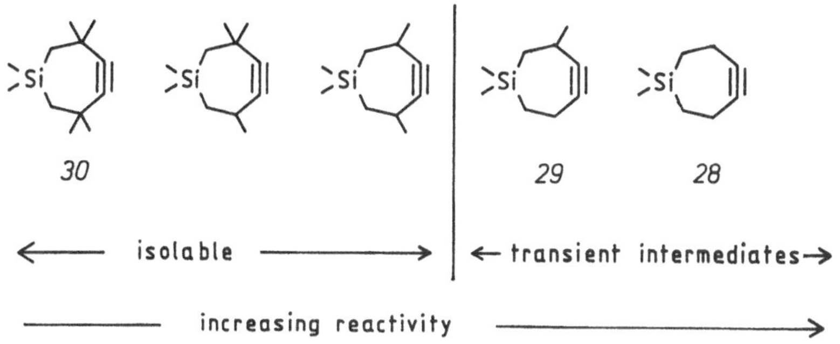

(10)

Within a given number of ring members, the heterocycloalkynes with the larger carbon-heteroatom bond length are more stable.

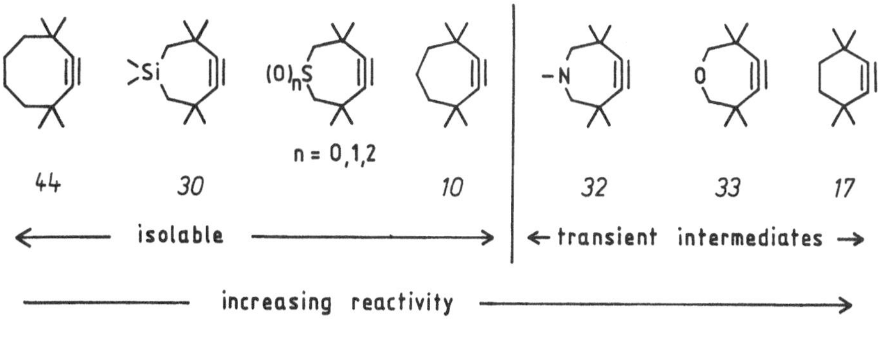

(11)

For example, in the series 1,1,3,3,6,6-hexamethyl-1-sila-4-cycloheptyne (30) [87], 3,3,6,6-tetramethyl-1-thia-4-cycloheptyne (31) [84, 85], (10) the reactivity increases significantly. This is probably due to the increase in $C—C≡C$-angle deformation with decreasing carbon-heteroatom bond length. This suggestion is further supported by the experimental fact, that 1,3,3,6,6-pentamethyl-1-aza-4-cycloheptyne (32) [63] and 3,3,6,6-tetramethyl-1-oxa-4-cycloheptyne (33) [62] cannot be isolated, since $C—N$ and $C—O$ are shorter than $C—C$ bond lengths.

Introduction of additional unsaturation in cyclooctyne leads to increasing ring strain and reactivity depending on the relative position and the number of double bonds, as is shown in the following scheme [106, 144].

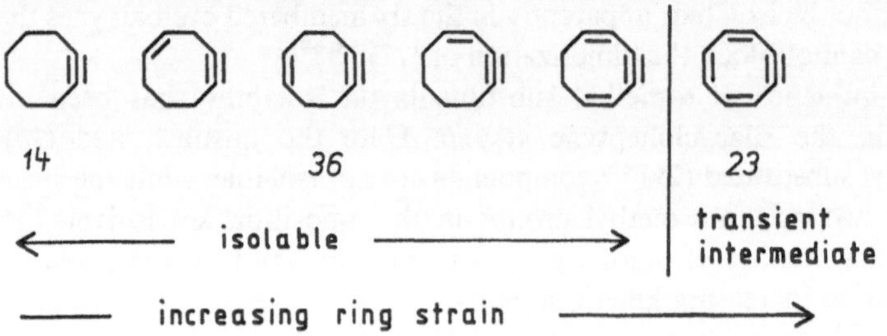

$$(12)$$

14   36   23

$\longleftarrow$ ——— isolable ———— $\longrightarrow$ transient intermediate

——— increasing ring strain ——— $\longrightarrow$

The 1,3,5-cyclooctatrien-7-yne (1,2-didehydrocyclooctatetraene) (23) is not isolable but has a half-life of about 12 minutes in $5 \cdot 10^{-2}$ molar solution at 25 °C [16].

In the benzoannellated compounds two opposing effects on isolability are to be expected. Annellation of a benzene ring to cyclooctyne should increase ring strain as does introduction of an additional double bond. On the other hand, the size of the benzene ring should, like a bulky substituent, impede oligomerization reactions and thereby kinetically stabilize the angle strained cycloalkyne.

$$(13)$$

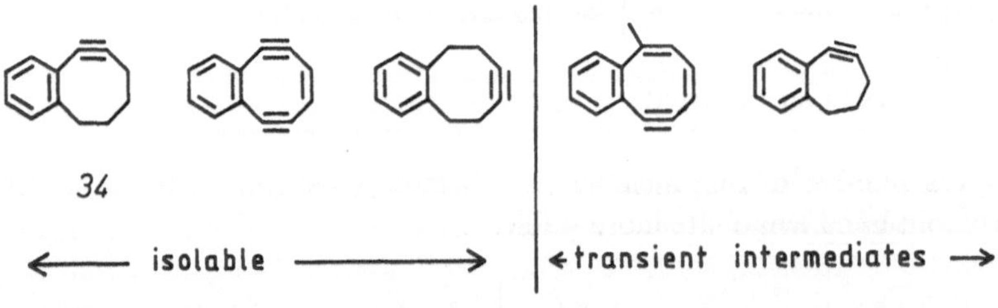

34

$\longleftarrow$ —— isolable ——— $\longrightarrow$ | $\longleftarrow$ transient intermediates $\longrightarrow$

——— increasing reactivity ——— $\longrightarrow$

The steric effect of *one* annellated benzene ring seems not sufficient to provide significant kinetic stabilization, particularly if the triple bond is not in an adjacent position as in (34).

However, two annellated benzene rings can, at least in the eight-membered systems, kinetically stabilize the angle strained cycloalkynes compared to the corresponding compounds with $C—C$ single bonds in place of the benzene rings, although the effect is not very large.

An interesting feature in this series, as well as in the corresponding unsubstituted 1,5-cyclooctadiyne (35) and 1-cycloocten-5-yne (36) is the higher kinetic stability of the diynes (35) [103, 104] and (37) [122, 123] compared to the enynes (36) [107] and (38) [122, 123], resp. This is probably due to the larger deformation of the angles at the triple bond in (38) [157] compared to (37) [158, 159] as shown by X-ray analyses.

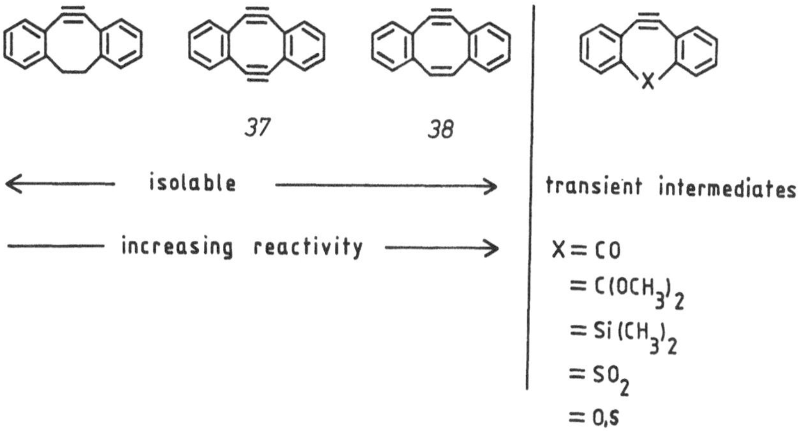

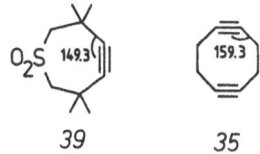

37   38   (14)

←————— isolable ————————————→ | transient intermediates

————— increasing reactivity ————→ | X = CO
| $= C(OCH_3)_2$
| $= Si(CH_3)_2$
| $= SO_2$
| $= O, S$

5 Molecular Structure of Angle Strained Cycloalkynes

5.1 X-Ray Analyses and Electron Diffraction

A few structural analyses of angle strained cycloalkynes have been carried out.
    Electron diffraction data show that the $C—C\equiv C$ bond angle is decreasing in the series (6) [160], (14) [161], (31) [162], although the difference between (6) and (14) is surprisingly small.

160.2   158.5   S 145.8   (15)

6   14   31

    The experimental data for (6) are consistent with a conformation of C_s symmetry for the lowest energy conformer, but other conformers may also be present in the equilibrium mixture, as is calculated by force field calculations [153, 154].
    Cyclooctyne (14) has C_2 symmetry. Interestingly, the $C\equiv C$-bond length in (14) is significantly larger (1.23 Å) [161] than in (6) (1.21 Å) [160] or in (31) (1.21 Å). [162]
    The seven-membered cycloalkyne (31) shows the smallest $C\equiv C—C$ angle (145.8°)

O_2S 149.3   159.3

39   35

205

yet observed [162]. A similar angle is observed in the X-ray analysis of the structurally related 3,3,6,6-tetramethyl-1-thia-4-cycloheptyne-1,1-dioxide (39), where two mole-cules of (39) with differing C—C≡C bond angles (151.7° and 147°, resp.) were found [163] in the unit cell.

The cyclic diyne (35) has C—C≡C angles of 159.3° as determined by an X-ray analysis, which is close to the value found in cyclooctyne by electron diffraction [161] (see above). The carbon framework of the molecule is almost planar with the two CH_2—CH_2 units in an almost eclipsed arrangement. This and the transannular repulsion between the π-moieties leads to the large value of 1.57 Å for the C_3—C_4 single bond. All other structural parameters seem to be normal. [164]

A smaller C—C≡C bond angle (155.9°) is observed in (37), which can be considered as a dibenzo derivative of (35). The molecule is completely planar. An interesting feature is the large bond length for the C—C bonds (1.426 Å) which are common to the eight membered and the benzene rings [158,159]; this is also observed in (38) [157].

(17)

The corresponding olefin (38) shows an even smaller C—C≡C bond angle than (37) of 154° [157]. Here, as in (39), two independent molecules with slightly different geometries were observed. In addition to the large angle deformation at the triple bond a significant widening of the C—C=C bond angles to almost 145° takes place [157]. Whereas one of the molecules of (38) is nearly planar, the other deviates from planarity by as much as 0.174 Å.

(18)

X-ray crystallographic studies have shown that the originally published structure of (40) [165] is not correct, but must be revised to (41) [166].

X-ray analyses of the two twelve membered tetraacetylenes (42) [167] and (43) [168] have been carried out. The bond angles at the triple bonds must be bent by about 10–15°, but the exact structural parameters of both compounds have not yet been published.

(19)

5.2 Force Field and Quantum Mechanical Calculations

Since for most angle strained cycloalkynes, particularly for those which are not isolable, structural data is lacking, calculations have been carried out. The structural parameters and strain energies of cyclononyne (6) [152–154, 160], cyclooctyne (14) [152, 154, 106, 144], cycloheptyne (16) [152, 154], cyclohexyne (20) [154] and cyclopentyne (18) [154] were calculated. Some values for strain energies and C—C≡C bond angles are given in Tables 3 and 4.

Table 3. Calculated Strain Energies of Some Angle Strained Cycloalkynes

Compound	Strain Energy [kcal/mol]				
	Ref.				
	152)	144)	154)	107)	153)
Cyclononyne (6)	16.4		11.3		15.2
Cyclooctyne (14)	20.8	11.8	19.2		
Cycloheptyne (16)	31.1		26.8		
Cyclohexyne (20)			45.7		
Cyclopentyne (18)			79.2		
1,5-Cyclooctadiyne (35)				29.0	
1-Cycloocten-5-yne (36)		18.2		20.3	
1-Cycloocten-4-yne		16.1			
1-Cycloocten-3-yne		20.9			

Table 4. Calculated C—C≡C Bond Angles in Some Angle Strained Cycloalkynes [°]

Compound	Ref.				
	152)	144)	154)	107)	153)
Cyclononyne (6)	169.7		167.4		164
Cyclooctyne (14)	158.5	163	158.8		
Cycloheptyne (16)	150.4		145.0		
Cyclohexyne (20)			130.6		
Cyclopentyne (18)			116.2		
1,5-Cyclooctadiyne (35)				159	
1-Cycloocten-5-yne (36)		162		156	
1-Cycloocten-4-yne		161			
1-Cycloocten-3-yne		154			
		162			

The results for cyclooctyne and cyclononyne are only in moderate accord with the electron diffraction data [160–161].

In addition to these calculations, the influence of an additional C=C double bond in the cyclooctyne system was investigated by MINDO/2 calculations which led to the results discussed on p. 204 [106, 144]. The calculated strain energies are in good agreement with the observed gradation of reactivity [106–108].

6 Thermochemical Data

The heats of hydrogenation of some angle strained cycloalkynes have been determined in acetic acid [150] or cyclohexane [151]. The experimental values including the heats of hydrogenation of some open-chain reference substances are given in Table 5.

Table 5. Enthalpies of Hydrogenation ($-\Delta H^{25\,°C}$) of Angle Strained Cycloalkynes, their Corresponding Alkenes and Open-chain Reference Compounds

Compound	$-\Delta H^{25\,°C}$ [kcal/mole]	Ref.
Cyclononyne	61.9[a]	[150]
Cyclooctyne	70.8[b]	[151]
3,3,8,8-Tetramethyl-cyclooctyne	74.3[b]	[151]
4,4,7,7-Tetramethyl-cyclooctyne	71.8[b]	[151]
1,1,3,3,6,6-Hexamethyl-1-sila-4-cycloheptyne	76.3[b]	[151]
3,3,6,6-Tetramethyl-1-thiacycloheptyne	86.2[b]	[151]
1,5-Cyclooctadiyne	154.0[b]	[151]
cis-Cyclooctene	22.4[a]	[169]
3,3,8,8-Tetramethyl-cis-cyclooctene	33.4[b]	[151]
4,4,7,7-Tetramethyl-cis-cyclooctene	24.7[b]	[151]
1,1,3,3,6,6-Hexamethyl-1-sila-4-cis-cycloheptene	31.4[b]	[151]
3,3,6,6-Tetramethyl-1-thia-cis-cycloheptene	34.1[b]	[151]
1,5-cis,cis-Cyclooctadiene	53.7[a]	[170]
4-Octyne	62.8[a]	[171]
4-cis-Octene	27.4[a]	[171]
2,2,7,7-Tetramethyl-4-octyne	63.8[a]	[151]
2,2,7,7-Tetramethyl-cis-4-octene	26.9[a]	[150]
2,2,5,5-Tetramethyl-3-hexyne	67.7	[171]
2,2,5,5-Tetramethyl-cis-3-hexene	36.2	[171]

a) in acetic acid   b) in cyclohexane

These values show that ring strain increases in the following order

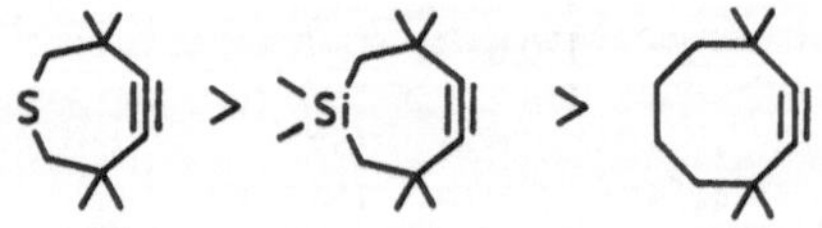

$$\text{(20)}$$

$$31 \qquad 30 \qquad 44$$

as already concluded from reactivity data (see p. 203). If one defines angle strain by the following equation

Angle Strain = [ΔH ⬡ − ΔH ⬡] − [ΔH ⫶ − ΔH ⫶]   Eq. (1)

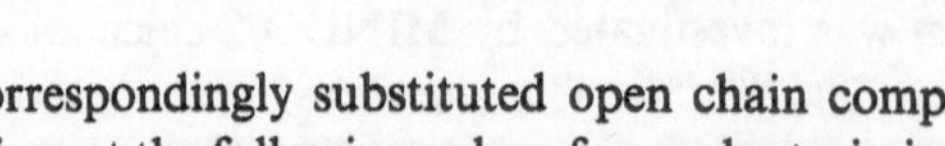

and uses the correspondingly substituted open chain compounds as reference substances, one arrives at the following values for angle strain in these cycloalkynes:

Table 6. Values of Angle Strain Calculated According to Equation (1) from Enthalpies of Hydrogenation [kcal/mole]

Cyclononyne (6)	2.9
Cyclooctyne (14)	13.0
3,3,8,8-Tetramethylcyclooctyne (44)	9.4
4,4,7,7-Tetramethylcyclooctyne (46)	10.2
1,1,3,3,6,6-Hexamethyl-1-silacycloheptyne (30)	13.4
3,3,6,6-Tetramethyl-1-thiacycloheptyne (31)	20.6
1,5-Cyclooctadiyne (35)	30.5 (15.3 per C≡C bond)

This table shows that the silacycloheptyne (30) is, as expected, more strained than the corresponding eight-membered system (44), but its value of 13.4 kcal/mole is closer to that of (44) than to the corresponding thiacycloheptyne (31). This is also reflected in the much lower reactivity of (30) compared to (31) [65]. These values of "angle strain" are not directly comparable with the values in Table 3.

With decreasing ring size the isomeric cycloallenes become more stable than the corresponding cycloalkynes, because an allene requires only three colinear carbon atoms (see p. 191–192). This is shown by comparing the equilibrium constants for the cycloalkyne/1,2-cycloalkadiene equilibria at 100 °C in tert.butanol/potassium-tert.-butoxide (Table 7) [4].

Table 7. The Equilibrium Composition of Cycloalkyne/1,2-Cycloalkadiene Mixtures [4]

Cycloalkyne	% Cycloalkyne in the Mixture
Cycloundecyne	74
Cyclodecyne	35
Cyclononyne	7

It has been shown that the cycloalkyne/1,2-cycloalkadiene equilibrium ratios depend markedly on the medium used for the equilibrations [4]. One reason may be that additional isomerization to the more stable 1,3-cycloalkadienes may occur which would impede the measurement of exact equilibrium constants. Therefore, for an exact comparison, heats of hydrogenation of the 1,2-cycloalkadienes would be very useful. A comparison of cyclooctyne and 1,2-cyclooctadiene is not possible, since 1,2-cyclooctadiene cannot be isolated because of its fast dimerization [173, 174]. This shows clearly the difference between thermodynamic and kinetic stability. Although the 1,2-cyclooctadiene may be thermodynamically more stable than cyclooctyne, its fast rate of dimerization makes it kinetically less stable. Uncatalyzed dimerization of cycloalkynes is apparently impossible, since it would lead to the energetically unfavorable cyclobutadiene system.

7 Spectroscopic Properties

The angle strain in cycloalkynes has a profound effect on several spectroscopic properties, as already pointed out on p. 202.

7.1 Photoelectron (PE) and Electron Transmission (ET) Spectroscopy

From a simple point of view one would expect that "in plane" cis-bending at the triple bond would lead to less efficient overlap in the "in plane" π-bond, while the π-bond perpendicular to this plane would not be affected very much by in plane bending. However, substantial σ-π-mixing must necessarily occur as a consequence of angle bending.

Model calculations which show the influence of cis-bending on the orbital structure of acetylene and 2-butyne have been performed using the Extended Hückel Theory [175], the MINDO/2 and SPINDO procedures [149, 176, 177] and ab initio methods [178, 179]; they all come to the conclusion that cis-bending up to 30° does not change significantly the energy of highest occupied molecular orbital (HOMO) (an example is given in Fig. 2). The influence of bending is larger in the semiempirical methods than in ab initio calculations [177, 179].

In contrast, according to ab initio [179] and Extended Hückel calculations [180] the energy of the lowest unoccupied molecular orbital (LUMO) is lowered dramatically in the cis-bent models.

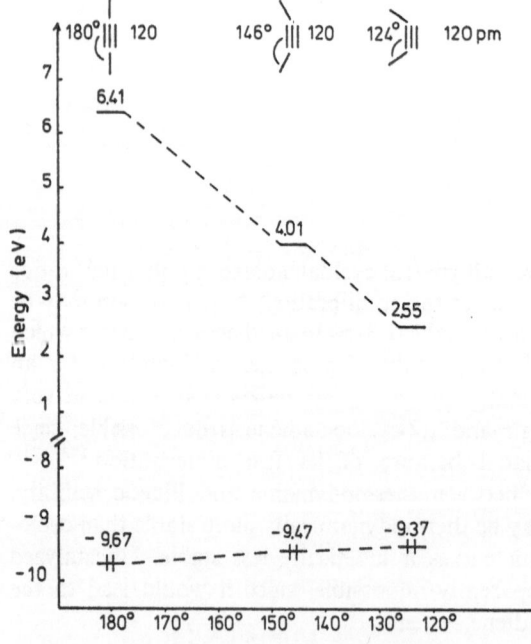

Fig. 2. Frontier Orbitals and Energies (eV). Calculated for 2-Butyne Models (4-31G) [179] vs. C—C≡C bond angles. The angles of 146° and 124° correspond to (31) and 1,2-benzyne, resp.

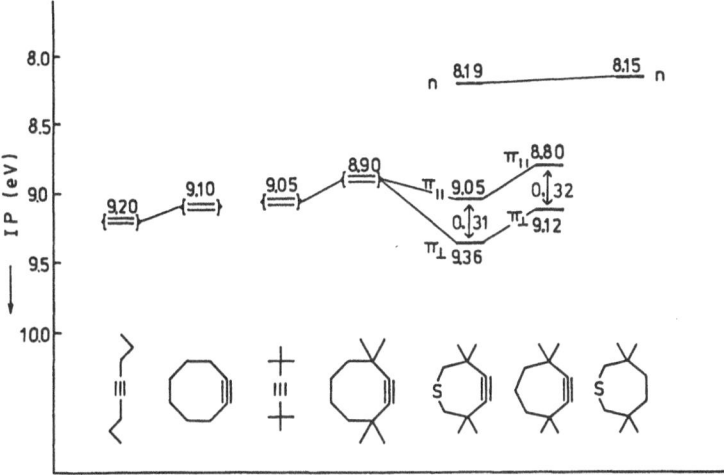

Fig. 3. PE Ionization Potentials (IP) of the HOMO's in Some Angle Strained and Open-Chain Alkynes [149, 181]

Experimental investigations confirm these predictions. PE spectroscopic results show that cis-bending of the $C—C \equiv C$ angles in angle strained cycloalkynes does not change the HOMO energy very much, but a splitting of the otherwise degenerate π-orbitals is observed in seven-membered cycloalkynes (*31*) and (*10*) (Fig. 3). This splitting cannot be observed yet in eight-membered cycloalkynes [149, 177].

Recent investigations by ET spectroscopy show that the expected lowering of the LUMO energies is indeed observed (Table 8 and Fig. 4) [146, 147]. The splitting of the two π^* orbitals in (*31*) amounts to 1.66 eV, which is in good accord with the results of the ab initio calculations. Fig. 4 shows the observed vertical ionization potentials and electron affinities of some angle strained cycloalkynes and some open-chain alkynes of analogous substitution pattern. By extrapolation of these results to an angle of 124°, one can conclude that the 1,2-benzyne anion will be stable.

Table 8. Vertical Electron Affinities (eV) of Some Alkynes as Determined by Electron Transmission Spectroscopy [146, 147]

2-Butyne	−3.43
2,2,5,5-Tetramethyl-3-hexyne	−3.10
Cyclooctyne (*14*)	−2.18, −3.28
3,3,6,6-Tetramethyl-1-thiacycloheptyne (*31*)	−1.16, −2.82

From the ET spectra vertical electron affinities are obtained. However, for several reasons it would be interesting to know the adiabatic electron affinities, particularly since there might be a large difference between the vertical and the adiabatic electron affinities in these cycloalkynes. Therefore, electrochemical investigations were carried out to find additional evidence for the increased electron affinity.

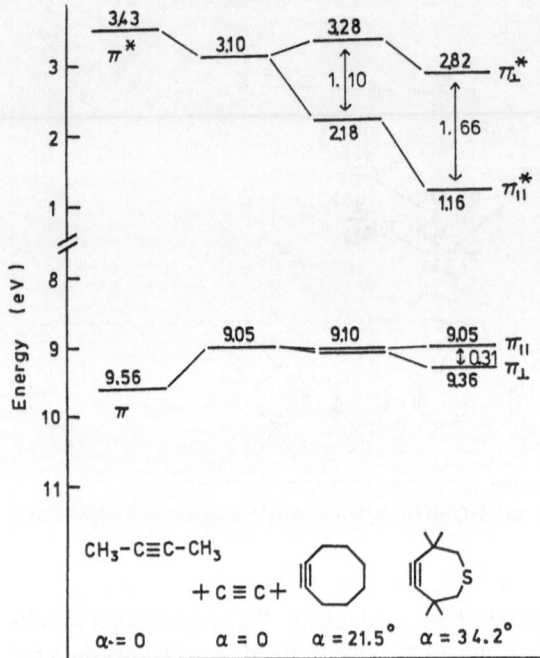

Fig. 4. Correlation Diagram for the Vertical Electron Affinities and Ionization Potentials of Some Linear Alkynes and Some Angle Strained Cycloalkynes [146, 147]

(α = Deviation of C—C≡C bond angles from 180°)

Dialkylacetylenes usually cannot be reduced electrochemically up to —3.0 V which is the limit in most solvents suitable for these experiments. However, the three seven-membered cycloalkynes, (31) and its corresponding S-oxide and S,S-dioxide, exhibit irreversible reduction waves at —2.93, —2.78 and —2.83 V vs. saturated calomel electrode, although the corresponding cyclooctynes are not reduced up to —3.0 V [182].

It was shown by reduction experiments with the corresponding thiacycloalkanes and cis-cycloalkenes that reduction starts at the triple bond and not at the sulfur. Thus, increasing deformation at the triple bond leads to larger electron affinity even in solution [182]. The implications of these results also support the idea that linear alkynes are more electrophilic than related alkenes, because they can easily adopt a bent structure in reaction transition states.

PE spectra of the two angle strained cyclopolyynes 1,5-cyclooctadiyne (35) [177] and the twelve membered tetraacetylene (43) [88] have been measured and interpreted with the aid of semiempirical and ab initio molecular orbital calculations. In both compounds evidence for through-space and through-bond interactions between the acetylene moieties has been found [88, 177]. The compound (43) has been described on the basis of these results as weakly antiaromatic [88].

7.2 ¹H- and ¹³C-NMR-Spectroscopy

For the obvious reason that the carbon atoms of a C≡C bond are not directly bonded to protons, ¹H-NMR chemical shifts are not of much importance as a probe for ring strain. There have been some attempts to get an idea of the C≡C—C bond angle from the ¹H-chemical shifts of the α-CH$_2$-groups, but the observed effects are small and no quantitative correlation is possible [100]. However, temperature dependent ^{1}H-NMR-spectroscopy has been used to study the intra-molecular mobility of angle strained cycloalkynes (see below).

Interesting effects are observed in ^{13}C-NMR-spectroscopy (see Table 9).

Table 9. ^{13}C-Chemical Shifts of the Acetylenic Carbons in Angle Strained Cycloalkynes and Some Reference Compounds (CDCl$_3$, rel. to TMS)

Ring members	Compound	δ	Ref.
7	3,3,7,7-Tetramethylcycloheptyne (*10*)	109.8	99)
	3,3,5,5,7,7-Hexamethylcycloheptyne	107.3	9)
	3,3,6,6-Tetramethyl-1-thia-4-cycloheptyne (*31*)	108.6	85)
	3,3,6,6-Tetramethyl-1-thia-4-cycloheptyne-1-oxide	104.6	99)
	3,3,6,6-Tetramethyl-1-thia-4-cycloheptyne-1,1-dioxide (*39*)	101.7	183)
	1,1,3,3,6,6-Hexamethyl-1-sila-4-cycloheptyne (*30*)	100.7	87)
	1,1,3,3,6-Pentamethyl-1-sila-4-cycloheptyne	103.2	65)
		98.2	
	1,1,3,6-Tetramethyl-1-silacycloheptyne	100.7	87)
8	Cyclooctyne (*14*)	94.4	85)
	3-Methylcyclooctyne	98.6	65)
		92.9	
	3,8-Dimethylcyclooctyne	96.8	98)
	3,3-Dimethylcyclooctyne	100.5	98)
		91.4	
	3,3,8-Trimethylcyclooctyne	98.7	98)
		95.3	
	3,3,8,8-Tetramethylcyclooctyne (*44*)	98.2	99)
	3,3,7,7-Tetramethylcyclooctyne (*46*)	98.1	102)
		91.3	
	5-Chlorcyclooctyne	95.2	184)
		93.3	
	5-Hydroxycyclooctyne	94.7	184)
		94.3	
	5-Oxocyclooctyne	94.1	184)
		91.9	
	1-Cycloocten-3-yne	114.5	144)
		94.5	
	1-Cycloocten-4-yne	96.0	144)
		95.8	
	1-Cycloocten-5-yne	100.4	107)
	1,5-Cyclooctadiyne	95.8	104)
	5,6-Didehydro-11,12-dihydrodibenzo[a.e]cyclooctene (*45*)	111.6	70, 185)
	5,6-Didehydrodibenzo[a.e]cyclooctene (*38*)	108.5	186)
	5,6,11,12-Tetradehydrodibenzo[a.e]cyclooctene (*37*)	109.3	186)
9	Cyclononyne	87.5	85)
10	Cyclodecyne	83	144)
12	Cyclododecyne	81.5	144)
12	1-Cyclododecen-3-yne	94.5	144)
		78.6	
—	4-Octyne	80.3	85)
—	2,2,5,5-Tetramethyl-hexyne	87.0	85)
—	Diphenylacetylene	89.5	70)

213

Table 9 demonstrates that the ^{13}C chemical shift of the acetylenic carbon is a sensitive probe of ring strain. Increasing ring strain displaces the ^{13}C signals to lower field. Thus, among the systems without an additional C=C bond in the ring (10) has the highest δ value of 109.8, which is a shift of about 22.8 ppm from the corresponding carbon signal of the unstrained open-chain reference compound 2,2,5,5-tetramethyl-3-hexyne or even 11.6 ppm from that of (44). (31) and the corresponding sila compound (30) have lower δ values of 108.6 and 100.7, resp. which correlate well with the diminished reactivity of these compounds and the heats of hydrogenation (see p. 208). However, a comparison of the ^{13}C shifts of (31) and its corresponding S-oxide and S,S-dioxide (δ = 108.6, 104.6 and 101.7, resp.) reveals that other influences, such as inductive effects, cannot be neglected.

Therefore, to assess ring strain effects on chemical shifts it is important to use good reference compounds.

Additional C=C bonds increase the ring strain in cyclooctynes as was shown by a thorough investigation of the ^{13}C chemical shifts of the cyclooctenynes (see p. 213), which were correlated with calculated and experimental values of C—C≡C bond angles [144]. The same type of shift of 19–22 ppm relative to diphenylacetylene is observed in the dibenzoannellated eight membered systems (45), (38) and (37) [70, 185, 186]. The observed shift due to ring strain is explained as a consequence of the changed hybridization of the acetylenic carbons in the angle strained cycloalkynes [144].

In order to test this interpretation the $^{13}C-^{13}C$-coupling constants of the two strained cycloalkynes, 3-methylcyclooctyne and 1,1,3,3,6-pentamethyl-1-sila-cycloheptyne were determined, and compared to the corresponding coupling constants in 2,2,5-trimethyl-3-hexyne (Fig. 5). [145]

There is a significant decrease of about 10 Hz in the $^{13}C-^{13}C$-coupling constants on going from the open-chain reference compound to the seven-membered sila-cycloheptyne system; however, considering a difference of 83 Hz between acetylene

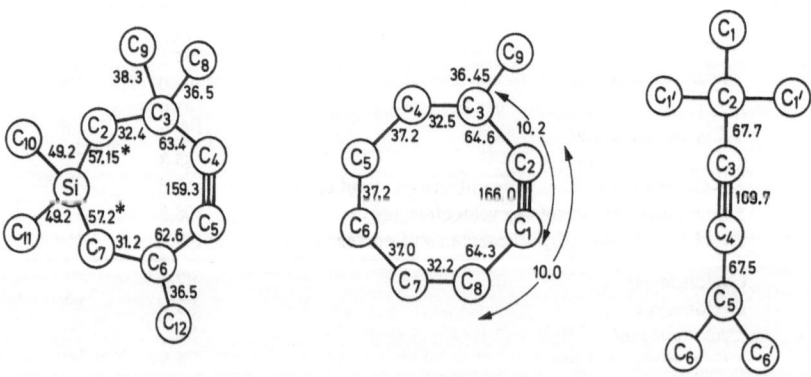

Fig. 5. $^{13}C-^{13}C$—NMR Coupling Constants in two Angle Strained Cycloalkynes and an Unstrained Reference Compound

and ethylene in $^{13}C-^{13}C$-coupling constants a change of about 10 Hz is not much [187, 188]. Probably a more detailed knowledge of the factors, which influence $^{13}C-^{13}C$ coupling constants and additional experimental results are necessary to interpret the observed effects correctly.

Cyclooctyne (14) and the dibenzoannellated system (45) possess only C_2-symmetry [100, 118, 185] and are therefore chiral molecules. Models suggest that ring strain should increase the barrier to racemization compared to unstrained compounds.

$$(21)$$

44   45   46

However, (46) exhibits only 3 singlets in the ^{1}H-NMR-spectrum at $+35\,°C$ for methyl, α- and γ-methylene groups, resp. At lower temperatures ($-70\,°C$, 60 MHz) the expected splitting is observed and from the coalescence temperatures a value $\Delta G^{\neq} = 12$ kcal/mole can be obtained for the inversion process. Thus, ring strain increases the barrier for inversion in the cyclooctyne system, but the increase is not sufficiently l rge to allow separation of enantiomers at or near room temperature.

Annellati n of two benzene rings increases the barrier for the racemization process considerabl and the ^{1}H-NMR-spectrum of (45) shows the expected splitting at room temperature [120]. Coalescence of the AA′BB′ system for the CH_2-CH_2 moiety occurs at 128 °C (60 MHz) which corresponds to $\Delta G^{\neq} = 19.7$ kcal/mole [118, 185]. By chromatography of (45) on a chiral phase (TAPA on silica gel) at $-28\,°C$ a partial separation of the enantiomers was achieved.

Ring strain in (31) increases also the barrier to inversion compared to analogous unstrained compounds; here a barrier $\Delta G^{\neq} = 9.3$ kcal/mole is observed [85, 189].

7.3 IR- and Raman-Spectroscopy

Angle strained cycloalkynes exhibit usually a weak band for the $C\equiv C$ stretching mode in the IR spectrum and a strong band in the Raman spectrum.

Table 10 summarizes some wave numbers for angle strained cycloalkynes.

Table 10. $C\equiv C$-Stretching Bands in Angle Strained Cycloalkynes

Ring Size	Compound	Wave Number (cm^{-1})	Method (IR, Raman/Solvent)	Ref.
6	1,2-Benzyne	2085	IR/Ar matrix	190)
	3,3,6,6-Tetramethyl-cyclohexyne (17)	2108	IR/Ar matrix	12)
7	3,3,7,7-Tetramethyl-cycloheptyne (10)	2190, 2170, 2210, 2179, 2158	IR/CCl$_4$ Raman/neat	8)
	3,3,5,5,7,7-Hexamethyl-cycloheptyne (79)	2180, 2165	IR/CCl$_4$	9)

Table 10. (continued)

Ring Size	Compound	Wave Number (cm^{-1})	Method (IR, Raman/Solvent)	Ref.
7	3,3,6,6-Tetramethyl-1-thia-4-cycloheptyne (*31*)	2200, 2170, 2172, 2145	IR/CCl$_4$ Raman/neat	[85]
	3,3,6,6-Tetramethyl-1-thia-4-cycloheptyne-1-oxide	2180	IR/KBr	[85]
	3,3,6,6-Tetramethyl-1-thia-4-cycloheptyne-1,1-dioxide (*39*)	2177	IR/KBr	[86]
	1,1,3,3,6,6-Hexamethyl-1-sila-4-cycloheptyne (*30*)	2210, 2190	IR/neat	[87]
	1,1,3,6-Tetramethyl-1-sila-4-cycloheptyne	2190	IR/neat	[87]
	1,1,3,3,6-Pentamethyl-1-sila-4-cycloheptyne	2190	IR/neat	[65]
8	Cyclooctyne (*14*)	2260, 2206	IR/neat	[185]
	3-Methylcyclooctyne	2220	IR/neat	[65]
	3,8-Dimethylcyclooctyne	2240, 2210	IR/neat	[98]
	3,3-Dimethylcyclooctyne	2210	IR/neat	[98]
	3,3,8-Trimethylcyclooctyne	2215	IR/neat	[98]
	3,3,8,8-Tetramethyl-cyclooctyne (*44*)	2230	IR/neat	[99]
	4,4,7,7-Tetramethyl-cyclooctyne (*46*)	2230	IR/neat	[100]
	3,3,7,7-Tetramethylcyclooctyne	2220	IR/neat	[102]
		2222	Raman/neat	[102]
	1-Cycloocten-3-yne	2180	IR/neat	[144]
		2180.5	Raman/neat	[144]
	1-Cycloocten-4-yne	2280, 2220	IR/neat	[144]
	1,5-Cyclooctadiyne	2237, 2175	IR/CCl$_4$	[104]
	7,8,9,10-Tetrahydro-5,6-didehydrobenzocyclooctene	2185, 2150	IR/CCl$_4$	[185]
	6,6,9,9-Tetramethyl-5,6,9,10-tetrahydro-7,8-didehydro-benzocyclooctene	2225	IR/CCl$_4$	[118]
	5,6,9,10-Tetradehydro-benzocyclooctene	2100	IR/CCl$_4$	[119]
	5,6-Didehydro-11,12-dihydrodibenzo[a.e]cyclooctene (*45*)	2170, 2140 2153	IR/CCl$_4$ Raman	[70, 120, 185] [120]
	5,6,11,12-Tetrahydro-dibenzo[a.e]cyclooctene	2180	IR/KBr	[122]
9	Cyclononyne	2230	IR	[191]

From these data it follows that increasing ring strain, i.e. smaller C—C≡C angles, generally leads to a shift to lower wave numbers. A linear correlation between wave numbers and experimental or calculated C—C≡C bond angles has been proposed [144]; however some compounds, like 1-cycloocten-4-yne, deviate significantly from the linear correlation.

In some cases two or three bands show up in the region between 2100 and 2300 cm^{-1}; this was previously observed in open-chain alkynes and was ascribed to Fermi resonance with overtones or combination bands. The shift to lower wave numbers may be a consequence of the smaller $C-C\equiv C$ angles. This deformation can influence the $C\equiv C$ stretching frequency in two ways.

1. The hybridization may be changed from sp towards sp^2, which would lead to a lower force constant of the $C\equiv C$ bond stretch and a lower frequency.
2. Angle deformation can affect coupling of the $C\equiv C$ bond stretch with the adjacent $C-C$ bonds; this interaction was used to explain the dependence of IR bands on the bond angles in cycloalkenes [192]. This effect would also shift the $C\equiv C$ stretching band to lower wave numbers.

7.4 Electronic Absorption Spectra

The wave length and the intensity of the absorptions in the electronic spectra of the dibenzocycloalkynes (37) and (38) has been used as an argument for the planarity of these systems [186].

8 Reactions

8.1 General Remarks

Angle strained cycloalkynes exhibit enhanced reactivity, particularly in addition reactions to the triple bond, over analogous open chain alkynes. Therefore many reactions can be carried out with angle strained cycloalkynes which are not feasible or at least cannot be carried out under such mild conditions with analogous open chain alkynes. This review concentrates on the reactions typical for angle strained cycloalkynes, while those typical also for "normal" alkynes are not discussed in detail.

The cycloalkynes (10), (31) and related compounds offer an interesting combination. They combine high reactivity of the triple bond with steric protection of the resulting addition product; therefore they are used for the synthesis of new systems and for the realization of new reaction pathways.

8.2 Isomerization

It was shown that eleven- to nine-membered cycloalkynes under basic conditions rearrange to the isomeric allenes with the equilibrium shifted towards the allene as the ring size decreases (see p. 209) [4]; the allenes can subsequently isomerize to the more stable 1,3-dienes. Should the observed trend continue to smaller rings, the short-lived cycloallenes should be considerably more stable than the corresponding cyclo-alkynes. The intermediate occurrence of 1,2-cyclooctadiene, 1,2-cycloheptadiene and 1,2-cyclohexadiene has been proved by trapping reactions and labeling experiments [155, 156, 172, 173, 174]. However, the rearrangement of a short-lived cycloalkyne has not yet been demonstrated; the reactive cycloalkyne usually undergoes other faster reactions, such as oligomerizations.

8.3 Oligomerization

Cyclooctyne (14) trimerizes on heating to give the benzene derivative (47) [18,93]; the trimerization can be catalyzed by many transition metal salts or complexes, such as Ni(CN)$_2$ [193], TiCl$_4$ [194], Al$_2$O$_3$ [195] and some transition metal carbonyls [196].

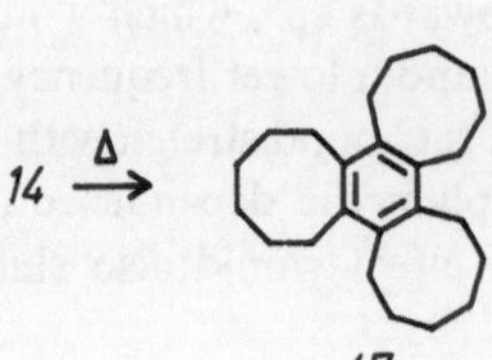

$$14 \xrightarrow{\Delta} \qquad (22)$$

47

The polymerization of cyclooctyne can be initiated by some tungsten carbene complexes [197]. A rather unique reaction is the dimerization of the isolable (10) to give (11) [9] (see p. 192); the same type of product is observed in the attempted preparation of (17) [38, 39]. The mechanism of this dimerization is not clear but it shows that in these strained cycloalkynes cleavage of ring C—C bonds is possible under mild conditions. The methyl groups in the α-position of (10) and (17) must have an important influence; for unsubstituted cycloheptyne (16) and cyclohexyne (20), if generated from aminotriazoles yield two hexamers [14] and a tetramer (48) [41, 42], resp. as the main products, the corresponding trimers are only by-products. If (16) is generated at 250 °C by pyrolysis of cycloheptenocyclopropenone only the trimer (49) is found [56, 57].

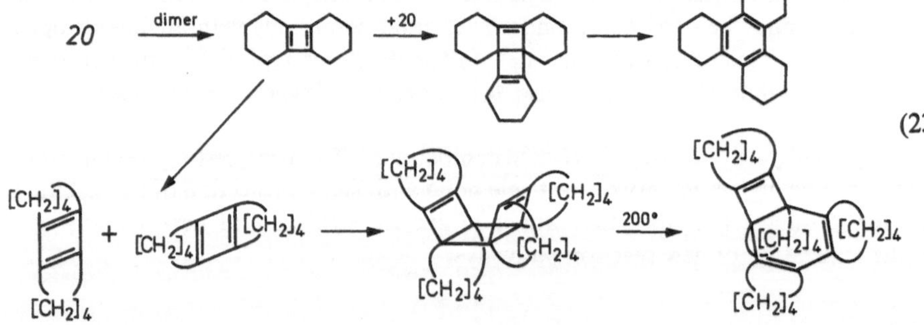

$$(23)$$

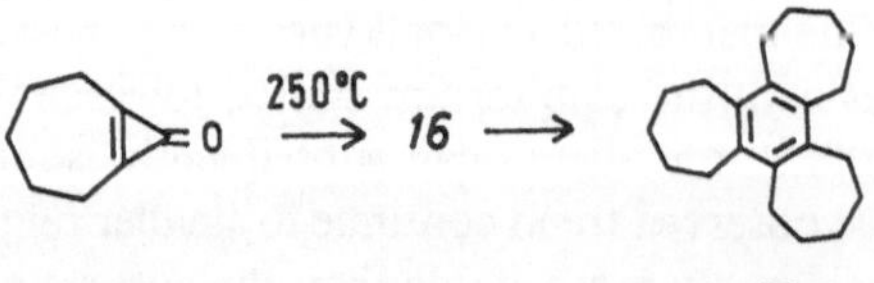

$$\xrightarrow{250\,°C} 16 \longrightarrow \qquad (24)$$

49

The formation of hexamers and tetramers is best explained by the intermediate occurence of cyclobutadienes, which then trimerize or dimerize [14,41,42], resp. It is important to note that the stable cycloheptynes (10) and (31) can only be converted to the corresponding cyclobutadienes via the cyclobutadiene-PdCl$_2$-complexes [9,198,199]. The formation of (49) from (16) at 250 °C suggests that only at low temperatures sufficient cyclobutadiene can accumulate for formation of a hexamer.

Another reaction which may proceed via a cyclobutadiene intermediate and subsequent proton shifts is the dimerization of (23) to yield cycloocta[b]naphthalin (50) [15].

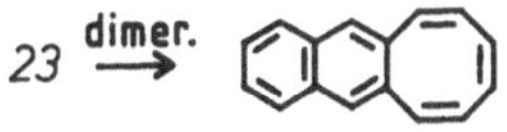

$$23 \xrightarrow{\text{dimer.}}$$

50

(25)

Particularly interesting is the dimer formed from didehydrobullvalene (51) in which the two fluctuating bullvalene systems are fused to each other by a four-membered ring. According to ^{1}H-NMR data (52), with the three-membered rings in the anti-configuration, is the thermodynamically most stable of the 17 possible valence isomers, whereas the cyclobutadiene structure (53) is not present in the equilibrium mixture [200]; in addition to these dimers, a trimer is also obtained [200].

51   52   53

(26)

8.4 Electrophilic Additions

Angle strained cycloalkynes undergo electrophilic additions typical for open-chain dialkylacetylenes, at a faster rate.

Cyclooctyne (14) adds bromine to give 1,2-dibromo-cis-cyclooctene [93]; this is presently the only route to prepare a pure 1,2-dihalocyclooctene. The addition of bromine and iodine to (31) leads to the unexpected products (54); apparently trans-annular attack of the cationic intermediate (55) by the nucleophilic sulfur produces a strained sulfonium ion which stabilizes itself by ring opening to (54) [201].

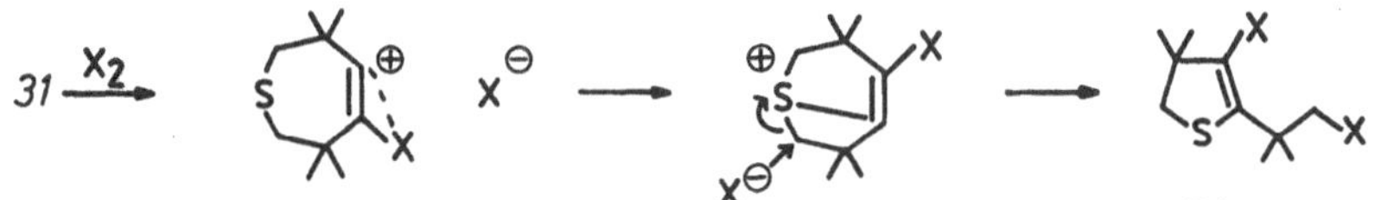

$$31 \xrightarrow{X_2}$$

55

54

X = Br, I

(27)

In contrast, addition of bromine and iodine to the cyclic sulfone (39) yields the expected 1,2-dihalocycloalkenes (57) [86,201].

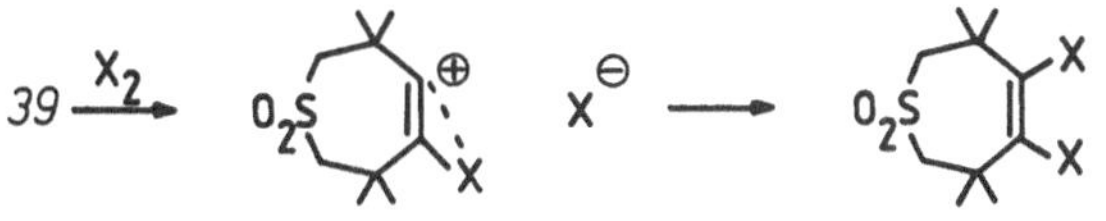

$$39 \xrightarrow{X_2}$$

58   57

X = Br, I

(28)

Since (57) had the cis-configuration a detailed NMR spectroscopic investigation of the bromine addition was carried out. In the ^{13}C- and ^{1}H-NMR spectra at low temperatures as intermediate was found whose signals can be assigned to the unsymmetrically bridged structure (58). However, the corresponding 1,2-dihalo-trans-cycloalkene could not be observed [86].

Since sulfenylhalides exhibit high anti-selectivity in addition reactions to alkynes [202] and since a bulky substituent at the sulfur should stabilize the trans-cycloalkene relative to the cis-alkene, trichloromethylsulfenylchloride was used as an electrophile.

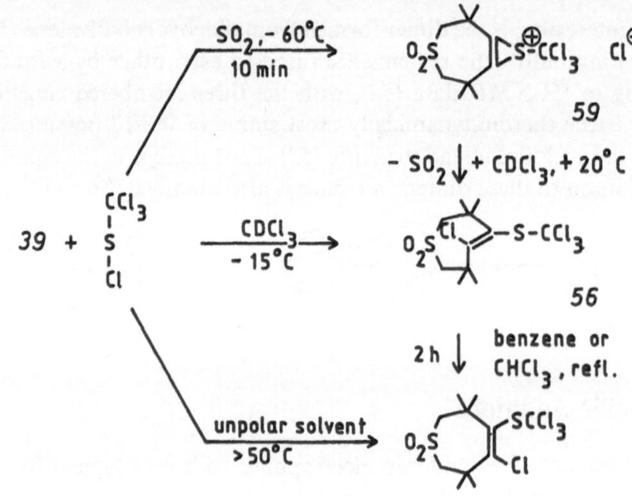

$$(29)$$

If the reaction was carried out in liquid SO_2 at $-60\,°C$ the thiirenium ion (59) was formed, which in CDCl$_3$ at 20 °C yielded (56). This represents the first seven-membered trans-cycloalkene, which is isolable at room temperature. In refluxing chloroform (56) isomerized to the corresponding cis-cycloalkene; the activation parameters of this trans-cis-isomerization were determined [86].

Uncatalyzed addition of acetic acid yields the corresponding enolacetates. Since this reaction proceeds with reasonable rates for all isolable eight- and seven-membered cycloalkynes, it was used for a quantitative determination of the influence of ring

Table 11. Rates of Addition of Acetic Acid to Some Angle Strained Cycloalkynes (40 °C, CDCl$_3$)

Rate constants $(M^{-1} \cdot sec^{-1})$	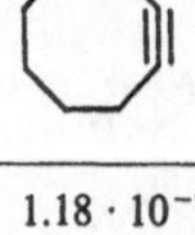 $1.18 \cdot 10^{-9}$	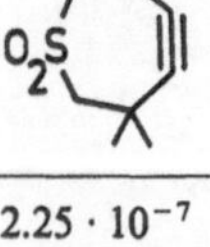 $2.25 \cdot 10^{-7}$	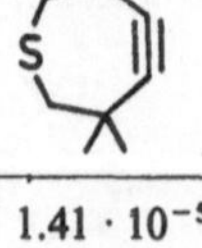 $1.41 \cdot 10^{-5}$	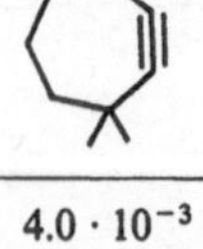 $4.0 \cdot 10^{-3}$
Relative Rate (Cyclooctyne = 1)	1	190	11950	3390000

strain on reactivity. The rate constants and relative rates (cyclooctyne = 1) are shown in Table 11 [203].

This demonstrates the enormous increase in the rates of electrophilic additions with increasing ring strain.

Even phenol, being less acidic than acetic acid, adds to (31) at 50 °C. However, the final product is not the enolether (60) but an o-substituted phenol (61), which is probably formed by a rearrangement of (60) [204].

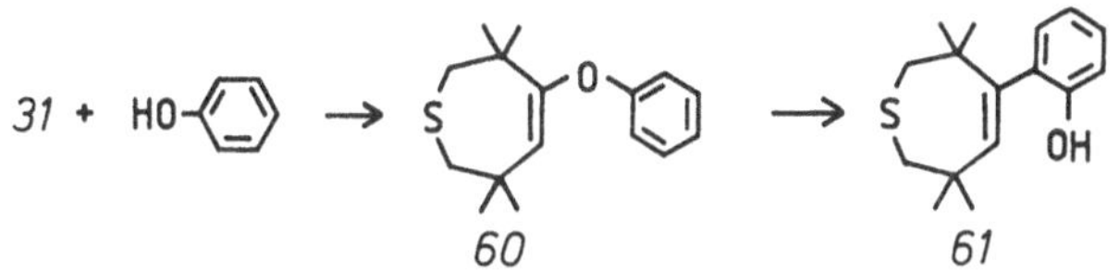

$$31 + HO \text{—} \quad \rightarrow \quad 60 \quad \rightarrow \quad 61 \tag{30}$$

8.5 Nucleophilic Additions

Although there is ample evidence for nucleophilic additions to benzyne [1a] and some other unstable angle strained cycloalkyne intermediates [15,27,31,205–207], only a few addition reactions to isolable angle strained cycloalkynes are known which can be classified as nucleophilic. Hydroxylamine and hydrazine add to (31) to yield the corresponding oxime and hydrazone, resp. [208].

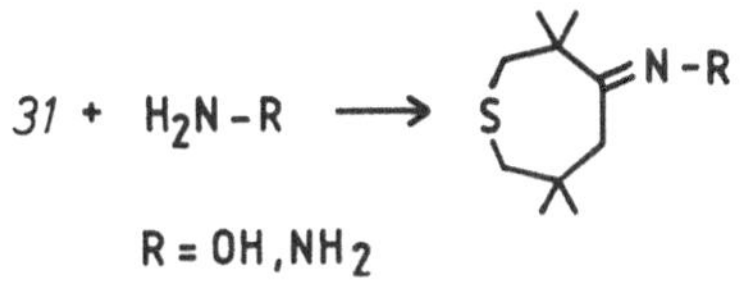

$$31 + H_2N\text{–}R \quad \longrightarrow \quad \tag{31}$$

$$R = OH, NH_2$$

The very strong bases, butyllithium and tert. butyllithium add to (31) to give the corresponding alkenes in moderate yields [208,209]. The addition of butyllithium has also been observed with cyclooctyne [19] and cyclooctenynes [108].

As a consequence of the drastic lowering of the LUMO energy on C–C≡C bending in angle strained cycloalkynes a stronger increase is expected for nucleophilic than for electrophilic additions [178,179]. This prediction could not be proved so far for isolable cycloalkynes [209,210]; however, there is some, although not conclusive evidence in benzyne chemistry for such a trend [179].

8.6 Radical Addition Reactions

The reactions of (31) and (39) with molecular oxygen has been investigated in some detail [208,211]. The reaction products for the autoxidation and rose bengale sensitized oxidation are shown below.

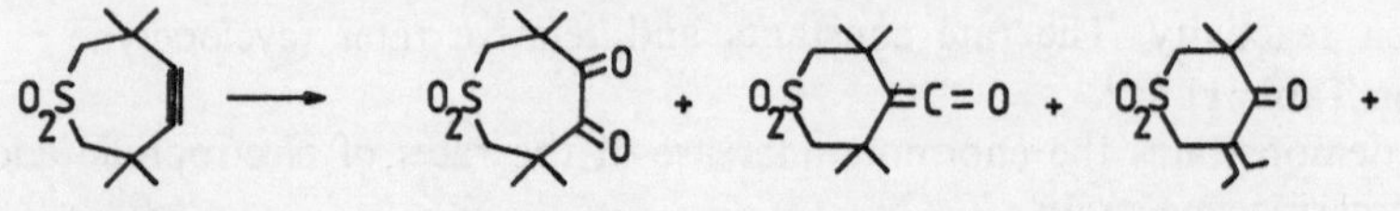

39

Yield	Autoxidation	20-24 %	8-9 %	4-7 %
	Photooxidation (Rose Bengale, -30°C, CH₃OH)	49 %	1 %	4 %

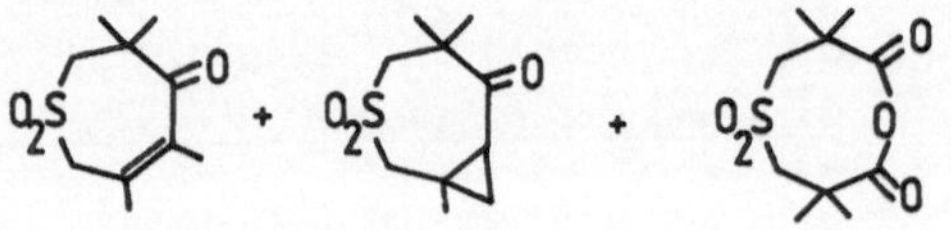

Yield

Autoxidation	4-7 %	5-10 %	* 8 %
Photooxidation	5 %	3 %	12 %

(32)

It has beeen shown by inhibition experiments that more than 90% of the oxidation reaction proceed via a radical pathway. The formation of products can be explained by the following scheme:

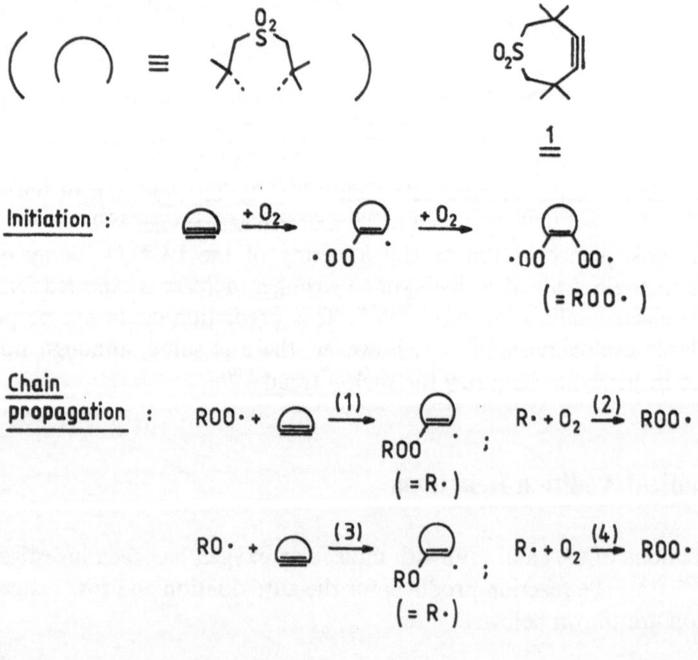

$$\left(\bigcap \equiv \begin{array}{c} O_2 \\ S \end{array} \right)$$

1

Initiation :

$\cdot OO$ $(=ROO\cdot)$

Chain propagation :

$ROO\cdot +$ (1) ROO $(=R\cdot)$; $R\cdot + O_2$ (2) $ROO\cdot$

$RO\cdot +$ (3) RO $(=R\cdot)$; $R\cdot + O_2$ (4) $ROO\cdot$

222

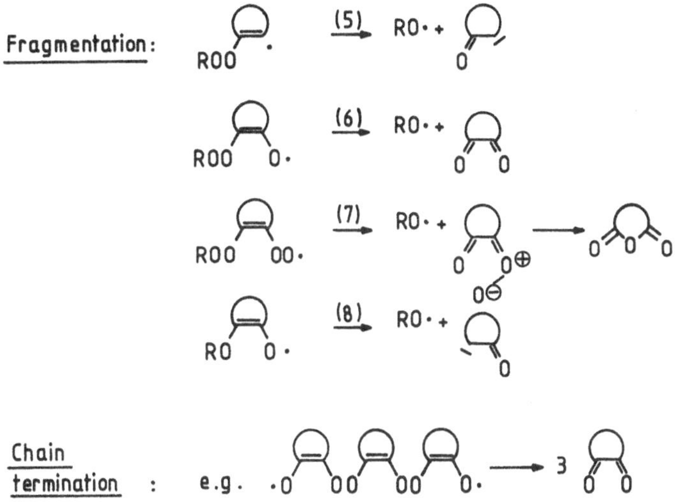

Fragmentation:

$$(5) \quad RO\cdot + $$

$$(6) \quad RO\cdot + $$

$$(7) \quad RO\cdot + $$

$$(8) \quad RO\cdot + $$

Chain termination : e.g.

$$\longrightarrow 3 \qquad (33)$$

However, the chemiluminescent reaction in the autoxidation of (31) is not inhibited by radical scavengers [208]; it may proceed via a 1,2-dioxete (62). The proposed [211] thermal generation of singlet oxygen in the presence of (31) could not be confirmed in various experiments [208].

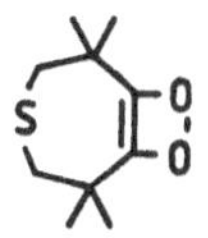

$$(34)$$

62

8.7 Hydrogen Transfer Reactions

Isolable angle strained cycloalkynes can be hydrogenated in the presence of a catalyst and this reaction has been used to determine the heats of hydrogenation (see Chapter 6).

More typical for angle strained cycloalkynes is the dehydrogenation of alcohols, amines and thiols [208]. The reaction of (31) with methanol to yield the alkene (63) and formaldehyde was investigated in detail. On the basis of rate laws and H/D kinetic isotope effects the following reaction mechanism was proposed [208, 212].

Table 12. Half-Lives $t_{1/2}$ of Hydrogenation of (31) at 44.3 °C, excess of reagent >20, without solvent

	Reagent	$t_{1/2}$
31 → 63	$C_3H_7NH_2$	54 h
	CH_3OH	3.4 h
	C_2H_5SH	~ sec.
	$C_6H_5-CH_2NH_2$	2.5 h
	$C_6H_5-CH_2OH$	<3 min

223

The rate of the hydrogen transfer reaction depends on the structure of the alcohol, amine or thiol as well as on the structure of the cycloalkyne. Cyclooctyne (*14*) dehydrogenates ethanol, but the reaction is 10^2–10^3 times slower than with (*31*) [213]. The half-lives of some dehydrogenation reactions are given in Table 12 [208].

The dehydrogenation of thiols is not a radical reaction; this was proved by inhibition experiments and isotope studies [208].

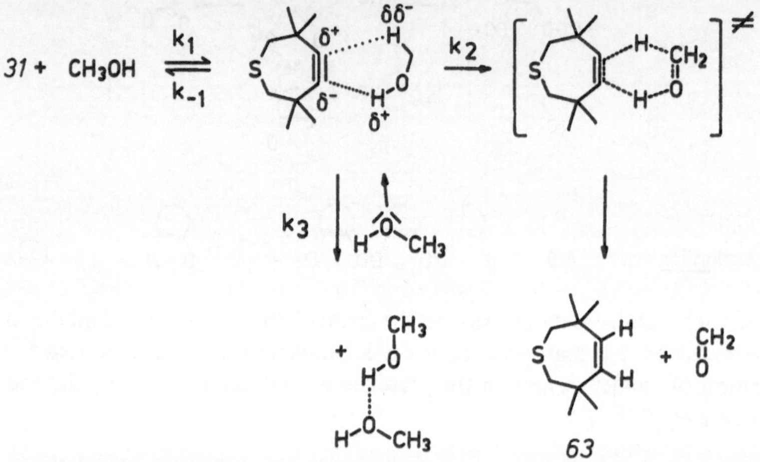

$$(35)$$

8.8 Cycloadditions

Cycloaddition reactions, particularly the Diels-Alder and 1,3-dipolar cycloaddition, have been used extensively in trapping unstable cycloalkynes. In addition, some typical cycloaddition reactions leading to interesting products are observed only with angle strained cycloalkynes.

8.8.1 [2+1]-Cycloadditions

Carbenes add to cyclooctyne (*14*) to give the corresponding cyclopropenes or compounds, which can be considered as rearrangement products of these cyclopropenes [214]. The addition of isonitriles is typical for seven membered cycloalkynes [8,215]; the rate of formation of the cyclopropenimines depends on the electrophilicity of the isonitriles [215]. Electron deficient isonitriles, such as p-nitrophenylisonitrile, add much faster than alkylisonitriles.

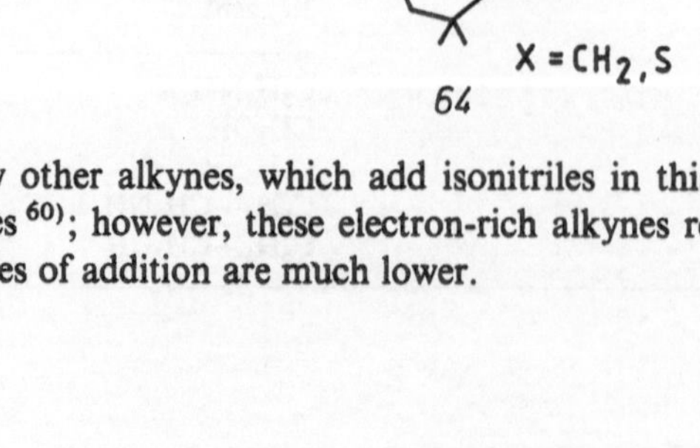

$$(36)$$

The only other alkynes, which add isonitriles in this way are ynamines [216] and ynediamines [60]; however, these electron-rich alkynes react only with arylisonitriles and the rates of addition are much lower.

8.8.2 [2+2]-Cycloadditions

Dichloroketene adds to (*31*) [85], (*37*) [186] and (*38*) [186] to give the corresponding dichlorocyclobutenones. The first azetinone (*65*), whose structure is sufficiently proven, was synthesized by the reaction of (*31*) with tosylisocyanate [208]. The mechanism of its formation is given below.

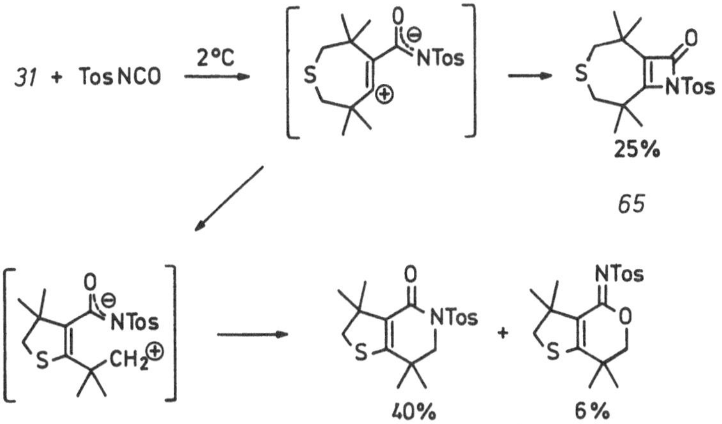

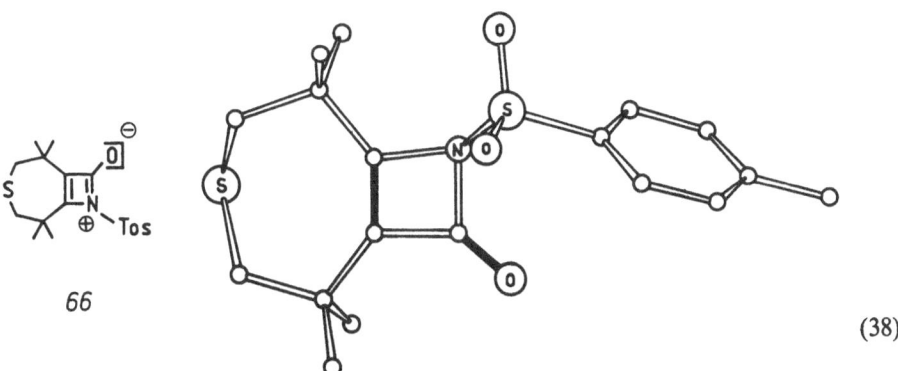

(37)

25%

65

40%   6%

The X-ray structural analysis of (*65*) reveals some interesting features. The length of the C—N amide bond is 1.482 Å compared to 1.41 Å in an azetidinone. In addition the nitrogen has a tetrahedral structure, it is even more tetrahedral than ammonia. These parameters indicate that the conceivable azacyclobutadiene structure (*66*) does not contribute to (*65*), apparently as a consequence of an antiaromatic destabilization of (*66*) [217].

66

(38)

Length of amide bond: 148.2 pm (normal: 141 pm)
Sum of all bond angles at N: 324°
Torsional angle in the four membered ring: 8.4°

Sulfur adds to (*31*) and (*39*) at elevated temperatures to give the 1,2-dithietes (*67*); these represent the first alkylsubstituted 1,2-dithietes [218].

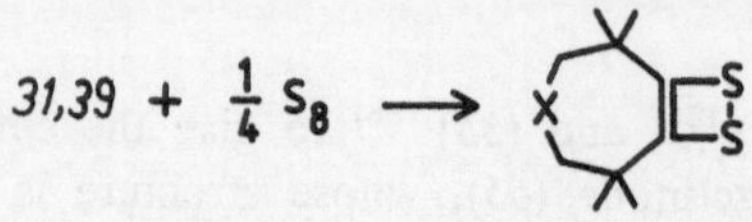

$$31,39 + \frac{1}{4} S_8 \longrightarrow \qquad (39)$$

$$X = S$$
$$= SO_2$$
$$67$$

An interesting entry into the heptalene system starting from 1,3,5-tri-tert.butyl-pentalene was achieved via a [2 + 2]-cycloaddition of cyclooctyne (14) [225].

8.8.3 [3+2]-Cycloadditions

Phenylazide has often been used to trap unstable cycloalkynes [15,18,25,62] and to characterize angle strained cycloalkynes [18,89]. The addition of suitable diazoalkanes to cyclooctyne (14) was applied as the first step in the synthesis of threemembered ring spirenes [223,224].

A reaction typical of angle strained cycloalkynes is the addition of CS_2 leading to tetrathiafulvalenes [8,70,85−87,102,65,107]. The formation of the tetrathiafulvalenes is explained by the intermediate occurrence of the carbene (70), which has been trapped by reaction with methanol in the case of (10) and (31) [85,219].

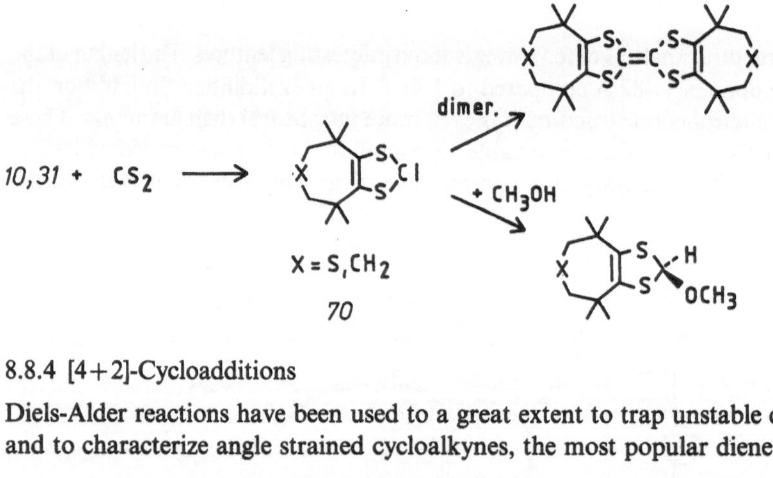

$$10,31 + CS_2 \longrightarrow \qquad (40)$$

$$X = S, CH_2$$
$$70$$

8.8.4 [4+2]-Cycloadditions

Diels-Alder reactions have been used to a great extent to trap unstable cycloalkynes and to characterize angle strained cycloalkynes, the most popular diene being tetra-

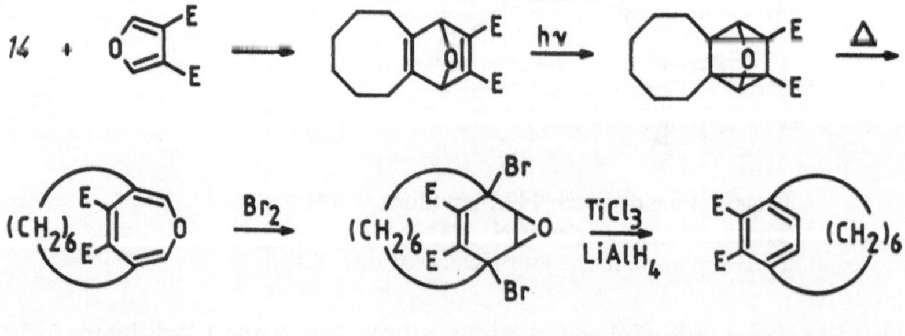

$$14 + \qquad (41)$$

$$E = COOC_2H_5 \qquad\qquad 68$$

phenylcyclopentadienone. An example for such a trapping reaction is given on p. 194 [15].

Benzo-phosphabarrelenes have been obtained from the addition of cyclooctyne (*14*) to 2-phenyl-1-phosphanaphthalene [226]. Cyclooctyne (*14*) has recently been used to synthesize a [6]-paracyclophane (*68*) the first step being a Diels-Alder addition of (*14*) to a substituted furan [220-222].

8.9 Reactions with Metal Salts or Metal Complexes

The first synthesis of an isolable cyclobutadiene (*71*) kinetically stabilized by bulky substituents was achieved starting from the angle strained cycloalkyne (*31*) via the $PdCl_2$ complex (*69*) [9, 198, 199]. Here again the high reactivities of (*10*), (*31*) and (*79*)

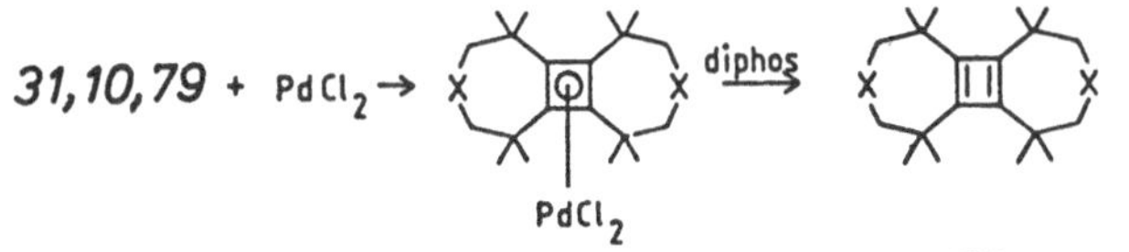

$$31, 10, 79 + PdCl_2 \rightarrow \quad \xrightarrow{diphos}$$

$$\begin{array}{cc} 69 & 71 \\ X = S & X = S \\ = CH_2 & = CH_2 \\ = C(CH_3)_2 & = C(CH_3)_2 \end{array}$$ (42)

are essential for the success of the syntheses. Di-tert.butylacetylene having the same substitution pattern as (*10*), (*31*) and (*79*) with $PdCl_2$ gives only the alkyne-$PdCl_2$ complex [227]. On the other hand, (*31*) forms with $CpCo(CO)_2$ at low temperatures an alkyne complex (*72*) in almost quantitative yield [204].

$$72 \qquad\qquad 79$$ (43)

Complexes of this type have been assumed as intermediates in reactions of $CpCo(CO)_2$ with other alkynes [228], but (*72*) is the first isolated example of this

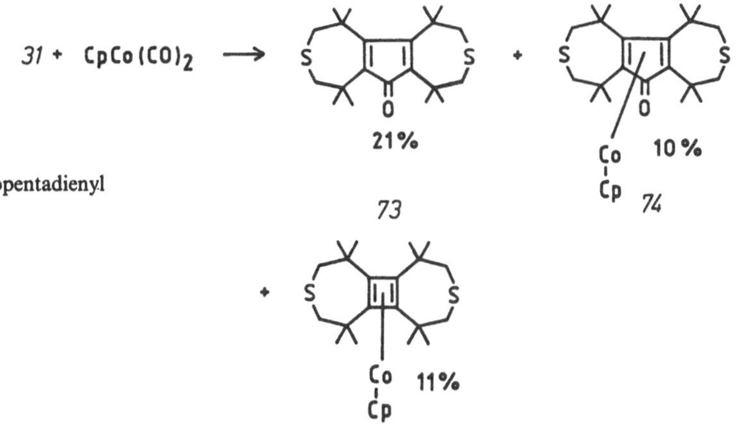

$$31 + CpCo(CO)_2 \rightarrow$$

21%
73

10%
74

Cp = Cyclopentadienyl

11%
75

(43)

type. At elevated temperatures from (31) and $CpCo(CO)_2$ the cyclopentadienone (73), the cyclopentadienone complex (74) and the cyclobutadiene complex (75) have been isolated in good yields [204]. Complex (72) seems to be an intermediate in the formation of these products.

Thus far, it has not been possible to prepare the cyclobutadiene (71) and related cyclobutadienes from metal complexes other than the $PdCl_2$-complexes [199,229,230].

The reaction of cyclooctyne (14) with $NiBr_2$ and NiI_2 gave the corresponding cyclobutadiene-complexes in low yield, while with $Ni(CO)_4$ the nickel(0) cyclopentadienone complex (76) was formed, whose thermal decomposition yielded the cyclopentadienone (77). In this case again a cyclooctyne Ni(0) complex (78) was postulated as an intermediate, but could not be isolated due to its instability [193].

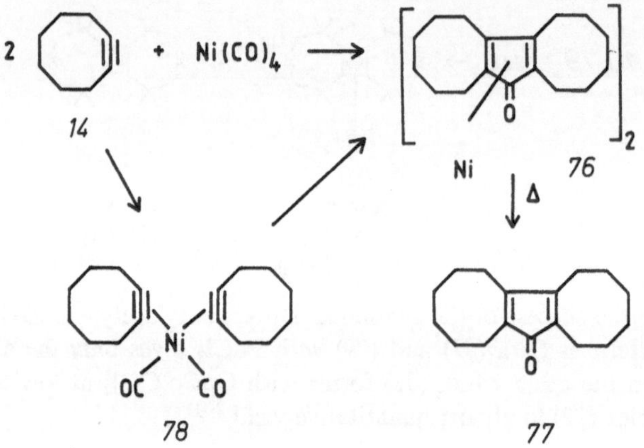

(44)

A stabilization of the unstable cyclohexyne (20) and cycloheptyne (16) has been achieved through zerovalent platinum complexes [46,47]; X-ray analyses of these complexes have been published [231].

9 Conclusions and Outlook

The chemistry of angle strained cycloalkynes has brought about as well novel insights into structural and spectroscopic problems as possibilities for the synthesis of new interesting systems.

There are still many more applications of angle strained cycloalkynes conceivable, which have not yet been pursued.

10 References

1. The literature is covered until June 1982, in few cases later references are included. Previous reviews on this topic or parts of it are: a) R. W. Hoffmann: Dehydrobenzene and Cycloalkynes, Verlag Chemie, Weinheim 1967; b) A. Krebs: Cyclic Acetylenes, in: Chemistry of Acetylenes, (Ed. H. G. Viehe), Marcel Dekker, New York 1969, p. 987; c) M. Nakagawa: Cyclic Acetylenes, in: The Chemistry of the carbon-carbon triple bond, (Ed. S. Patai), J. Wiley & Sons, New Aork 1978, p. 635; d) H. Meier, Synthesis 1972, 235.

2. Isolability, as used throughout this article, means that the respective compound can be isolated at or near 20 °C in neat form. However, in some cases it is not evident from the respective publications, if these conditions are met.
3. P. Saxe, H. F. Schaefer III: J. Am. Chem. Soc. *102*, 3239 (1980)
4. W. R. Moore, H. R. Ward: ibid. *85*, 86 (1963)
5. W. J. Ball, S. R. Landor: Proc. Chem. Soc. *1961*, 143
6. A. T. Blomquist, L. H. Liu, J. C. Bohrer: J. Am. Chem. Soc. *74*, 3643 (1952)
7. H. Meier, E. Voigt: Tetrahedron *28*, 187 (1972)
8. A. Krebs, H. Kimling: Angew. Chem. *83*, 540 (1971); Angew. Chem. Int. Edit. *10*, 509 (1971)
9. J. Pocklington: Ph. D. Thesis Hamburg, 1979
10. L. K. Montgomery, L. E. Applegate: J. Am. Chem. Soc. *89*, 5305 (1967)
11. L. Brandsma, H. D. Verkruijsse: Synthesis *1978*, 290
12. A. Krebs, W. Cholcha: unpublished results
13. O. L. Chapman, J. Gano, P. R. West, M. Regitz, G. Maas: J. Am. Chem. Soc. *103*, 7033 (1981)
14. G. Wittig, I. Meske-Schüller: Liebigs Ann. Chem. *711*, 65 (1968)
15. A. Krebs: Angew. Chem. *77*, 966 (1965); Angew. Chem. Int. Ed. Engl. *4*, 953 (1965); A. Krebs, D. Byrd: Lieb. Ann. Chem. *707*, 66 (1967)
16. A. S. Lankey, M. A. Ogliaruso: J. Org. Chem. *36*, 3339 (1971)
17. G. Wittig, A. Krebs, R. Pohlke: Angew. Chem. *72*, 324 (1960)
18. G. Wittig, A. Krebs: Chem. Ber. *94*, 3260 (1961)
19. G. Wittig, R. Pohlke: ibid. *94*, 3276 (1961)
20. G. Wittig, J. Weinlich, E. R. Wilson: ibid. *98*, 458 (1965)
21. G. Wittig, J. Heyn: Liebigs Ann. Chem. *726*, 57 (1969)
22. G. Wittig, J. Heyn: ibid. *756*, 1 (1972)
23. L. K. Montgomery, J. D. Roberts: J. Am. Chem. Soc. *82*, 4750 (1960)
24. L. K. Montgomery, F. Scardiglia, J. D. Roberts: ibid. *87*, 1917 (1965)
25. J. M. Bolster, R. M. Kellogg: ibid. *103*, 2868 (1981)
26. G. Wittig, H. Heyn: Chem. Ber. *97*, 1609 (1964)
27. K. L. Erickson, J. Wolinsky: J. Am. Chem. Soc. *87*, 1142 (1965)
28. K. L. Erickson, B. E. Vanderwaart, J. Wolinsky: J.C.S. Chem. Commun. *1968*, 1031
29. L. Fitjer, U. Kliebisch, D. Wehle, S. Modaressi: Tetrahedron Lett. *1982*, 1661
30. P. G. Gassman, T. J. Atkins: ibid. *1975*, 3035
31. P. G. Gassman, J. J. Valcho: J. Am. Chem. Soc. *97*, 4768 (1975)
32. P. G. Gassman, I. Gennick: ibid. *102*, 6863 (1980)
33. J. Nakayama, T. Segiri, R. Ohya, M. Hoshino: J.C.S. Chem. Comm. *1980*, 791
34. F. Scardiglia, J. D. Roberts: Tetrahedron *1*, 343 (1957)
35. A. T. Bottini, F. P. Corson, R. Fitzgerald, K. A. Frost II: ibid. *28*, 4883 (1972)
36. P. Caubère, J. J. Brunet: ibid. *29*, 4835 (1972)
37. B. Fixari, J. J. Brunet, P. Caubère: ibid. *32*, 927 (1976)
38. D. E. Applequist, P. A. Gebauer, D. E. Gwynn, L. H. O'Connor: J. Am. Chem. Soc. *94*, 4272 (1972)
39. C. N. Bush, D. E. Applequist: J. Org. Chem. *42*, 1076 (1977)
40. G. Wittig, U. Mayer: Chem. Ber. *96*, 329 (1963)
41. G. Wittig, U. Mayer: ibid. *96*, 342 (1963)
42. G. Wittig, J. Weinlich: ibid. *98*, 471 (1965)
43. J. Wolinsky: J. Org. Chem. *26*, 704 (1961)
44. J. J. Brunet, B. Fixari, P. Caubère: Tetrahedron *30*, 2931 (1974)
45. A. T. Bottini, B. R. Anderson, V. Dev, K. A. Frost jr.: ibid. *32*, 1613 (1976)
46. M. A. Bennett, G. B. Robertson, P. O. Whimp, T. Yoshida: J. Am. Chem. Soc. *93*, 3797 (1971)
47. M. A. Bennett, T. Yoshida: ibid. *100*, 1750 (1978)
48. C. F. Huebner, R. T. Puckett, M. Brzechffa, S. L. Schwartz: Tetrahedron Lett. *1970*, 359
49. H. Hart, S. Shamouilian, Y. Takehira: J. Org. Chem. *46*, 4427 (1981)
50. J. G. Duboudin, B. Jousseaume, M. Pinet-Vallier: J. Organomet. Chem. *172*, 1 (1979)
51. E. Müller, G. Odenigbo: Chemiker-Ztg. *97*, 662 (1973)
52. E. Müller, G. Odenigbo: Liebigs Ann. Chem. *1975*, 1435
53. A. Faworsky, W. Boshowsky: ibid. *390*, 122 (1912)
54. A. Favorsky, N. A. Domnine, M. F. Chestakowski: Bull. Soc. Chim. France [5] *3*, 1727 (1936)

55. A. T. Bottini, K. A. Frost II, B. R. Anderson, V. Dev.: Tetrahedron 29, 1975 (1973)
56. R. Breslow, J. Posner, A. Krebs: J. Am. Chem. Soc. 85, 234 (1963)
57. R. Breslow, L. J. Altman, A. Krebs, E. Mohacsi, I. Murata, R. A. Peterson, J. Posner: ibid. 87, 1326 (1965)
58. H. Meier, M. Layer, W. Combrink, S. Schniepp: Chem. Ber. 109, 1650 (1976)
59. F. G. Willey: Angew. Chem. 76, 144 (1964); Angew. Chem. Int. Ed. 3, 138 (1964)
60. A. Güntner: Diplom Thesis, Hamburg 1982
61. H. Meier, T. Molz, U. Merkle, T. Echter, M. Lorch: Liebigs Ann. Chem. 1982, 914
62. A. Krebs, G. Burgdörfer: Tetrahedron Lett. 1973, 2063
63. G. Burgdörfer: Ph. D. Thesis, Heidelberg 1975
64. S. F. Karaev, A. Krebs: Tetrahedron Lett. 1973, 2853
65. H. J. Hohlt: Ph. D. Thesis, Hamburg 1981
66. W. Tochtermann: Angew. Chem. 74, 432 (1962)
67. W. Tochtermann, K. Oppenländer, U. Walter: Chem. Ber. 97, 1318, 1329 (1964)
68. W. Tochtermann, K. Oppenländer, M. N. D. Hoang: Liebigs Ann. Chem. 701, 117 (1967)
69. M. Lorch, H. Meier: Chem. Ber. 114, 2382 (1981)
70. J. Odenthal: Ph. D. Thesis, Heidelberg 1975
71. J. A. Elix, M. V. Sargent: J. Am. Chem. Soc. 91, 4734 (1969)
72. J. A. Elix, M. V. Sargent, F. Sondheimer: J.C.S. Chem. Commun. 1966, 509
73. J. F. M. Oth, R. Merenyi, T. Martini, G. Schröder: Tetrahedron Lett. 1966, 3087
74. J. Gasteiger, G. E. Gream, R. Huisgen, W. E. Konz, U. Schnegg: Chem. Ber. 104, 2412 (1971)
75. H. N. C. Wong, F. Sondheimer: Tetrahedron Lett. 1980, 983
76. H. N. C. Wong, F. Sondheimer: J. Org. Chem. 45, 2438 (1980)
77. H. Gugel, H. Meier: Chem. Ber. 113, 1431 (1980)
78. H. N. C. Wong: Ph. D. Thesis, Univ. College London, 1976 cited in 79)
79. N. Z. Huang, F. Sondheimer: Acc. Chem. Res. 15, 96 (1982)
80. P. E. Eaton, C. E. Stubbs: J. Am. Chem. Soc. 89, 3722 (1967)
81. R. Wolovsky, F. Sondheimer: ibid. 88, 1525 (1966)
82. W. A. Böll: Angew. Chem. 78, 755 (1966); Angew. Chem. Int. Ed. 5, 733 (1966)
83. M. Psiorz, H. Hopf: Angew. Chem. 94, 639 (1982); Angew. Chem. Int. Ed. Engl. 21, 623 (1982)
84. A. Krebs, H. Kimling: Tetrahedron Lett. 1970, 761
85. A. Krebs, H. Kimling: Liebigs Ann. Chem. 1974, 2074
86. U. Höpfner: Ph. D. Thesis, Heidelberg 1979
87. R. Neubauer: Ph. D. Thesis, Heidelberg 1977
88. C. Santiago, K. N. Houk, G. J. De Cicco, L. T. Scott: J. Am. Chem. Soc. 100, 692 (1978)
89. A. T. Blomquist, L. H. Liu: ibid. 75, 2153 (1953)
90. G. Schröder, J. F. M. Oth: Angew. Chem. 79, 458 (1967); Angew. Chem. Int. Ed. Engl. 6, 414 (1967)
91. G. Schröder, H. Röttele, R. Merenyi, J. F. M. Oth: Chem. Ber. 100, 3527 (1967)
92. J. F. M. Oth, R. Merenyi, H. Röttele, G. Schröder: ibid. 100, 3538 (1967)
93. G. Wittig, H.-L. Dorsch: Liebigs Ann. Chem. 711, 46 (1968)
94. Eu. Müller, H. Meier: ibid. 716, 11 (1968)
95. H. Meier, I. Menzel: J.C.S. Chem. Commun. 1971, 1059
96. E. V. Dehmlow, M. Lissel: Liebigs Ann. Chem. 1980, 1
97. H. Meier, I. Menzel: Synthesis 1971, 215
98. A. Krebs, K. D. Seidel: unpublished results
99. A. Krebs: unpublished results
100. A. Krebs: Tetrahedron Lett. 1968, 4511
101. A. Krebs, U. Joerss: unpublished results
102. J. Deutsch: Ph. D. Thesis, Heidelberg 1979
103. E. Kloster-Jensen, J. Wirz: Angew. Chem. 85, 723 (1973); Angew. Chem. Int. Ed. Engl. 12, 671 (1973)
104. E. Kloster-Jensen, J. Wirz: Helv. Chim. Acta 58, 162 (1975)
105. H. Meier, C. Schulz-Popitz, H. Petersen: Angew. Chem. 93, 286 (1981); Angew. Chem. Int. Ed. Engl. 20, 270 (1981)
106. H. Petersen, H. Kolshorn, H. Meier: Angew. Chem. 90, 483 (1978); Angew. Chem. Int. Ed. Engl. 17, 461 (1978)

107. W. Leupin, J. Wirz: Helv. Chim. Acta *61*, 1663 (1978)
108. H. Petersen, H. Meier: Chem. Ber. *113*, 2383 (1980)
109. H. Meier, T. Echter, H. Petersen: Angew. Chem. *90*, 997 (1978); Angew. Chem. Int. Ed. Engl. *17*, 942 (1978)
110. H. Bühl, H. Gugel, H. Kolshorn, H. Meier: Synthesis *1978*, 536
111. H. Meier, N. Hanold, H. Kolshorn: Angew. Chem. *94*, 67 (1982); Angew. Chem. Int. Ed. Engl. *21*, 66 (1982)
112. H. Meier, T. Echter: Angew. Chem. *94*, 68 (1982); Angew. Chem. Int. Ed. Engl. *21*, 67 (1982)
113. C. B. Reese, A. Shaw: J.C.S. Chem. Commun. *1970*, 1172
114. C. B. Reese, A. Shaw: J.C.S. Perkin I *1975*, 2422
115. H. Meier, H. Petersen: Chem. Ber. *111*, 3423 (1978)
116. H. Hanack, W. Spang: ibid. *113*, 2015 (1980)
117. H. Meier, M. Layer, A. Zetzsche: Chemiker-Ztg. *98*, 460 (1974)
118. A. Krebs, J. Oldenthal, H. Kimling: Tetrahedron Lett. *1975*, 4663
119. H. N. C. Wong, F. Sondheimer: Angew. Chem. *88*, 126 (1976); Angew. Chem. Int. Ed. Engl. *15*, 117 (1976)
120. G. Seitz, L. Pohl, R. Pohlke: Angew. Chem. *81*, 427 (1969); Angew. Chem. Int. Ed. Engl. *8*, 447 (1969)
121. H. Meier, H. Gugel: Synthesis *1976*, 338
122. H. N. C. Wong, P. J. Garratt, F. Sondheimer: J. Am. Chem. Soc. *96*, 5604 (1974)
123. H. N. C. Wong, F. Sondheimer: Tetrahedron *37*, 99 (1981)
124. C. B. Reese, A. Shaw: J.C.S. Chem. Commun. *1972*, 787
125. C. B. Reese, H. P. Sanders: Synthesis *1981*, 276
126. V. Prelog, K. Schenker, W. Küng: Helv. Chim. Acta *36*, 471 (1953)
127. A. C. Cope, M. W. Fordice: J. Am. Chem. Soc. *89*, 6187 (1967)
128. R. D. Miller: Ph.D. Thesis, Cornell Univ., Ithaca, N.Y. 1968
129. K. Grohmann, F. Sondheimer: Tetrahedron Lett. *1967*, 3121
130. A. J. Hubert, J. Dale: J. Chem. Soc. *1963*, 86
131. R. Wolovsky, F. Sondheimer: J. Am. Chem. Soc. *84*, 2844 (1962)
132. R. Wolovsky, F. Sondheimer: ibid. *87*, 5720 (1965)
133. F. Sondheimer, R. Wolovsky, P. J. Garratt, I. C. Calder: ibid. *88*, 2610 (1966)
134. L. T. Scott, G. J. DeCicco: Tetrahedron Lett. *1976*, 2663
135. G. Büchi, K. C. T. Luk: J. Org. Chem. *43*, 168 (1978)
136. O. M. Behr, G. Eglington, R. A. Raphael: Chem. and Ind. *1959*, 699
137. O. M. Behr, G. Eglington, A. R. Galbraith, R. A. Raphael: J. Chem. Soc. *1960*, 3614
138. O. M. Behr, G. Eglington, I. A. Lardy, R. A. Raphael: ibid. *1964*, 1151
139. G. M. Pilling, F. Sondheimer: J. Am. Chem. Soc. *93*, 1970 (1971)
140. T. Matsuoka, T. Negi, T. Otsubo, Y. Sakata, S. Misumi: Bull. Chem. Soc. Japan *45*, 1825 (1972)
141. T. Matsuoka, T. Negi, S. Misumi: Synthetic Commun. 2, 87 (1972)
142. G. L. Lange, T. W. Hall: J. Org. Chem. *39*, 3819 (1974)
143. Many aspects of angle strain and other kinds of strain are discussed in: A. Greenberg, J. F. Liebman: Strained Organic Molecules, Academic Press, New York 1978
144. H. Meier, H. Petersen, H. Kolshorn: Chem. Ber. *113*, 2398 (1980)
145. H. J. Hohlt, A. Krebs, R. Machinek, W. Lüttke, V. Sinnwell: unpublished results
146. A. Krebs, W. Rüger, L. Ng, K. D. Jordan: Bull. Soc. Chim. Belg. *91*, 363 (1982)
147. L. Ng, K. D. Jordan, A. Krebs, W. Rüger: J. Am. Chem. Soc. *104*, 7414 (1982)
148. C. Batich, O. Ermer, E. Heilbronner, J. R. Wiseman: Angew. Chem. *85*, 302 (1973); Angew. Chem. Int. Ed. Engl. *12*, 312 (1973)
149. H. Schmidt, A. Schweig, A. Krebs: Tetrahedron Lett. *1974*, 1471
150. R. B. Turner, A. D. Jarrett, P. Goebel, B. J. Mallon: J. Am. Chem. Soc. *95*, 790 (1973)
151. W. R. Roth, H. W. Lennartz: unpublished results
152. N. L. Allinger, A. Y. Meyer: Tetrahedron *31*, 1807 (1975)
153. F. A. L. Anet, I. Yavari: ibid. *34*, 2879 (1978)
154. D. W. Garrett: Ph.D. Thesis, Indiana Univ., Bloomington 1974
155. G. Wittig, P. Fritze: Liebigs Ann. Chem. *711*, 82 (1968)
156. G. Wittig, J. Meske-Schüller: ibid. *711*, 76 (1968)

157. R. A. G. de Graff, S. Gortner, C. Romers, H. N. C. Wong, F. Sondheimer: J. Chem. Soc. Perkin. Trans. II *1981*, 478
158. R. Destro, T. Pilati, M. Simonetta: J. Am. Chem. Soc. *97*, 658 (1975)
159. R. Destro, T. Pilati, M. Simonetta: Acta Cryst. Sect. B *B33*, 447 (1977)
160. V. Typke, J. Haase, A. Krebs: J. Mol. Struct. *56*, 77 (1979)
161. J. Haase, A. Krebs: Z. Naturforsch. *26a*, 1190 (1971)
162. J. Haase, A. Krebs: ibid. *27a*, 624 (1972)
163. H. H. Bartsch, H. Colberg, A. Krebs: Z. Kristallogr. *156*, 10 (1981)
164. C. Römming: unpublished results, cited in 104)
165. M. K. Au, T. W. Siu, T. C. W. Mak, T. L. Chan: Tetrahedron Lett. *1978*, 4269
166. C. Romers, T. L. Chan: unpublished results, cited in 79)
167. W. K. Grant, J. C. Speakman: Proc. Chem. Soc. *1959*, 231
168. R. Weiss: Ph.D. Thesis, Univ. of California, Los Angeles 1976; cited in [88]
169. D. W. Rogers, F. J. McLafferty: Tetrahedron *27*, 3765 (1971)
170. R. B. Turner, B. J. Mallon, M. Tichy, W. von E. Doering, W. R. Roth, G. Schröder: J. Am. Chem. Soc. *95*, 8605 (1973)
171. R. B. Turner, D. E. Nettleton, jr., M. Perelman: ibid. *80*, 1430 (1958)
172. G. Wittig, P. Fritze: Angew. Chem. *78*, 905 (1966); Angew. Chem. Int. Ed. Engl. *5*, 846 (1966)
173. E. T. Marquis, P. D. Gardner: Tetrahedron Lett. *1966*, 2793
174. J. P. Visser, J. E. Ramakers: J.C.S. Chem. Commun. *1972*, 178
175. R. Hoffmann, A. Imamura, W. J. Hehre: J. Am. Chem. Soc. *90*, 1499 (1968)
176. E. Haselbach: Helv. Chim. Acta *54*, 1981 (1971)
177. G. Bieri, E. Heilbronner, E. Kloster-Jensen, A. Schmelzer, J. Wirz: ibid. *57*, 1265 (1974)
178. R. W. Strozier, P. Caramella, K. N. Houk: J. Am. Chem. Soc. *101*, 1340 (1979)
179. N. G. Rondan, L. N. Domelsmith, K. N. Houk, A. T. Bowne, R. H. Levin: Tetrahedron Lett. *1979*, 3237
180. D. M. Hoffman, R. Hoffmann, C. R. Fisel: J. Am. Chem. Soc. *104*, 3858 (1982)
181. P. Carlier, J. E. Dubois, P. Masclet, G. Mouvier: J. Electron Spectrosc. *7*, 55 (1975)
182. A. Krebs, W. Rüger: unpublished results
183. U. Höpfner: Diplom Thesis, Heidelberg 1976
184. H. Meier, H. Petersen: Synthesis *1978*, 596
185. H. Meier, H. Gugel, H. Kolshorn: Z. Naturforsch. *31b*, 1270 (1976)
186. H. N. C. Wong, F. Sondheimer: Tetrahedron *37*, 99 (1981)
187. R. M. Lynden-Bell, N. Sheppard: Proc. Roy. Soc. (London) *A 269*, 385 (1962)
188. D. M. Graham, C. E. Holloway: Can. J. Chem. *41*, 2114 (1963)
189. J. Ulmen: Ph.D. Thesis, Köln 1973
190. O. L. Chapman, C.-C. Chang, J. Kolc, N. R. Rosenquist, H. Tomioka: J. Am. Chem. Soc. *97*, 6586 (1975)
191. A. T. Blomquist, R. E. Burge, jr., L. H. Liu, J. C. Bohrer, A. C. Sucsy, J. Kleis: ibid. *73*, 5510 (1951)
192. L. J. Bellamy: Advances in Infrared Group Frequencies, 1ˢᵗ Ed., Methuen, London 1968, p. 21–26
193. G. Wittig, P. Fritze: Liebigs Ann. Chem. *712*, 79 (1968)
194. H. Meier, J. Heiss, H. Suhr, Eu. Müller: Tetrahedron *24*, 2307 (1968)
195. V. Franzen, H.-I. Joschek: Liebigs Ann. Chem. *703*, 90 (1967)
196. H. Kolshorn, H. Meier, Eu. Müller: Tetrahedron Lett. *1972*, 1589
197. T. J. Katz, S. J. Lee: J. Am. Chem. Soc. *102*, 422 (1980)
198. H. Kimling, A. Krebs: Angew. Chem. *84*, 952 (1972); Angew. Chem. Int. Ed. Engl. *11*, 932 (1972)
199. A. Krebs, H. Kimling, R. Kemper: Liebigs Ann. Chem. *1978*, 431
200. R. J. Böttcher, H. Röttele, G. Schröder, J. F. M. Oth: Tetrahedron Lett. *1968*, 3935
201. P. Puster: Ph.D. Thesis, Heidelberg 1976
202. G. H. Schmid: Thiiranium and Thiirenium Ions, in: Topics in Sulfur Chemistry, Vol. 3, 101, Thieme-Verlag, Stuttgart 1977
203. A. Krebs, H. Colberg: unpublished results
204. A. Krebs, B. Jessel: unpublished results
205. T. Nozoe, Y. Kitahara: Proc. Japan Acad. *30*, 204 (1954); C.A. *49*, 13200 (1955)

206. T. Toda: Bull. Chem. Soc. Japan *40*, 588 (1967)
207. T. Yamatani, M. Yasunami, K. Takase: Tetrahedron Lett. *1970*, 1725
208. H. Colberg: Ph.D. Thesis, Hamburg 1980
209. A. Krebs, W. Rüger: unpublished results
210. A. Krebs, H. Colberg: unpublished results
211. N. J. Turro, V. Ramamurthy, K.-C. Liu, A. Krebs, R. Kemper: J. Am. Chem. Soc. *98*, 6758 (1976)
212. A. Krebs, H. Colberg: Chem. Ber. *113*, 2007 (1980)
213. W. Spang, M. Hanack: ibid. *113*, 2025 (1980)
214. G. Wittig, J. J. Hutchison: Liebigs Ann. Chem. *741*, 79 (1970)
215. A. Krebs, H. Kimling: Angew. Chem. *83*, 401 (1971); Angew. Chem. Int. Ed. Engl. *10*, 409 (1971)
216. Y. Tjung: Ph.D. Thesis, Heidelberg 1975
217. K. H. Klaska, H. Colberg: unpublished results
218. A. Krebs, H. Colberg, U. Höpfner, H. Kimling, J. Odenthal: Heterocycles *12*, 1153 (1979)
219. A. Krebs, K. Schütz: unpublished results
220. W. Tochtermann, P. Rösner: Tetrahedron Lett. *1980*, 4905
221. W. Tochtermann, P. Rösner: Chem. Ber. *114*, 3725 (1981)
222. J. Liebe, C. Wolff, W. Tochtermann: Tetrahedron Lett. *1982*, 171
223. H. Dürr, B. Weiss: Angew. Chem. *87*, 674 (1975); Angew. Chem. Int. Ed. Engl. *14*, 646 (1975)
224. H. Dürr, A. C. Ranade, I. Halberstadt: Synthesis *1974*, 878
225. K. Hafner, H. Diehl, H. U. Süss: Angew. Chem. *88*, 121 (1976); Angew. Chem. Int. Ed. Engl. *15*, 104 (1976)
226. G. Märkl, K. H. Heier: Tetrahedron Lett. *1974*, 4369
227. T. Hosokawa, I. Moritani, S. Nishioka: ibid. *1969*, 3833
228. W.-S. Lee, H. H. Brintzinger: J. Organometal. Chem. *127*, 87 (1977)
229. K. W. Lienert: Ph.D. Thesis, Hamburg 1981
230. J. Wilke: Ph.D. Thesis, Hamburg 1982
231. G. B. Robertson, P. O. Whimp: ibid. *97*, 1051 (1975)

Author Index Volumes 101–109

Contents of Vols. 50–100 see Vol. 100
Author and Subject Index Vols. 26–50 see Vol. 50

The volume numbers are printed in italics

Ashe, III, A. J.: The Group 5 Heterobenzenes Arsabenzene, Stibabenzene and Bismabenzene. *105*, 125–156 (1982).

Bestmann, H. J., Vostrowsky, O.: Selected Topics of the Wittig Reaction in the Synthesis of Natural Products. *109*, 85–163 (1983).
Bourdin, E., see Fauchais, P.: *107*, 59–183 (1983).

Chivers, T., and Oakley, R. T.: Sulfur-Nitrogen Anions and Related Compounds. *102*, 117–147 (1982).
Consiglio, G., and Pino, P.: Asymmetrie Hydroformylation. *105*, 77–124 (1982).
Coudert, J. F., see Fauchais, P.: *107*, 59–183 (1983).

Edmondson, D. E., and Tollin, G.: Semiquinone Formation in Flavo- and Metalloflavoproteins. *108*, 109–138 (1983).
Eliel, E. L.: Prostereoisomerism (Prochirality). *105*, 1–76 (1982).

Fauchais, P., Bordin, E., Coudert, F., and MacPherson, R.: High Pressure Plasmas and Their Application to Ceramic Technology. *107*, 59–183 (1983).

Gielen, M.: Chirality, Static and Dynamic Stereochemistry of Organotin Compounds. *104*, 57–105 (1982).
Groeseneken, D. R., see Lontie, D. R.: *108*, 1–33 (1983).

Hellwinkel, D.: Penta- and Hexaorganyl Derivatives of the Main Group Five Elements. *109*, 1–63 (1983).
Hilgenfeld, R., and Saenger, W.: Structural Chemistry of Natural and Synthetic Ionophores and their Complexes with Cations. *101*, 3–82 (1982).

Keat, R.: Phosphorus(III)-Nitrogen Ring Compounds. *102*, 89–116 (1982).
Kellogg, R. M.: Bioorganic Modelling — Stereoselective Reactions with Chiral Neutral Ligand Complexes as Model Systems for Enzyme Catalysis. *101*, 111–145 (1982).
Krebs, S., Wilke, J.: Angle Strained Cycloalkynes. *109*, 189–233 (1983).

Labarre, J.-F.: Up to-date Improvements in Inorganic Ring Systems as Anticancer Agents. *102*, 1–87 (1982).
Laitinen, R., see Steudel, R.: *102*, 177–197 (1982).
Landini, S., see Montanari, F.: *101*, 111–145 (1982).
Lavrent'yev, V. I., see Voronkov, M. G.: *102*, 199–236 (1982).
Lontie, R. A., and Groeseneken, D. R.: Recent Developments with Copper Proteins. *108*, 1–33 (1983).
Lynch, R. E.: The Metabolism of Superoxide Anion and Its Progeny in Blood Cells. *108*, 35–70 (1983).

McPherson, R., see Fauchais, P.: *107*, 59–183 (1983).
Majestic, V. K., see Newkome, G. R.: *106*, 79–118 (1982).
Margaretha, P.: Preparative Organic Photochemistry. *103*, 1–89 (1982).
Montanari, F., Landini, D., and Rolla, F.: Phase-Transfer Catalyzed Reactions. *101*, 149–200 (1982).
Müller, F.: The Flavin Redox-System and Its Biological Function. *108*, 71–107 (1983).
Mutter, M., and Pillai, V. N. R.: New Perspectives in Polymer-Supported Peptide Synthesis. *106*, 119–175 (1982).

Newkome, G. R., and Majestic, V. K.: Pyridinophanes, Pyridinocrowns, and Pyridinycryptands. *106*, 79–118 (1982).

Oakley, R. T., see Chivers, T.: *102*, 117–147 (1982).

Painter, R., and Pressman, B. C.: Dynamics Aspects of Ionophore Mediated Membrane Transport. *101*, 84–110 (1982).
Pillai, V. N. R., see Mutter, M.: *106*, 119–175 (1982).
Pino, P., see Consiglio, G.: *105*, 77–124 (1982).
Pommer, H., Thieme, P. C.: Industrial Applications of the Wittig Reaction. *109*, 165–188 (1983).
Pressman, B. C., see Painter, R.: *101*, 84–110 (1982).

Recktenwald, O., see Veith, M.: *104*, 1–55 (1982).
Reetz, M. T.: Organotitanium Reagents in Organic Synthesis. A Simple Means to Adjust Reactivity and Selectivity of Carbanions. *106*, 1–53 (1982).
Rolla, R., see Montanari, F.: *101*, 111–145 (1982).
Rzaev, Z. M. O.: Coordination Effects in Formation and Cross-Linking Reactions of Organotin Macromolecules. *104*, 107–136 (1982).

Saenger, W., see Hilgenfeld, R.: *101*, 3–82 (1982).
Schöllkopf, U.: Enantioselective Synthesis of Nonproteinogenic Amino Acids. *109*, 65–84 (1983).
Siegel, H.: Lithium Halocarbenoids Carbanions of High Synthetic Versatility. *106*, 55–78 (1982).
Steudel, R.: Homocyclic Sulfur Molecules. *102*, 149–176 (1982).
Steudel, R., and Laitinen, R.: Cyclic Selenium Sulfides. *102*, 177–197 (1982).

Thieme, P. C., see Pommer, H.: *109*, 165–188 (1983).
Tollin, G., see Edmondson, D. E.: *108*, 109–138 (1983).

Veith, M., and Recktenwald, O.: Structure and Reactivity of Monomeric, Molecular Tin(II) Compounds. *104*, 1–55 (1982).
Venugopalan, M., and Vepřek, S.: Kinetics and Catalysis in Plasma Chemistry. *107*, 1–58 (1982).
Vepřek, S., see Venugopalan, M.: *107*, 1–58 (1983).
Vostrowsky, O., see Bestmann, H. J.: *109*, 85–163 (1983).
Voronkov, M. G., and Lavrent'yev, V. I.: Polyhedral Oligosilsequioxanes and Their Homo Derivatives. *102*, 199–236 (1982).

Wilke, J., see Krebs, S.: *109*, 189–233 (1983).

Topics in Current Chemistry

Fortschritte der Chemischen Forschung
Managing Editor: F.L. Boschke

Volume 99
Cosmo- and Geochemistry

1981. 39 figures, 18 tables. VII, 133 pages. ISBN 3-540-10920-X

Volume 100
New Trends in Chemistry

1982. 95 figures, 22 tables. VII, 213 pages. ISBN 3-540-11287-1

Volume 101
Host Guest Complex Chemistry II
Editor: F. Vögtle
With contributions by numerous experts
1982. 140 figures, 40 tables. VIII, 203 pages. ISBN 3-540-11103-4

Volume 102
Inorganic Ring Systems

1982. 111 figures, 39 tables. VII, 237 pages. ISBN 3-540-11345-2

Volume 103
P. Margaretha
Preparative Organic Photochemistry

Editor: J.-M. Lehn
1982. 9 figures. X, 89 pages. ISBN 3-540-11388-6

Volume 104
Organotin Compounds

1982. 34 figures, 22 tables. VII, 137 pages. ISBN 3-540-11542-0

Volume 105
Organic Chemistry

1982. 99 figures, 22 tables. VII, 158 pages. ISBN 3-540-11636-2

Volume 106
Synthetic and Structural Problems

1982. 3 figures, 13 tables. VII, 178 pages. ISBN 3-540-11766-0

Volume 107
Plasma Chemistry IV

Guest Editors: S. Veprek, M. Venugopalan
1983. 122 figures, 26 tables. VIII, 188 pages. ISBN 3-540-11828-4

Volume 108
Radicals in Biochemistry

1983. 13 figures, 13 tables. VIII, 144 pages. ISBN 3-540-11846-2

Springer-Verlag
Berlin
Heidelberg
New York

Heidelberger Taschenbücher

Band 72: M. Becke-Goehring, H. Hoffmann

Vorlesungen über Anorganische Chemie: Komplexchemie

Teilweise mitbearbeitet von K.Chr. Buschbeck

1970. 104 Abbildungen. VIII, 245 Seiten.
ISBN 3-540-04873-1

Band 131: W. Bähr, H. Theobald

Organische Stereochemie

Begriffe und Definitionen

1973. XV, 122 Seiten. ISBN 3-540-06339-0

Band 135: D. Hellwinkel

Die systematische Nomenklatur der Organischen Chemie

Eine Gebrauchsanweisung

3. Auflage. 1982. VIII, 170 Seiten.
ISBN 3-540-11764-4

Aus dem Vorwort zur dritten Auflage: „Nachdem auch die zweite Auflage schnell vergriffen war, wurde bei dieser neu durchgesehenen, korrigierten Auflage insbesondere der neueren Entwicklung bei der Heterocyclen-Nomenklatur Rechnung getragen."
Der Verfasser

Band 148: J. Schurz

Physikalische Chemie der Hochpolymeren

Eine Einführung

1974. 76 Abbildungen. VIII, 196 Seiten.
ISBN 3-540-06708-6

Band 161: H. Preuß, F.L. Boschke

Die chemische Bindung

Eine verständliche Einführung

1975. 32 Abbildungen. VIII, 82 Seiten.
ISBN 3-540-07041-9

Band 193: H.P. Latscha, H.A. Klein

Anorganische Chemie

Chemie – Basiswissen I

1978. 190 Abbildungen, 1 Ausklapptafel.
XIII, 424 Seiten. ISBN 3-540-08630-7

Inhaltsübersicht: Allgemeine Chemie: Chemische Elemente und chemische Grundgesetze. Aufbau der Atome. Periodensystem der Elemente. Moleküle, chemische Verbindungen, Reaktionsgleichungen und Stöchiometrie. Chemische Bindung. Komplexverbindungen. Zustandsformen der Materie. Mehrstoffsysteme. Redox-Systeme. Säure-Base-Systeme. Energetik chemischer Reaktionen (Grundlagen der Thermodynamik). Kinetik chemischer Reaktionen. Chemisches Gleichgewicht. – Spezielle Anorganische Chemie: Hauptgruppenelemente. Nebengruppenelemente. – Literaturauswahl und Quellennachweis. – Sachverzeichnis. – Abbildungsnachweis. – Maßeinheiten. – Ausklapptafel: Periodensystem der Elemente.

Band 211: H.P. Latscha, H.A. Klein

Organische Chemie

Chemie – Basiswissen II

1982. 121 Abbildungen, 56 Tabellen, 700 Formeln.
XXII, 554 Seiten. ISBN 3-540-10814-9

Inhaltsübersicht: Grundwissen der organischen Chemie. – Chemie und Biochemie von Naturstoffen. – Angewandte Chemie. – Trennmethoden und Spektroskopie. – Register und Nomenklatur.

Springer-Verlag
Berlin
Heidelberg
New York